Perspectives on the Human Controller

Henk Stassen

Perspectives on the Human Controller

Essays in honor of Henk G. Stassen

Edited by

Thomas B. Sheridan
Ton Van Lunteren

LEA LAWRENCE ERLBAUM ASSOCIATES, PUBLISHERS
1997   Mahwah, New Jersey

Lawrence Erlbaum Associates, Inc., Publishers
10 Industrial Avenue
Mahwah, New Jersey 07430

Library of Congress Cataloging-in-Publication Data

Perspectives on the human controller : essays in honor of Henk G.
  Stassen / edited by Thomas B. Sheridan, Ton Van Lunteren.
      p.   cm.
    Includes bibliographical references and index.
    ISBN 0-8058-2189-9(cloth). -- ISBN 0-8058-2190-2 (pbk.)
    1. Human-machine systems--Manual control. 2. Stassen, Henk G.
  I. Stassen, Henk G. II. Sheridan, Thomas B. III Van Lunteren,
  Ton.
  TA167.P46 1997
  620.8'2--dc21   96-37983
                                                      CIP

Books published by Lawrence Erlbaum Associates are printed
on acid-free paper, and their bindings are chosen
for strength and durability.

Printed in the United States of America

10 9 8 7 6 5 4 3 2 1

Contents

Contributors

Prof. Han Bakker MD
 Turfschip 39, 1186 XB Amstelveen, The Netherlands
Dr. Sheldon Baron
 BBN Systems and Technologies, 10 Moulton Street, Cambridge, MA 02138
Dr. ir. Paul Breedveld
 Laboratory for Measurement and Control, Dept of Mechanical Engineering and
 Marine Technology, Delft University of Technology, Mekelweg 2, 2628 CD
 Delft, The Netherlands
Prof. ir. Jan C. Cool
 Laboratory for Measurement and Control, Dept of Mechanical Engineering and
 Marine Technology, Delft University of Technology, Mekelweg 2, 2628 CD
 Delft, The Netherlands
Dr. ir. C.(Els) van Daalen
 Dept of Systems Engineering, Policy Analysis and Management,
 Delft University of Technology, Jaffalaan 5, 2628 BX Delft, The Netherlands
Dr. Jenny Dankelman
 Laboratory for Measurement and Control, Dept of Mechanical Engineering and
 Marine Technology, Delft University of Technology, Mekelweg 2, 2628 CD
 Delft, The Netherlands
Ir. Jan Goezinne
 Laboratory for Measurement and Control, Dept of Mechanical Engineering and
 Marine Technology, Delft University of Technology, Mekelweg 2, 2628 CD
 Delft, The Netherlands
Prof. dr. ir. C.A.(Kees) Grimbergen
 Dept of Medical Physics, Academic Medical Center, University of Amsterdam,
 PO Box 22700, 1100 DE Amsterdam, The Netherlands
Werneck E.C. van Haselen MD
 Revalidatiecentrum De Hoogstraat, Rembrandtkade 10, 3583 TM Utrecht, The
 Netherlands
Dr. Frans C.T. van der Helm
 Laboratory for Measurement and Control, Dept of Mechanical Engineering and
 Marine Technology, Delft University of Technology, Mekelweg 2, 2628 CD
 Delft, The Netherlands
Dr. ir. Rob B.M. Jaspers
 Unilever Research Laboratorium Vlaardingen, PO Box 114, 3130 AC
 Vlaardingen, The Netherlands
Henry Jex
 System Technology, Inc., 13766 S. Hawthorne Blvd., Hawthorne, CA 90250,
 U.S.A.
Prof. dr. ing. Gunnar Johannsen
 IMAT-MMS (Systems Engineering and Human-Machine Systems), University
 of Kassel, Moenchebergstr. 7, D-34109 Kassel, Germany
Dr. William Levison
 BBN Systems and Technologies, 10 Moulton Street, Cambridge, MA 02138,
 U.S.A.

Dr. ir. Antonie (Ton) van Lunteren
: Laboratory for Measurement and Control, Dept of Mechanical Engineering and Marine Technology, Delft University of Technology, Mekelweg 2, 2628 CD Delft, The Netherlands

G.H.M. (Erna) van Lunteren
: Nieuwe Plantage 105, 2611XV Delft, The Netherlands

Dr. Anil Macwan
: 1250 Chalet Road #103, Noperville, IL 60563, U.S.A.

Prof. Neville Moray
: LAMIH-PERCOTEC, Universiti de Valenciennes, B.P.311, Le Mont Houy, 59304 Valenciennes Cedex, France

Dr. ir. R. (Bob) Papenhuijzen
: Ministry of Transport, Public Works and Water Management, Transport Research Centre (AVV), P.O. Box 1031, 3000 BA Rotterdam, The Netherlands

Dr. William B. Rouse
: Search Technology, Inc., 4898 South Old Peachtree Rd., Norcross, GA 30071-4707, U.S.A.

Prof. Thomas B. Sheridan
: 3-346 MIT, Cambridge, MA 02139, U.S.A.

Prof. dr. Gerda J.F. Smets
: Dept of Industrial Design Engineering, Delft University of Technology, Jaffalaan 9, 2628 BX, Delft, The Netherlands

Prof. dr. ir. Jos. A.E. Spaan, Dept of Medical Physics, Academic Medical Center, University of Amsterdam, PO Box 22700, 1100 DE Amsterdam, The Netherlands

Prof. dr. ir. Henk G. Stassen
: Laboratory for Measurement and Control, Dept of Mechanical Engineering and Marine Technology, Delft University of Technology, Mekelweg 2, 2628 CD Delft, The Netherlands

Lenie A. de Vries
: Revalidatiecentrum De Hoogstraat , Rembrandtkade 10, 3583 TM Utrecht, The Netherlands

Ir. Ben J.M. Wenneker
: Laboratory for Measurement and Control, Dept of Mechanical Engineering and Marine Technology, Delft University of Technology, Mekelweg 2, 2628 CD Delft, The Netherlands

Dr. ir. Peter A. Wieringa
: Laboratory for Measurement and Control, Dept of Mechanical Engineering and Marine Technology, Delft University of Technology, Mekelweg 2, 2628 CD Delft, The Netherlands

Dr. ir. Ron A. van Wijk
: Shell International Petroleum Maatschappy B.V., Manufacturing, Dept Catalytic Cracking, PO Box 162, 2501 AN Den Haag, The Netherlands

INTRODUCTION AND OVERVIEW

Chapter 1

From Biomechanical Control to Large Scale Systems

Thomas B. Sheridan

1   Introduction

This book is about human control of mechanical things. That includes people controlling the mechanical movements of their own limbs, extensions of their limbs such as prostheses (limb replacements) or orthoses (limb braces), and hand tools or telemanipulators. It also includes people controlling the mechanical movements of vehicles that they ride in, such as aircraft, automobiles, and trains, and movements of discrete products through maufacturing plants, and of chemicals and other fluids through process plants, such as refineries and nuclear power stations.

Within academia and industry the control of limbs (or their substitutes), the control of vehicles, and the control of discrete manufacturing and continuous process plants are usually studied by very different research and engineering communities. The control of limbs is generally regarded as a subfield of biomechanics, and more generally of biomedical engineering, and the hospital or medical clinic is where most work is being carried out. Industry mostly has ignored the challenges of prosthetics and orthotics because there is not much money to be made in it; payers are typically third party insurers or the government itself. In contrast, the problems of control of vehicles (particularly aircraft and military vehicles) and industrial plants have driven work in the field of control, both in terms of its science and technology. In the field of robotics, where biomechanics is an obvious model, industry has expected too much too soon, and in some cases has overinvested and later been forced to withdraw in disappointment.

Curiously, when considering human control and the human-machine interface, the problems of biomechanics and biomedical control have played a critical role. From the beginning, the structure and operation of animals, and of the human body in particular, have provided the inspiration and standard references for designers of all kinds of machines.

2   Origins of feedback control

It is never easy to point to exactly where or when some basic ideas in science or technology began. Certainly Leonardo da Vinci contributed significantly to scientists' appreciation of human control of both their own limbs and limb extensions, such as machines to exert power and perform various mechanical tasks. D'Arcy Wentworth Thompson, in his landmark (1959) work, *On Growth and Form*, by focusing on the mechanisms of human muscle motion and locomotion, further inspired interest and even awe with respect to the human body's mechanisms of control.

As so often happens, the intuitive idea of feedback control was well known to technologists long before the science (one might say, the mathematics of control) were developed. The most evident example of this is the flyball governor on the steam engine, attributed to James Watt. The founding fathers of control recognized these phenomena and were keenly aware of the close analogies to driving cars, flying airplanes, and aiming guns. The science of control was developing, by chance, just as World War II was breaking out, and the MIT Radiation Laboratory published what is recognized as the first widely accepted treatise on control theory (James, et al., 1947). It presented a theory of servomechanisms (slave mechanisms) based on the notion of feedback, which is obvious to us today but was then a novel idea, particularly the idea that one could actually write equations to describe the effects of feedback. It also included a chapter on manual control.

For readers not familiar with the basic ideas of control, we present here the bare minimum of mathematics to represent these ideas. If one desired the output position **x** of some arm, vehicle, plant, or other controlled process **G** to follow some reference position **r**, one could measure the discrepancy of **x** from **r** (specifically **r** - **x**), amplify that signal by **K**, and use the result to drive **G** into conformity. The simple block diagram representation of this "negative feedback control" is shown in Figure 1. Thus the automobile driver observes the deviation of his car position (**x**) from the center of the lane (**r**) and turns the steering wheel in proportion to **r** - **x** to force **G** to the centerline. The more sensitive he is to the deviation (**r** - **x**) the larger the amplitude **K** and the more quickly the car returns to the centerline.

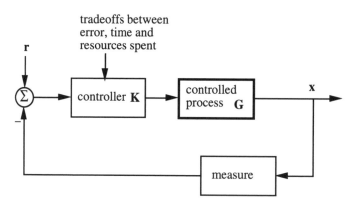

Figure 1. Simple feedback control system

Symbolically, in terms of Figure 1, **x** = **G** **K** (**r** - **x**). Solving for **x** in terms of **r**,

x/r = KG / (1 + KG)

(Mathematically, treating these terms algebraically is valid only if the corresponding time functions are linear differential equations and can be subjected to LaPlace or Fourier transforms, but for purposes of general discussion we will ignore this essential formality.) The point is that as **K** gets large, **x** follows **r** more and more precisely, provided that the denominator of the above equation never approaches zero (i.e., **KG** never approaches -1). In the latter case Figure 1 becomes a positive feedback system that continually sustains itself even if **r** = 0, which means the energy in the system simply accumulates until the system "blows up" or becomes unstable. Stated in another way, when there is even a minuscule and brief energy component of **r** that is periodic AND the dynamic properties of **KG** produce a phase lag of 180 degrees AND the net gain (absolute value of **KG**) is greater than 1 the system will go unstable, and it will go unstable more quickly the greater |**KG**| is.

These unhappy conditions are more than hypothetical. They occur commonly in control systems, since **G** is very likely to have dynamic lags. Therefore the science and art of control is much more than just having negative feedback and high gain; it is also a matter of avoiding instability and choosing the controller gain and dynamics to match both the controlled element dynamics **G** and the criteria of compromise between different attributes of system performance **x**, such as acceptable degree of oscillation, speed and overshoot in responding to sudden changes in **r**, ability to respond to **r** in spite of disturbances entering the system elsewhere, smoothness of steady-state response, and inaccuracies in measuring **x** in the feedback loop.

3   Birth of manual and biomechanical control

Working with the scientists and engineers at the MIT Radiation Lab, the mathematician Norbert Wiener was among the first to point out the close relations between what was being developed for gun laying and what apparently were the principles by which our nervous systems control the limbs and other organs of our bodies. Wiener coined the term *cybernetics* to mean "communication and control in the animal and the machine" and discussed his ideas in several books, including *Cybernetics* (1947), *The Human Use of Human Beings* (1950), and *God and Golem Inc.* (1964), as well as his own autobiographies. He was fascinated by the number and variety of control mechanisms in the human body, and how various diseases of the nervous system, such as Parkinson's disease, lead to tremors and instabilities much as were found in flying aircraft and aiming guns.

During the early phase of World War II, Britain was challenged to refine the understanding of human control of tanks and aircraft. The first engineering-oriented manual control models were probably those of Prof. Arnold Tustin in the United Kingdom applied to tank control, followed closely by models by J.P. North of Boulton-Paul Aircraft Co.

Just after the war, development of nuclear power was deemed a high priority and demanded robotic means by which to handle radioactive materials. Raymond Goertz (1954), at Argonne National Lab near Chicago, and Jean Vertut (1986), of the French atomic energy establishment, were pioneers in developing manually controlled remote manipulator arms.

In 1960 the International Federation of Automatic Control (IFAC) was founded, which was meant to provide a basis by which control scientists and engineers from all nations could share their ideas and applications. Dr. Harold Chestnut of General Electric Co. was elected the first president, and the first world congress was held in Moscow. This location was particularly striking because the Cold War was already in place and because at the particular time of that meeting the (until then secret) American U2 spy plane had just been shot down by the Soviets and was being displayed in parts in a Moscow park. Delegates to that conference, including the writer, had the experience of being ushered ahead of thousands of Soviet citizens waiting in line to see the exhibit. Meanwhile, the American government was denying that the Soviets had the aircraft. The tone of the meeting was nevertheless one of being hopeful about cooperation among scientists on both sides of the Iron Curtain at the scientific level.

One of the earliest applications of the new science of control was to understanding the basic physiological mechanisms of the human body, particularly neuromuscular control, with the aim of developing better rehabilitation treatment of the physically handicapped and of designing better prostheses and other technological aids. At that first IFAC congress, Dr. Aaron Kobrinskii of Moscow, at that time a renowned robotics engineer (but already being persecuted as a Jew), demonstrated what I regard as the first successful myographic-controlled prosthetic arm. Soon afterwards the Americans (principally at Rancho Los Amigos Hospital in Los Angeles and later at MIT), the Japanese (at Waseda University), the British, and others developed similar prosthetic arms, and these developments spurred interest in robotic control as well as biomechanics. Prof. Kato at Waseda University in Tokyo began developing a series of walking machines. The biomedical application of control engineering caught on and spawned myographic-controlled knees and other artificial joint articulations. Success in rehabilitation engineering further motivated researchers to explore the human body's amazingly intelligent control capabilities. It was at this exciting and hopeful time that Dr. Henk Stassen came into the field.

In 1967 Henk was awarded a Fullbright fellowship to study bioengineering at University of California, Los Angeles, under Prof. John Lyman. The writer, as it happened, had also studied with Lyman 12 years earlier, and it was partly through this contact that Stassen came to be a good friend. Both of us acknowledge a great debt to Lyman's mentorship and his enthusiasm for applied science.

Two years earlier, Duane McRuer (1965) of Systems Technology, Inc. and his colleagues had laid out the basic structure of a useable differential equation model for the human pilot. It was based on the idea that the human operator in a control loop tends to adopt an approximately linear input-output equation that compensates for the dynamics of the aircraft or other process being controlled so as to leave the combination of the human operator and controlled process (**G** in the earlier equations) to be the equivalent of a simple integrator plus an equivalent time delay of roughly 0.25 - 0.50 sec, with a gain coefficient **K** (adjustable by the human but hovering roughly around 10). The integrator was known to be a relatively ideal **G** for most control applications, while the time delay was undesirable, but also unavoidable due to finite delays between vision and corresponding muscle response. This so-called simple crossover model and its somewhat more complex variants have served well not only the aircraft engineers but also those interested in the handling qualities of other vehicles.

An important paper by Kleinman, et al. (1970) presented a different approach to modeling the human operator. It made use of the ideas of modern control theory, based on the fact that a model of the controlled process contained within the control logic (called a Kalman filter) permitted a best estimate of system state **x** despite disturbance or state measurement noise. Having this best estimate of **x** plus an assumed simple form of the objective function (an equation specifying degree of "goodness" of system performance), one could calculate the control that would maximize that goodness within the given constraints. They thus formulated an *optimal control model* (OCM) of the human. This model has also seen considerable application to human-machine systems. To invoke this model is to assume that the human operator possesses in some form an "internal model" of the external controlled process.

In the early 1970s within IFAC a group of control engineers, including Stassen, founded within the IFAC structure a Man-Machine Systems Committee (at this writing Stassen chairs that committee), which has sponsored a series of six triennial international meetings on analysis, design, and evaluation of human-machine systems of all kinds, including biomedical, vehicular, and process control.

4   Extensions to large-scale systems

Thus efforts proceeded to model biomechanical systems, and these extended to model human control of vehicles such as aircraft and ships, and to model human control of remote manipulators for space and other hostile environments. Experiments on both how the human controls himself and how he controls external systems were mutually supportive. Then gradually, as control technology became more sophisticated and some simpler control tasks were automated, the human was quite naturally relegated to more complex tasks. The simple crossover and OCM models no longer fit the situation. The human control modeler was quite naturally challenged to develop more ambitious models, embodying higher level and more cognitive capabilities by which to predict more complex human-machine system behavior and to engineer those systems.

The latter challenges led to emphasis on what is now called *supervisory control*, wherein the human (1) communicates instructions to a computer coded in high level language to specify goals, constraints, and criteria of goodness (the objective function, in whatever form the human can specify this) and (2) receives from the computer information about system performance at whatever level of detail is desired. The computer, in turn, closes the loop through the controlled process and its own artificial sensors and actuators. Supervisory control pushed the challenge of modeling well beyond the McRuer et al. crossover model and the Kleinman et al. OCM. The internal model of control engineering thus sought affinity with the "mental model" of the cognitive scientist, and the supervisory roles of planning, instructing the computer, monitoring automated actions and detecting abnormalities, intervening as necessary to revise instructions or to take over in emergencies, and finally learning from experience had to be viewed in engineering as well as in cognitive terms. However, as yet no comprehensive and commonly accepted model of supervisory control exists, although Chapter 19 suggests one approach.

5   Organization of the book

The next chapter by van Lunteren provides a second introduction, giving a more personal statement of Prof. Henk Stassen's activities from within the Man-Machine Systems Group at Delft University of Technology, where he has focused his career. It describes how Stassen and his students branched out from biomechanics and rehabilitation of the handicapped into ship control, chemical plant process control, manipulator control, and other fields. This is followed by Stassen's curriculum vitae.

This first introductory chapter concludes with a simple statement of the organization of the main body of the book. The chapters are organized into three sections: (A) control of body mechanisms, and rehabilitation and design of aids for the disabled; (B) human control of vehicles and manipulation; and (C) human control of large, complex systems. This organization helps to provide a historical perspective as well as to show the interrelations between these seemingly different applications.

References

Goertz, R.C. and Thompson, R.C. (1954). Electronically controlled manipulator. *Nucleonics*, No. 46–47

James, H.M., Nichols N.B., and Philips, R.S (1947). *Theory of Servomechanisms*. New York: McGraw Hill.

Kleinman, D.L., Baron, S., and Levison, W. H. (1970). An optimal control model of human response, Part I. *Automatica* 6, no. 3: 357–369.

McRuer, D.T., Graham, D., Krendel, E., and Reisener, W., Jr. (1965). Human pilot dynamics in compensatory systems. Air Force Flight Dynamics Laboratory, Wright Patterson AFB, OH, Report AFFDL–TR–65–15.

Thompson, D.W. (1959). *On Growth and Form*, Vols. I and II. Cambridge (U. K.): University Press.

Vertut, J. and Coiffet, P. (1986). *Robot Technology, Volume 3A: Teleoperation and Robotics: Evolution and Development, Volume 3B: Teleoperation and Robotics: Applications and Technology*, Englewood, NJ: Prentice Hall.

Wiener, N. (1947). *Cybernetics*. New York: Wiley.

Wiener, N. (1950). *The Human Use of Human Beings*. Boston: Houghton Mifflin Co.

Wiener, N. (1964). *God and Golem*, Inc. Cambridge, MA: MIT Press.

Chapter 2

Man-Machine Systems at Delft, an Inside Story

Ton van Lunteren

1   Origin of the Man-Machine Systems Group

Henk Stassen was a student of Roelof G. Boiten, who was the first-full time professor of control engineering in a Mechanical Engineering Department in the Netherlands. Boiten was a man with many ideas and a broad view. As Henk sometimes explained, Boiten planted his ideas among his collaborators like seedlings in a garden to see whether they were cared for and had some viability. The underlying idea was to let people do something they liked to do, because that was the best way to get good results. This is a very basic principle that Henk adopted from his tutor.

One of the seedlings in Boiten's garden was called Man-Machine Systems, a topic the importance of which had already been pointed out by the great mathematician and founder of the science of cybernetics, Norbert Wiener. Two application fields of man-machine systems had Boiten's interest. One was the medical field. For instance, he invented a syringe for blind diabetic patients. The other was the subject of humans as controllers of technical systems such as vehicles. In the medical field, contacts were established with the world of rehabilitation, especially focused on the control of arm prostheses. On the subject of vehicle control, he started a project on the design of a bicycle simulator to study human control behavior in a very common, though not trivial, control task.

While Henk Stassen was working on his Ph.D. thesis on random lateral motions of railway vehicles, which he finished in 1967, Boiten asked him whether he was willing to stay in the laboratory in order to further develop the topic of man-machine systems. When Henk agreed, it marked the beginning of the Man-Machine Systems Group in its present form. It is interesting to note that Boiten's two application fields have remained as his heritage and that we find them represented in the contents of this book.

The activities of the Laboratory for Measurement and Control were focused on four main application fields, that is, process control, hydraulic servo systems, instruments, and man-machine systems.

The Process Control Group, then headed by Frans J. Schijff, focused its activities on modelling and control of industrial systems, such as electrical power plants. Taco Viersma, who was head of the Hydraulics Servo Systems Group, was the inventor of the hydraulic actuator without Coulomb friction. This type of actuator is presently used in all moving base flight simulators. Presently his group is incorporated in the Process Control Group.

The Instruments Group, headed by Jan C. Cool, was involved with pneumatics, because in those days all control systems in the process industry were based on pneumatic instruments. One of the topics of study in this group was the development of miniature components for pneumatic logics. At that time externally powered prostheses and orthoses had pneumatic actuators. However, these devices were far from optimal. Because of that experience with pneumatics, the Instruments Group got involved in the design of prostheses and orthoses. The Man-Machine Systems Group has a close interaction with the Instruments Group in the field of rehabilitation. According to the plans of the department, these groups will be joined in the near future.

The following two sections give an overview of the research activities of the Man-machine Systems Group over the past 30 years, separated into the two application fields. In Section 4, the relation between the two fields is addressed. Finally, something is said about the philosophy behind the research program as a whole.

During the first 15 years of the existence of the Man-Machine Systems Group, four progress reports were written (Stassen & van Lunteren, 1969; Stassen et al., 1973; Stassen, 1977; Van Lunteren, 1983). Progress reports have been abandoned since they are very time consuming to produce. In a sense, the present edition could be considered to be a kind of fifth progress report.

2   Projects in the field of medical applications

About 1960, a number of children in some European countries were born with limb defects, due to the use of a new medicine for pregnant women. Although in the Netherlands this number was small, because there the medicine was not yet on the market officially, it was a point of concern for a number of rehabilitation centers.

In the previous years, in a number of countries new developments had taken place that resulted in the manufacture of components for pneumatically powered prostheses. Therefore the rehabilitation centers sought contact with a number of technical institutions, among which were the Medical Physics Institute (TNO-MFI) and the Laboratory for Measurement and Control of the Department of Mechanical Engineering of the Delft University of Technology. The latter were sought out for their know-how in the field of pneumatic control. These contacts led to the formation of a specialized rehabilitation team, the Working Group for Orthoses and Prostheses (WOP), in the rehabilitation center De Hoogstraat, then at Leersum. Initiators were the medical director Dr. A. Verkuyl and Ms. W.J. Luitse, who was one of the first occupational therapists in the Netherlands, perhaps even the first, and at that time head of the OT Department. Besides the members of the different disciplines within De Hoogstraat, the team also consisted of members of the cooperating technical institutes TNO-MFI and the DUT. Henk Stassen and Jan Cool joined the team from the latter institute. The practical problems of patients with upper limb defects were analyzed in the meetings of the WOP (Luitse and Stassen, 1974). The goals were not only to benefit the individual patient but also to find solutions for problems that were useful for another group with comparable problems (see Chapter 9). This formed the basis for a number of projects that were executed at the DUT as illustrated in Figure 1, and that will be elucidated in the following paragraphs.

In the beginning, much attention was paid to the design of orthoses for patients with an arm paralyzed due to a traffic accident. At the start of the WOP, pneumatic components were commercially available for externally powered orthoses, which, however, had a lot of friction and thus were not very efficient.

This was the starting point for the design activities of the Instruments Group, which finally resulted in the development of an elbow orthosis that did not need an external power source, and also in the development of a device to press the upper arm into the shoulder joint for people who have lost the function of the muscles that normally keep the arm in position. Further activities of this group later also led to the design of arm prostheses for children, an ongoing activity that still aims at improvements in comfort, cosmesis, and easy control (Chapter 8).

The activities of the Man-Machine Systems Group were primarily focused on the investigation of control principles, design criteria, and evaluation studies. In the case of the evaluation studies, a principal decision was made that designers perform only mechanical tests of components but that evaluation studies with clinical prototypes should be done by independent investigators in order to provide objective judgement in a comparison with commercially available devices (see Chapter 7).

The activities of the WOP also had their influence on the treatment of other patient categories. Specialized teams were started, such as the Working Group for Quadriplegics, headed by Han Bakker, of which Henk Stassen also became a member. Quadriplegics have a spinal cord injury leading to loss of motor functions of their lower limbs and in some cases the upper limbs as well, depending on the location of the injury. The study group consisted mainly of young men who had received their injuries through diving into shallow water. Their rehabilitation period varied between a few months and one and a half years, dependent on the location and the severity of the damage.

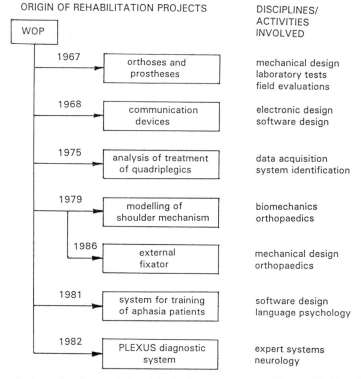

Figure 1. Projects that found their origin in the cooperation with the Working Group for Orthoses and Prostheses of rehabilitation center De Hoogstraat. The year when the project was started is also indicated.

In most cases, homes have to be adapted for these individuals to make them accessible to a wheelchair. Because this takes some time, the procedure of applying for financial support has to be started well ahead of time. In a number of cases, it was found that patients had to stay longer in the clinic because home adaptations were not yet ready, due to team members postponing submitting the application for support because they were not yet sure about the level of independence the patient could reach, and thus did not yet know what type of adaptations were required.

Therefore, in 1973 Henk Stassen proposed a project to develop a model that could predict the future state of quadriplegic patients. The necessary data were gathered weekly and concerned four fields of interest in medical state, psychosocial information, progress in the execution of daily life activities, and state of adaptations. In the course of the project, it was found that the activity of data gathering itself yielded a lot of information for the treatment team. In those days computing facilities were still not very common, so data processing was done at the central computer of the university. In the meantime a collaboration among a number of specialized rehabilitation centers in the Netherlands had led to the establishment of a working group among these centers, which now uses a modified version of the system. Finally, it should be mentioned that the introduction of such a system does not simply provide an additional facility for a treatment team but also can have other effects, such as resulting in a change in the method of treatment (see Chapter 12).

Another research area has its roots in 1968, when a wheelchair-bound patient who had brain stem damage due to a traffic accident was introduced to the WOP. This damage resulted in loss of speech and loss of motor control over a large part of the patient's body. The patient could make only head motions. The DUT was asked for advice on the choice of a communication device. At that time two recently developed commercially available systems existed, Possum and Pilot.

Possum was a device on which a matrix of characters was presented. After activation of the device, the rows of the matrix were successively indicated by a signal light. When the user gave a signal, for instance, by pressing a button or, in the case of a patient without hand function, by blowing into a little pipe, the lighted row was selected. Thereafter, the elements of the row were scanned in the same way. After selection of a character, it was printed on a typewriter.

Pilot also had a panel with characters, but each of its characters was accompanied by a light-sensitive element. The user had a head-mounted light source with which a character could be indicated. When the photosensitive element was lighted for a given period of time, the indicated character was sent to a typewriter. The time could be adjusted so that the light spot could move freely over the panel without activating the selection mechanism.

Possum was designed for persons who could only generate an on-off signal. Selection was paced by the machine. The user had to wait for the machine and then had to produce a relatively quick reaction. Moreover, the choice of a character consisted of two selections. Pilot was designed for users who still could generate a good control signal in two dimensions. In this case the machine was paced by the user, and a reasonable typing speed could be obtained. Therefore, Pilot was the more suitable system for the patient in question. However, a test in the laboratory showed that the reliability of the system was not very high. The group thought that a more reliable system could be designed. This marked the beginning of a project on communication devices for the motorically disabled which resulted in the present version of Lucy (see Chapter 10).

One of the initiatives of the WOP was an evaluation study with two purposes. On the one hand, the team wanted to know more about the acceptance of prostheses and orthoses, because the impression existed that some of these devices were no longer used. On the other hand, evaluation of the team approach of the WOP was also thought to be useful.

In the course of the project, it was found that the first goal was too broad, so the evaluation was limited to adult wearers of arm prostheses who had lost their arms in an accident. The subjects were recruited from the WOP and from a hospital which also had a large group of patients with an amputation. This study gave much insight in the factors which determined the usefulness of a prosthesis for arm amputees and also on the way they coped with their handicap (see Chapter 7).

Treatment of patients with a paralyzed arm due to a plexus lesion, that is, an injury to the nerve bundle between the spinal cord and arm, depends on the nature of the damage (see Chapter 9). If the sheath of the damaged nerve is still connected to the spinal cord or the connection has been restored surgically using a nerve graft, the nerve cells can grow back to their actuator or sensor at a speed of 1 mm/day. Hence, after a certain length of time, function will return. When the nerve is pulled out of the spinal cord, no restoration of function of the corresponding muscles is possible. If the muscles that keep the arm in the shoulder joint are involved, the patient loses control over the arm. In addition, the soft tissues including the blood vessels are overstretched and the shoulder is very painful.

A possible solution is to fuse the upper arm bone to the shoulder blade, an operation known as a shoulder arthrodesis. By using the muscles between the trunk and the shoulder blade, the patient again has control over his or her upper arm. By adding an elbow orthosis, the patient also can fix the lower arm in two positions, fully stretched or in a 90-degree flexed position. In this way, the pain is relieved, the blood supply is restored, and the patient again has some arm function. A disadvantage is that the operation is irreversible. This is a difficult decision for the patient, especially in the case in which a reliable diagnosis is not available. Furthermore, the amount of function restoration will depend on the fixation angles applied in the operation.

This problem led to a question about the optimal fixation angles, which in 1979 initiated a project on modelling the shoulder mechanism. One of the byproducts of this project was the development of a special external fixator for performing a shoulder arthrodesis. In the meantime, modelling activities have led to a number of other applications, for example, in the fields of orthopedics and ergonomics, resulting in establishment of the Dutch Shoulder Group, a cooperation among the DUT, the Department of Orthopaedics of Leiden University, and the Department of Human Movement Studies of the Free University of Amsterdam (see Chapter 4).

The evaluation of arm orthoses and the problem of timing the fixation of the shoulder was provided another impetus to start a field study with this group of WOP patients. As a first step, the files of the patients who were seen in the period 1967-1980 were studied with the idea that it would also be useful to create a database for this group like that created to describe the rehabilitation process for quadriplegics. Based on this study, an analysis could be made of the rehabilitation process of this category of patients. One of the problems encountered in this study was that in most cases the diagnosis with respect to the location and nature of the damage was not clearly known. This was also the reason why a patient often had to wait a long time to learn whether some muscle function would return or not, before the possibility of performing a shoulder arthrodesis could be considered. Furthermore, for some types of damage, neurosurgical treatment executed in an early stage could provide some restoration of function. These points motivated the investigation of the usefulness of establishing a medical decision support system. Due to the cooperation of two neurosurgeons who were specialists in the treatment of plexus lesions, such a system has been created and tested in an international evaluation study (see Chapter 5).

The introduction of inexpensive microprocessors led in 1980 to investigation of the question posed by one of the speech therapists at De Hoogstraat about developing possible training devices for aphasia patients. Aphasia is a language disorder caused by brain damage, in many cases resulting from a cerebro-vascular accident (CVA), an occlusion, or rupture of a blood vessel. These language disorders vary in gradation and nature, from not being able to recognize words to not being able to find the right words. Therapy aims at restoring communication abilities by using as many of the remaining communication capabilities as possible. The project resulted in the development of a system in which a speech therapist can generate a number of exercises that the patient can use on a personal computer (PC) at home. Results of the exercises are recorded so that the therapist can adjust the degree of difficulty of the next set of exercises to the level of the patient (see Chapter 11).

At the Delft University of Technology, the field of medical applications was not the exclusive domain of the MMS Group of the Department of Mechanical Engineering. Groups in several other departments were active, and there were a number of informal contacts among these groups. These contacts were formalized in 1971 with the foundation of the Center for Medical Technology. The main purpose of this center was to function as a communication channel, on the one hand for the medical world who sought solutions for their technical problems, and to provide contacts within the university. Henk Stassen was one of the founders of this center and became one of the members of its board.

This board represented the university in more formal contacts with medical departments of other universities. An important result of these contacts was that a number of universities allowed one of their professors in the field of medical physics to spend one day per week teaching and doing research at Delft, and the DUT provided the facilities. Over the years this collaboration was created in four departments; Electrical Engineering, Physics, Chemistry, and Mechanical Engineering. It began in the Department of Mechanical Engineering in 1976, when John Laird was appointed by Leiden University to work in the field of heart physiology. This was the beginning of work on the research theme of control of the blood supply to heart muscle. When John Laird left Leiden University in 1985, the medical activities of this research were shifted to the University of Amsterdam. There his collaborator Jos Spaan was appointed using the same model of cooperation with the DUT. This fruitful interaction led to the unfolding of increasing insight into the control mechanisms for oxygen supply to heart muscle. Besides providing a model of blood supply tissue on the capillary level, a model was also made on a macro level of blood supply to the heart muscle as a function of both blood pressure and oxygen consumption. In another study of this group, the influence of heart contraction on blood supply was also investigated (see Chapter 6).

The development of systems to measure and process physiological data combined with new imaging techniques has led to new diagnostic methods that make extensive use of physiological data. The technical staff of the Academic Hospital of the University of Amsterdam have played an important role in these developments. To encourage further developments, an extension of the collaboration with the MMS-Group of the DUT was thought to be a worthwhile investment. This led to the appointment in 1994 of Kees Grimbergen as a professor at Delft for one day per week, also financed by the University of Amsterdam. In addition to teaching measurement techniques, Professor Grimbergen's activities are focused on aiding the human operator in performing minimally invasive surgery by improving displays and controls. This project is being undertaken in cooperation with the Robotics Group of the Control Engineering Laboratory of the Department of Electrical Engineering (see Chapter 18).

3   Manual and supervisory control projects

The study of manual control originated during the Second World War. Towards the end of the 1950's, the development of advanced aircraft that posed higher demands on human pilots gave rise to increased attention on the part of investigators to the human operator as a controller, culminating in the famous crossover model (McRuer et al., 1965; McRuer & Jex, 1967), followed by the optimal control model of Kleinman, Baron, and Levison (1969, 1970a, b). During the 1970's, increasing automation in industry shifted attention to the role of the human operator as a supervisor of complex systems. Especially after a number of industrial accidents at the end of the 1970's, the importance of a well-designed human-machine interface was generally acknowledged.

As already mentioned, modelling human control behavior was a topic of study from the beginning of the MMS Group. This work started in the late 1960's with modelling the rider of a bicycle simulator using a describing function model. Based on that experience, the next step was to model the helmsman of a large ship, a study executed in the period 1971-1976. Two models were identified, a nonlinear model with a decision element that generated discrete rudder actions and a describing function model that generated a continuous approximation of the helmsman's output in accordance with predictions made by the crossover model. Nine years later the subject again attracted interest when increasing traffic demands of sea vessels led to the question of whether it was possible to predict if a given type of ship could safely maneuver in narrow waters along a given trajectory under all weather conditions. A solution to this question can be found by executing a great number of trials in a ship simulator. However, this is an expensive and time-consuming approach. A reliable model that includes the ship, helmsman, and navigator can predict the dynamic behavior of the ship, and hence can be used for assessing safety limits and for designing new harbors. This resulted in a project that was successfully completed in 1995 with two models, one based on fuzzy sets and the other on optimal control theory, (see Chapter 15).

Attention shifted from manual control tasks to supervisory control. Manual control tasks came into focus again from a different viewpoint, that of the context of prosthesis control. Here, the more fundamental question of what is known about the role of proprioception in human movement control in step tracking tasks was raised. For this purpose, a test setup was built consisting of a force-controlled hydraulic manipulator, with adjustable dynamics, which could be used with or without visual feedback. The project still continues and presently forms part of the shoulder research project (see Chapter 4).

The availability of this setup led to a new field of interest when Fokker Space and Systems, the main contractor for the European robot arm, became interested in studying the effects of a visual delay on a single joint control task. A small pilot study was executed, and this finally led to a project to design and evaluate human-machine interfaces for remotely controlled space manipulators (see Chapter 16). In the meantime, increasing attention was being paid to manipulation tasks in not easily attainable or hazardous environments. Different aspects of this are being studied in other departments of DUT, including Electrical Engineering and Industrial Design. When the aforementioned project on minimal invasive surgery was started, it formed the basis for a more regular contact among a number of these groups. The contribution of one of these groups from the Department of Industrial Design in the area of visualization in remote manipulation is given in Chapter 17.

Work in the field of supervisory control began in the 1970's with an attempt to model human supervisory behavior by means of a model based on optimal control theory. In the same period, the MMS Group participated in an evaluation study conducted by Hoogovens (the Dutch steel company) and the British Steel Corporation on the human-machine interaction in Hot Strip Mill 2. The objective of the study was to provide design specifications for future automation systems in rolling mills. Applied methods were task analyses, interviews, and observations. Another study was undertaken in cooperation with the TNO Laboratory for Ergonomics Psychology. This involved analysis of air traffic controllers. The methods used were task analysis and mental load measurements. An interesting bias in the measurements was discovered when Henk Stassen, who had been an air traffic controller during his own military service, discovered during a site visit that a difficult situation was sometimes deliberately created. This was due to air traffic controllers wanting to be paid at a higher salary scale and hoping that the study would show their desire to be justified by the demands of their work.

Serious industrial accidents or near accidents like the one at Three Mile Island brought a greater awareness of the importance of topics such as fault management, human reliability, task allocation between human operator and computer, human-machine interface design, and human operator support systems, all items that have been or are part of the research program of the Man-Machine Systems Group (see Chapters 20 and 23).

4   Connections between the two application fields

The previous overview shows that the two application fields, roughly denoted as medical and industrial, in which the Man-Machine Systems Group started working at the end of the 1960s, still exist. When viewed superficially, these application fields do not seem to have much in common, although there are certain connections, as shown in Figure 2.

Figure 2 shows all the research projects discussed over the history of the MMS Group and their interrelations. Considered from the methodological point of view, however, the fields have much in common, as is illustrated by the following examples of the research activities:

- Mechanical systems are considered in both application fields. To control a space manipulator, a model has to be made of its kinematics and dynamics in order to provide a realistic simulation for the human operator. To make a model of the shoulder mechanism for diagnostic and ergonomic simulations, similar mathematics are required, although the shoulder model is much more complex due to the facts that joints in general have more than one degree of freedom and that the system has a large number of actuators, some of which influence motions in more than one joint.

- Optimal control models are also applied in both fields, for example, in the navigator model and in the shoulder mechanism model.

- Modelling and identification of process dynamics is not unique to industrial systems. It is an important item in the study of the rehabilitation process of quadriplegics, and it is central in the study of control of oxygen supply to the heart muscle.

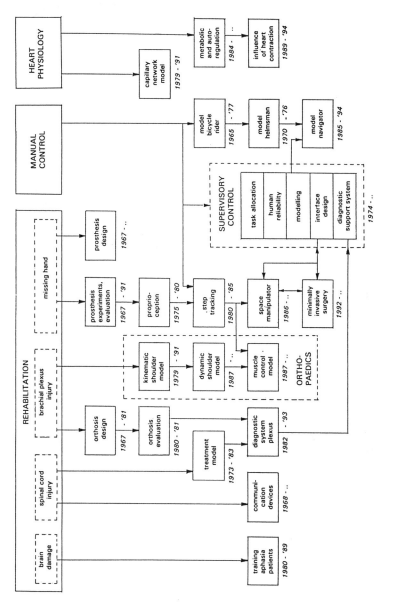

Figure 2: An overview of the research projects of the Delft Man-Machine Systems Group in time over the past 30 years. The origins and relations between the projects are indicated by arrows.

Mental load studies were executed to provide an indication of the workload of air traffic controllers and were also applied to compare different prosthesis types in a number of control tasks. The experience with interviews and data processing obtained in the study of the ergonomics of Hot Strip Mill 2 could be applied fruitfully in the evaluation study involving wearers of arm prostheses.

- The subject of human-machine interface design played a role not only in a comparison of different types of information presentation to the operator in an industrial control room and in different display-control configurations of a space manipulator, but also in the design of communication aids for severely motorically impaired users and the development of a therapy system for aphasia patients.

- The design of an alarm-monitoring system to detect and suppress false alarms does not differ much from that for either an industrial control room or a hospital intensive care unit.

- Human operator support systems to perform diagnostics were designed for both medical applications, the PLEXUS system, and for simplified simulation of a nuclear power plant.

5   Underlying philosophy

In all research projects executed over the past 30 years, the practical applicability of the results was always an important criterion, in addition to scientific value. In most cases, a project started with a practical question, which led to an assignment for an M.Sc. student, which sometimes was continued as a Ph.D. project. A working prototype resulted from a number of cases involving design activities. However, there still is a long way to go to arrive at a product that can be commercialized. Especially in the field of rehabilitation, the market is rather small, and producers have no interest in investing a lot of money and effort in the development of a product that probably will be better than the existing alternatives but is not very attractive from a commercial point of view. Henk Stassen, however, considered it an important social function of the university to do this type of research and development that otherwise would not be done. He felt that a good idea should not end with just an interesting publication but should benefit the people for whom it was meant, in this case the handicapped. Moreover, he succeeded in convincing government officials of his viewpoint, which resulted in their offering financial support to employ four conscientious objectors to fulfill their work alternative to military service. So, over a number of years two of these researchers were employed almost continuously in the design group working on arm prostheses and orthoses while two worked in the MMS group on rehabilitation projects. Without this support, products such as comfortable orthoses and prostheses, and LUCY, STAP, and the PLEXUS systems, developed at the DUT would never have been produced. Subsequently, versions of LUCY for a number of other languages, including French, German, and English, have also been made available.

In the industrial applications, much emphasis is placed on international contacts. As a European equivalent of the U.S. Annual Conference on Manual Control, in 1981 the initiative was taken to organize the First European Annual Conference on Human Decision Making and Manual Control, held at Delft. This conference is now an ongoing annual event held at different locations in Europe. Further, a number of related activities have been started on a European scale and have been supported by EEC grants. On a national level, support is obtained from government institutions, especially for safety items. In addition, contacts with industrial partners have led to a number of M.Sc. thesis topics.

As already mentioned, most projects start with an assignment for an M.Sc. student. This means that the problem should pose a scientific challenge for a future engineer.

Henk Stassen had the basic philosophy that we should not educate biomedical engineers, but rather mechanical engineers with a systems engineering background. One reason for this is a practical one, because the number of jobs available for biomedical engineers is very limited. The other, more fundamental reason is that he believed we should not educate engineers for a limited application field. The goal of our study program is not the acquisition of knowledge in a given field, but rather the ability to apply systems engineering methods to solving problems in mostly, but not necessarily, all mechanical engineering applications. This means that students have to learn to acquire knowledge in a formerly unknown problem field and to apply basic engineering methods, such as modelling, control, measurement techniques, data processing, and design.

Students always have to acquire some knowledge about humans in the field of human-machine systems. Depending on the type of application, they have to broaden their knowledge in anatomy, physiology, biomechanics, neurology, and psychology. An important aspect of this work involves using available measurement techniques. Many quantities are difficult to measure, such as the 3D coordinates of a moving shoulder blade, or have a low signal/noise ratio, such as with EEG signals. This requires knowledge of special techniques for measurement and data processing. Validation of hypotheses introduces the need for making models, from which hypotheses can be derived that make predictions of measurable quantities.

The design of aids for the handicapped poses extra problems, especially in the case of body-fixed appliances such as prostheses and orthoses. In this case, the fitting can be considered as a human-machine interface, which on the one hand, has an important function in information exchange, but, on the other, often has a function in the transfer of forces and energy. Furthermore, psychological factors are always involved in developing aids for the handicapped.

Over the course of time, MMS students have also contributed to cooperative projects with other institutions. Examples are model identification studies in physiological control mechanisms, such as the control of eye motions, a research topic of the Medical Department of the Erasmus University at Rotterdam, and the identification of a controller model for the respiratory system, a project of the Medical Department of the Leiden University.

It is our intention to continue this line of research in the future. A number of application fields can be distinguished, such as telemanipulation, biomechanics of the shoulder mechanism for application in orthopedics, design of aids for the handicapped, heart physiology, and human operation in large-scale systems.

References

Kleinman, D.L., Baron, S., and Levison, W. H. (1969). An optimal control model of human behavior. *Proeedings. 5th Annual Conference. on Manual Control.* NASA SP-215, pp. 343-365.

Kleinman, D.L., Baron S., and Levison, W. H. (1970a). An optimal control model of human response. Part I. *Automatica* 5: 357-369.

Kleinman D.L,, Baron S.,and Levison, W. H. (1970b). An optimal control model of human response. Part II. *Automatica* 6: 371-383.

Luitse, W.J. and Stassen, H.G. (1974). A multi-disciplinary team for the treatment of patients with defective upper extremities: a need for optimal rehabilitation. Delft University of Technology, Department of Mechanical Engineering. Report WTHD 67.

McRuer, D.T., Graham, D., Krendel, E. S., and Reisener, W. (1965). Human pilot dynamics in compensatory systems. Report AFFDL-TR-65-15.

McRuer, D.T., and Jex, H.R. (1967). A review of quasi-linear pilot models. *IEEE Transactions on Human Factors in Electronics*, HFE-8(3):231-249.

Stassen, H.G., and Van Lunteren, A. (1969). Annual Report 1969 of the Man-Machine Systems Group. Delft University of Technology, Dept. of Mechanical Engineering. Report WTHD 21.

Stassen, H.G. et al. (1973). Progress Report 1970-1973 of the Man-Machine Systems Group. Delft University of Technology, Dept. of Mechanical Engineering. Report WTHD 55.

Stassen, H.G. (Ed.) (1977). Progress Report 1973-1976 of the Man-Machine Systems Group. Delft University of Technology, Dept. of Mechanical Engineering. Report WTHD 95.

Van Lunteren, A. (Ed.) (1983). Fourth Progress Report of the Man-Machine Systems Group Over the Period 1976-1982. Delft University of Technology, Dept. of Mechanical Engineering. Report WTHD 161.

Chapter 3

Hendrik G. Stassen: Curriculum Vitae

Hendrik G. Stassen was born in Goes, The Netherlands, on September 29, 1935, where he lived for only two years. Then the family moved to Zeist, where he was educated in the public schools. Stassen attended the College for Manufacturing and Production Technology in Utrecht and received his BSc degree in 1961. He was then drafted for military service, which, after a one-year training program at Gilze-Rijen and Breda, he completed at Soesterberg Air Force Base, the U.S. base of "Camp New Amsterdam." During this period Stassen was an air traffic control officer, where he developed a strong interest in human-machine communication and airplanes. Due to his friends in the field, he became more and more interested in medicine, and in particular dental medicine. As a consequence, Stassen worked for some time as a dental technician, manufacturing dental prostheses and inlays. However, his desire to get a better education motivated him to apply to study mechanical engineering at the Delft University of Technology.

Stassen earned his Master's Degree with honor in Mechanical Engineering, with a specialization in Control Engineering, under the supervision of Prof. ir. Roelof G. Boiten in 1964. During this period Boiten taught him the pleasure of doing scientific research and showed him the power of using a multidisciplinary approach in a wide area of application. Boiten then asked Stassen to join the university in order to continue his study of railway dynamics. In 1967 Stassen obtained his Ph.D. degree with honors with his thesis "Random Lateral Motions of Railway Vehicles" which was supervised by Boiten and Prof. dr .ir. Anton de Pater, a specialist in railway dynamics.

During Stassen's Ph.D. research in Europe, a disaster took place. Several thousand babies with severe dysfunctions of the extremities were born, the birth defects caused by the use of the sleeping drug thalidomide taken during pregnancy. Boiten was thus inspired to start a biomedical research program on externally powered arm prostheses and orthoses, which attracted Stassen's attention. Consequently, directly after finishing his Ph.D. work, Stassen joined Prof. John Lyman's Biotechnology Laboratory at the University of California, Los Angeles as a Fullbright doctoral scholar. In 1967 he became Associate Professor in Control Engineering at the Department of Mechanical Engineering of the Delft University of Technology, where he taught systems theory, signal processing, and stochastic processes, as well as rehabilitation engineering.

As a member of the faculty committee on education, Stassen was heavily involved in the development of a new bachelor's degree program in mechanical engineering. The major shift to be made was the introduction of a series of courses on systems and control theory. In that period he also started a research group on human-machine systems. In 1976 he served as a visiting faculty member at the Massachusetts Institute of Technology with Tom Sheridan's Man-Machine Systems Laboratory. In 1977 he became a Delft University professor in human-machine systems, and started a masters program in this field. Henk started courses on mathematical models of human-machine systems, rehabilitation engineering, ergonomics in design, and control engineering. He also taught several courses on system ergonomics at the colleges of occupational therapy. In 1987 he served as a visiting faculty member at Stanford and the Veterans Administration in Palo Alto.

Stassen is the head of the Man-Machine Systems Group of the Delft University of Technology. His research started in the field of manual control, ergonomic aspects of assistive devices such as arm prostheses, orthoses, and communication aids for the disabled, and modeling the entire multidisciplinary rehabilitation treatment process. In the early years of the research group, large projects on the treatment of quadriplegics and the training of aphasia patients were carried out. However, gradually the research program turned from rehabilitation to orthopedics, and from manual to supervisory control, including human-machine interfaces for teleoperation and human reliability studies in nuclear power systems.

Around 1980 simulation studies on coronary circulation were started in close cooperation with Prof. Dr. ir. Jos A.E. Spaan of the Medical School of the University of Amsterdam. In addition, minimal invasive surgery became a topic of interest. The research is presently embodied in the research schools Dutch Institute of System and Control and Institute of Bio-Medical Engineering, both research schools acknowledged by the Royal Netherlands Academy of Science and Arts. In particular, the research activities related to the masters degree program, and later on the PhD program, pleased Stassen the most because he felt it always a challenge and stimulation to work with students, who often have very original ideas and positive attitudes. Since 1968 more than 220 students have obtained their MSc. degree in Mechanical Engineering, and 30 students have obtained their Doctorate degrees.

Stassen has held many administrative positions within the university. For more than 10 years he was a member of the Board of the Centre for Medical Technology, chairman and member of the Committee of Research of the Department, and vice-dean of the General Board of the Faculty of Mechanical Engineering and Marine Technology, where he also served for a period of 4 years as dean. Consequently he was a member of the Board of Deans of the university. For a relatively short period, Stassen was a member of the Departmental Educational Committee for the master degree program. Now, he is a member of the Committee on Ethics and Technology and the Center for International Cooperation and Appropriate Technology. Finally, he served for a long period as Director of the Laboratory for Measurement and Control.

Stassen has contributed to many activities outside the university community. Many of these activities are related to his special interest, the rehabilitation of severely motor disabled persons. For 4 years he was on the Scientific Board of the Foundation of Biophysics of the Organization of Pure Scientific Research. He was chairman and member of the Scientific Board of the Institute of Rehabilitation Research, and is chairman of the Foundation for the Development of Communication Aids for Rehabilitation. Stassen was a member of the Program Committee on Assistive Devices for the Handicapped of the Ministry of Education and Sciences, and of Economic Affairs, and he is active in the General Board of the Foundation on Handicap and Study. He was one of the founders of the Dutch working group on the treatment of quadriplegics.

Besides his interest in medical research, he also showed interest in the application of ergonomics to industry. As such, he is a member of the Royal Institute of Engineers, where he served for 4 years as a member of the board of the section on control engineering. He served as a member of the Nuclear Safety Council and the Railway Safety Council. In 1970 Stassen was involved in the founding of the Netherlands Technology Foundation. He is a member of the Netherlands Academy for Technology and Sciences, and was also appointed as a member of the Royal Netherlands Academy of Sciences and Arts, in both the physics and in medicine sections. Finally, Stassen is a member of the Board of Governors of the Royal Navy Institute. He is a senior member of the IEEE and actively works for the International Federation of Automatic Control. For many years he was member and chairman of the Technical Committee T6 on Man-Machine Systems.

Together with Prof. dr. dipl..ing. Gunnar Johannsen, Stassen started the European Annual Conference on Human Decision Making and Manual Control. He organized several conferences in the field of human-machine systems. In 1995 he obtained an Honorary Doctor's Degree of the University of Craiova, Romania, for to his contribution of the integration of medicine and technology. Stassen served on the editorial boards of *Human Movement Science*, the *Journal of Rehabilitation Sciences* and the *International Journal of Information, Education and Research in Robotics and Artificial Intelligence (Robotica)*. He is a regular member of International Program Committees of the IFAC Symposium on Analyses Design and Evaluation of Man-Machine Systems, of the Symposium on Cognitive Science Approaches to Process Control and the Man-Machine session of the IFAC world congresses.

Stassen is listed in *Who's Who in America*. In two different areas he has acted as a consultant. For many years he advised the three rehabilitation teams for the treatment of quadriplegics, of adults using arm prostheses and orthoses, and of children with assistive devices. He was also a consultant of the medical director of the rehabilitation centre "De Hoogstraat" for the introduction of multidisciplinary treatment teams. For many years he advised the Association of Netherlands Rehabilitation Centres on the introduction of a computer-aided treatment support system. For a couple of years he was a consultant to two industrial organizations, the International System Development and Support Organization, involved in the development of an industrial tunnel building method, and the Netherlands Institute for Naval Research and Development.

Stassen was married in 1964 to the former Maria Wilhelmina Petronella van Zutphen, and they have three children. It is because of her, he claims, that he has been able to have such a beautiful time with the students at the Delft University of Technology. In fact, he still describes his professional work as a well-paid hobby, as the best work environment he can imagine.

Publications

Books

Stassen, H.G. and Thijs W.L.T. (Eds.) (1981). *Proceedings of the 1st European Annual Conf. on Human Decision Making and Manual Control.* Delft, DUT.

Stassen, H.G. (Ed.) (1993). *Proceedings of the 5th IFAC/IFIP/IFORS/IEA Symposium on Analysis, Design and Evaluation of Man-Machine Systems.* IFAC Symposia Series N.5.

Stassen, H.G. and Wieringa, P.A. (Eds) (1995). *Proceedings of the 14th European Annual Conf. on Human Decision Making and Manual Control.* Delft, DUT.

Chapters in books

Soede, M., Stassen, H.G., van Lunteren, A. and Luitse, W.J. (1974). A lightspot operated typewriter (LOT). In K. Copeland. *Aids for the severely handicapped.* London: Sector Publications, pp.42-54.

Luitse, W. J. and Stassen, H.G. (1975). A multidisciplinary team for the treatment of patients with defect upper extermities: A need for optimal rehabilitation. In R.M. Pickett and T. J. Triggs. *Human Factors in Health Care.* Lext. Books, pp. 257-271.

Stassen, H.G. (1975). Man-machine systems: Manual and supervisory control (in Dutch). In: J.F. Michon. *Handbook for Psychonomy.* Zwolle; Kluwer, pp. 585- 607.

Stassen, H.G. and Levelt, W.J.M. (1975). Systems, automata and grammars. (In Dutch). Chapter 4. In J.F. Michon. *Handbook for Psychonomy*, Zwolle, Kluwer, 100-126.

Stassen, H.G. (1976). Man as a controller. In K.F. Kraiss and J. Moraal. *Introduction to Human Engineering.* Köln (FRG); Verlag TUV Rheinland GmbH, pp. 61-84.

Sheridan, T.B. and Stassen, H.G. (1979). Definitions, models and measures of human workload. In N. Moray. *Mental Workload: Its Theory and Measurement.* NATO-SAD, New York, Plenum Press, 219-234.

Stassen, H.G., Curry, R., Jex, H. and Levison, W. (1979). Final Report of Control Engineering Group on Mental Workload. In: N. Moray. *Mental Workload: Its Theory and Measurement.* NATO-SAD, New York; Plenum Press, pp. 235-252.

Stassen, H.G. (1981). Review and discussion overview on the session "System design and operator support". In J. Rasmussen and W.B. Rouse. *Human Detection and Diagnosis of System Failures.* NATO Conference Series, New York Plenum Press, pp. 339-353.

Wieringa, P.A., Stassen, H.G., Laird, J.D. and Spaan, J.A.E. (1985). Heterogeneous PO_2 distribution as a consequence of the capillary network. In F. Kreuzer et al. *Oxygen Transport to Tissue* VII. Plenum, 183-192.

Stassen, H.G. (1985). Decision demands and task requirements in work environments. What can be learned from human operator modelling? In E. Hollnagel, G. Mancini and D.D. Woods, *Intelligent Decision Support in Process Environments.* Berlin: Springer Verlag, NATO Scientific Affairs Division, San Miniato (Pisa), pp. 293-306.

Spaan, J.A.E., Bruinsma, P., Vergroesen, I., Dankelman, J., and Stassen, H.G. (1987). Distensibility of microvasculature and its consequences on coronary arterial and venous flow. In S. Sideman and R. Beyar. *Activation, metabolism and Perfusion of the Heart.* Nijhoff, pp. 389- 407.

Spaan, J.A.E., Vergroesen, I., Dankelman, J. and Stassen, H.G. (1987). Local control of coronary flow. In: J.A.E. Spaan, A.V.G. Bruschke and A.C. Gitterberg-de Groot. *Coronary circulation, from basic mechanisms to clinical implications.* Dordrecht, Martinus Nijhof, 45-58.

Salski, A., Noback, H. and Stassen, H.G. (1988). A model of the navigator's behaviour based on fuzzy set theory. In: J. Patrick and K.D. Duncan. *Training, Human Decision Making and Control,* Elsevier Science Publications, 205-222.

Stassen, H.G. (1988). On the modeling of manual control tasks. In: G.R. McMillan, D. Beevis, E. Salas, M.H. Strub, R. Sutton and L. van Breda. *Applications of Human Performance Models to System Design.* N.Y., Plenum Press, NATO Defence Research Series Vol.2, Orlando (FL), USA 107-122.

Stassen, H.G. (1989). The Rehabilitation of Severely Disabled Persons. A Man-Machine System Approach. In: W.B. Rouse (Ed.). *Advances in Man-Machine Systems Research,* Vol.5, JAI Press Inc., ISBN 1-55938-011, 153-227.

Dankelman, J., Stassen, H.G. and Spaan J.A.E. (1990). Coronary Circulation Mechanics. In: F. Kajiva, G.A. Klassen, J.A.E. Spaan and J.I.E. Hoffman (Eds.) *Coronary Circulation, Basic Mechanism and Clinical Relevance.* Berlin, FRG, Springer Verlag, 75-87.

Stassen, H.G. (1990). Supervisory control behavior modeling. The challenge and necessity. In: N. Moray, W.R. Ferrell and W.B. Rouse (Eds). *Robotics, Control and Society.* Taylor and Francis, NY, ISBN 0-85066-850-6, 105-117.

Stassen, H.G. (1992). Hoe complex is een industrieel proces voor een procesoperator. In: M.J.A. Alkemade. *Inspelen op complexiteit.* Samson Bedrijfsinformatie BV, Alphen aan de Rijn, ISBN 90-14-03883-6, 183-194.

Wieringa, P.A. and Stassen, H.G. (1993). Assessment of Complexity. In: J.A.Wise, V.D.Hopkin and P. Stager (Eds). *Verification and Validation of Complex Systems: Human Factors Issues.* NATO ASI Series F, Computers and Systems Sciences, Vol.110, ISBN 3-540-56574-4, 173-180.

Papers in refereed journals

Bommel, P. van and Stassen, H.G. (1965). Détermination de quelques caracteristiques stochastiques des déviations des files de rails, en vue de l'étude du mouvement de lacet, *Revue Française de Mécanique* 13, 65-69.

Stassen, H.G. (1969). The polarity coincidence correlation technique: A useful tool in the analysis of human-operator dynamics. *IEEE Transactions of Man-Machine Systems.* MMS-10(1), 34-39.

Bossers, G.Th.M., Stassen, H.G. and Verkuyl, A. (1970). Pneumatically powered arm prosthesis and orthosis (In Dutch). *Nederlands Tijdschrift voor Geneeskunde.* 114(47), 1954-1958.

Soede, M., Coeterier, J.F. and Stassen, H.G. (1971). Time analyses of the task of approach controllers in Air Traffic Control. *Ergonomics* 14(5), 591-601.

Stassen, H.G. (1972). Cybernetic aspects of prosthesis control (In Dutch). *Nederlands Tijdschrift voor Geneeskunde* 116(31), 1369-1370.

Soede, M. and Stassen, H.G. (1973). A Lightspot Operated Typewriter for severely disabled patients. Technical Note: Medical and Biological Engineering 11(9), 641-644.

Soede, M., Stassen, H.G., Lunteren, A. van and Luitse, W.J. (1973). A Lightspot Operated Typewriter for physically handicapped patients. *Ergonomics* 16(6), 829-844.

Veldhuyzen, W. and Stassen, H.G. (1977). The Internal Model concept: An application to modelling human control of large ships. 19(4), 367-380.

Stassen, H.G. and Schneider, H.W. (1978). Human manual and supervisory control models. Journal A 19(3), 160-172.

Beaverstock, M.C., Stassen, H.G. and Schneider, H.W. (1979). Modelling the operator: A needed approach to interface design. *Instrumentation Technology* 26(4), 49-51.

Kok, J.J. and Stassen, H.G. (1980). Human operator control of slowly responding systems: supervisory control. *Journal of Cybernetics and Information Sciences.* Special Issue on Man-Machine Systems 3(1-4), 123-174.

Wieringa, P.A., Spaan, J.A.E., Stassen, H.G.; and Laird, J.D. (1982). Heterogeneous flow distribution in a three dimensional network simulation of the myocardial microcirculation: A hypothesis. *Microcirculation* 2(2), 195-216.

Stassen, H.G., Soede, M. and Bakker, H. (1982). The Lightspot Operated Typewriter: A communication aid for severely bodily handicapped patients. *Scandinavian Journal of Rehabilitation Medicine* 14, 159-161.

Lunteren, A. van, Lunteren-Gerritsen, G.H.M. van, Stassen, H.G. and Zuithoff, M.J. (1983). A field evaluation of arm prostheses for unilateral amputees. *Prosthetics and Orthotics* 7(3), 141-151.

Stassen, H.G. (1987). Human Supervisor Modelling: Some New Developments. *Int. Journal of Man-Machine Studies,* 27, pp. 613-618.

Wieringa, P.A., Stassen, H.G., Laird, J.D. and Spaan, J.A.E. (1988). Quantification of arteriolar density and embolization by microspheres in rat myocardium. *American Journal of Physiology,* 254, Heart Circ. Physiol. 23, H636-H650.

Dankelman, J., Spaan, J.A.E., Stassen, H.G. and Vergroesen, I. (1989). Dynamics of Coronary adjustment to a Change in Heart Rate in the Anaesthetized Goat. *J. Physiology,* 408, 295-312.

Dankelman, J., Stassen, H.G. and Spaan, J.A.E. (1990). System analysis of the dynamic response of the coronary circulation to a sudden change in heart rate. *Medical and Biological Engineering and Computing* 28, 139-148.

Stassen, H.G., Johannsen, G. and Moray, N. (1990). Internal Representation, Internal Model, Human Performance Models and Mental Load . *Automatica* 26 (4), 811-820.

Heslinga, G., Stassen, H.G. and Spaan, J.A.E. (1991). Evaluation of the control function of interstitial osmolarity in coronary autoregulation. *Medical and Biological Engineering and Computing* 29, 212-216.

Heslinga, G. and Stassen, H.G. (1992). The Prediction of Human Performance Safety with Event Trees. *IEEE Transactions on Systems Man and Cybernetics* 22 (5), 1178-1182.

Stassen, H.G., Goezinne, J. and Visse, B. (1993). Communication Aids for Children and Adults. *Journal of Rehabilitation Sciences* 6 (1), 20-24.

Daalen C.van, Stassen, H.G., Thomeer, R.T.W.M and Slooff, A.C.J. (1993). Computer assisted Diagnosis and treatment planning of Brachial Plexus Injuries. *Clinical Neurology and Neurosurgery* 95, 550-555.

Wieringa, P.A., Stassen, H.G., Kan, J.J.I.M. van and Spaan, J.A.E. (1993). Oxygen Diffusion in a Network Model of the Myocardial Micro circulation. *Int.J. of Microcirculation: Clinical and Experimental* 13, 137-169.

Johannsen, G., Levis, A.H. and Stassen, H.G. (1994). Theoretical problems in Man-Machine Systems and their Experimental Validation. *Automatica* 30 (2), 217-231.

Vermeer, B. and Stassen, H.G. (1994). De mens komt er vaak bekaaid van af. *CA Techniek*, tijdschrift voor industriële automatisering 13 (6), 22-24.

Ploeg C.P.B. van der, Dankelman, J, Stassen, H.G. and Spaan, J.A.E. (1995). Comparison of different oxygen exchange models. *Medical and Biological Engineering and Computing* 33, 661-668.

Bos J.F.T., Stassen H.G. and Van Lunteren, A. (1995). Aiding the operator in the manual control of a space manipulator. *Control Engineering Practice* 3(2):223-230.

Conference proceedings

Lunteren, A. van and Stassen, H.G. (1967). Investigations on the characteristics of a human operator stabilising a bicycle model. *Proc. Int. Symp. on Ergonomics in Machine Design,* Prague, pp. 349-370.

Stassen, H.G. (1967). On the interaction between track and railway vehicle, in particular with respect to the hunting problem. *Proc. 4th Conf. on Non-Linear Oscillations,* Prague, pp. 481-493.

Stassen, H.G. (1969). Analyse statistique de la dynamique du bogie ferroviaire. *Proc. 5th Int. Coll. on Railway Vehicle Technology,* Vienna, pp. 1-33.

Lunteren, A. van and Stassen, H.G. (1969). On-line parameter estimation of the human transfer function in a man-bicycle system. *Proc. 4th IFAC Congress, Warszawa,* Session 70, pp. 41-53.

Aldrich, J.W., Lyman, J. and Stassen, H.G. (1969). A formal model for arm motion during target approach. *Proc. 5th Annual NASA-University Conf. on Manual Control,* Cambridge (Mass), pp. 581-608.

Lunteren, A. van, Stassen, H.G. and Schlemper, M.S.H. (1970). On the influence of drugs on the behavior of a bicycle rider. *Proc. 6th Annual Conf. on Manual Control,* Wright-Patterson AFB, Dayton (Ohio), pp. 419-437.

Lunteren, A. van and Stassen, H.G. (1970). On the variance of the bicycle rider's behavior. *Proc. 6th Annual Conf. on Manual Control,* Wright-Patterson AFB, Dayton (Ohio), pp. 701-722.

Stassen, H.G., Meyer, A.W.A. and Lunteren, A. van. (1970). On the possibilities of tactile information transmission for the use in arm prostheses. *Proc. 6th Annual Conf. on Manual Control,* Wright-Patterson AFB, Dayton (Ohio), pp. 513-534.

Veldhuyzen, W., Lunteren, A. van and Stassen, H.G. (1972). Modelling the helmsman of a supertanker: Some preliminary experiments. *Proc. 8th Annual Conf. on Manual Control,* Ann Arbor (Mich), pp. 485-501.

Soede, M. and Stassen, H.G. (1972). Lightspot Operated Typewriter for quadriplegic patients. *Proc. 8th Annual Conf. on Manual Control,* Ann Arbor (Mich), pp. 89-100.

Veldhuyzen, W. and Stassen, H.G. (1973). Modelling the behavior of the helmsman steering a ship. *Proc. 9th Annual Conf. on Manual Control,* Cambridge (Mass), pp. 639-658.

Lunteren, A. van and Stassen, H.G. (1973). Parameter estimation in linear models of the human operator in a closed loop with application of deterministic test signals. *Proc. 9th Annual Conf. on Manual Control,* Cambridge (Mass), pp. 503-520.

Veldhuyzen, W. and Stassen, H.G. (1975). Simulation of ship manoeuvring under human control. *Proc. 4th Symp. on Ship Manoeuvrability,* The Hague, KIM, Vol.6, pp. 146-163.

Stassen, H.G., Dieten, J.S.M.J. van and Soede, M. (1975). On the mental load in relation to the acceptance of arm prostheses. *Proc. 6th IFAC-Congress,* Cambridge MA, MIT, Session 40, 10 p.

Veldhuyzen, W. and Stassen, H.G. (1976). The Internal Model: What does it mean in human control. Preprints Int. NATO-Symposium on Monitoring Behavior and Supervisory Control, Berchtesgaden, pp. 109-123.

Beaverstock, M.C., Stassen, H.G. and Williams, R.A. (1977). Interface design in the process industries. *Proc. 13th Annual Conference on Manual Control,* Cambridge (Mass), MIT, pp. 258-265.

Beaverstock, M.C., Stassen, H.G. and Schneider, H.W. (1978). Control loops that include the operator - A new approach to interface design. *Proc. National ISA-78 Conference and Exhibit,* Philadelphia, pp. 373-378.

Stassen, H.G. (1979). Systems theory, and its relevance to language. *Proc. Workshop on Empirical Consequences of Learnability Theory with Special Relevance to Language,* Association Europeenne de Psycholinguistique, Paris, 21 p.

Stassen, H.G., Lunteren, A. van, Hoogendoorn, R., Kolk, G.J. van der, Balk, P., Morsink, G. and Schuurman, J.C. (1980). A computer model as an aid in the treatment of patients with injuries of the spinal cord. *Proc. of the ICCS,* Cambridge, Mass., IEEE, pp. 385-390.

Stassen, H.G., Lunteren, A. van, Hoogendoorn, R., Kolk, G.J. van der, Balk, P., Morsink, G. and Schuurman, J.C. (1981).: Computer aided treatment of patients with injuries of the spinal cord. *Proc. 1st European Annual Conference on Human Decision Making and Manual Control,* Delft, DUT, pp. 376-389.

Schneider, H.W., Veldt, R.J. van der and Stassen, H.G. (1982). The role of overview displays in human supervisory control. *Proc. 2nd European Annual Conf. on Human Decision Making and Manual Control,* Bonn, FAT, pp. 250-266.

Stassen, H.G. and Veldt, R.J. van der (1982). Human operator models: A useful tool in Man-Machine Interface design? *Proc. Workshop on European Research in Process Control: Human Interactions with Computer Based Systems,* 1982 Annual Human Factors Society Meeting, Seattle, Section 6, pp. 1-27.

Laird, J.D., Wieringa, P.A., Stassen, H.G. and Spaan, J.A.E. (1982). Inhomogenities in myocardial perfusion: A simple consequence of complex vascular micro-anatomy. *Proc. 5th International Conference of the Cardiovascular System Dynamics Society,* Oxford.

Stassen, H.G. (1983). Human operator models: A tool in design and experimentation with supervisory control systems. *Proc. Workshop in Research and Modelling of Supervisory Control Behavior,* Washington/DC (USA), pp. 59-78.

Stassen, H.G. (1984). Man-Machine guidelines: A need in the design of Man-Machine Interfaces. *Proc. European Seminar on Industrial Software Engineering and the EWICS,* Freiburg, pp. 181-188.

Thijs, W.L.Th., Stassen, H.G., Kok, J.J. and Veldt, R.J. van der (1984). Supervisory control and fault management. Voyage of discovery through the estate of theory and model creation. *Proc. 9th IFAC-Congress,* Budapest, Paper 11.4/A-3, pp. 209-215.

Stassen, H.G. (1984). The Ewics Man-Machine guidelines. *Proc. 4th European Annual Conference on Human Decision Making and Manual Control,* Soesterberg, IZF-TNO, pp. 277-284.

Stassen, H.G., Kok, J.J., Veldt, R.J. van der and Heslinga, G. (1985). Modelling human operator performance. Possibilities and limitations. *Proc. 2nd IFAC/IFIP/IFORS/IEA Conf. on Analysis, Design and Evaluation of Man-Machine Systems,* Varese (Italy), ISPRA(CEC)JRC, pp. 101-106.

Stassen, H.G. (1987). Supervisory control and rehabilitation: Are there common interests? In: *Proc. 10th IFAC Congress,* München, Vol. 7, Area 12.5-1, pp. 322-330.

Papenhuijzen, R. and Stassen, H.G. (1987). On the modelling of the behaviour of a navigator. In: *Proc. 8th Ship Control Systems Symp.,* The Hague (Neth.), Vol. 2, pp. 238-255.

Stassen, H.G., Johannsen, G. and Moray, N. (1988). Internal Representation, Internal Model, Human Performance Model and Mental Load. In: *Proc. 3rd IFAC/IFIP/IEA/IFORS Symp. on Analysis, Design and Evaluation of Man-Machine Systems.* Oulu, Univ. of Oulu (Finland), pp. 23-32. ISBN 0-08-036226-5.

Loos, H.F.M. van der, Michalowski, S.J.;,Hammel, J., Leifer, L.J. and Stassen, H.G. (1988). Assessing the Impact of Robotic Devices for the Severely Physically Disabled. *First Int. Workshop on Robotic Applications in Medicine and Health Care,* Ottawa, pp. 1.1-1.6.

Wieringa, P.A., Stassen, H.G. and Spaan, J.A.E. (1989). One-dimensional model to illustrate the importance of intercapillary O2 exchange with capillary flow heterogeneities. In: *Proc. ISOTT 88,* Ottowa, Canada, pp. 311-316.

Heslinga, G. and Stassen, H.G. (1989). The Prediction of Human Performance Safety with Event Trees. In: *Proc. 4th IFAC/IFIP/IEA/IFORS Conf. on Analysis, Design and Evaluation of Man-Machine Systems.* Xi'an, ISBN 0-08-035743-1. pp. 153-158.

Stassen, H.G., Steele, R.D. and Lyman, J. (1989). A man-machine System Approach in the Field of Rehabilitation: A Challenge or a Necessity. In: *Proc. 4th IFAC/IFIP/IEA/IFORS Conf. on Analysis, Design and Evaluation of Man-Machine Systems,* Xi'an, ISBN 0-08-035743-1, pp. 96-104.

Papenhuijzen, R. and Stassen, H.G. (1989). On the modelling of planning and supervisory behaviour of the navigator. An example of the application of linear optimal control theory. In: *Proc. 4th IFAC/IFIP/IEA/IFORS Conf. on Analysis, Design and Evaluation of Man-Machine Systems,* Xi'an, (1989), ISBN 0-08-035743-1, pp.33-38.

Wenneker, B.J.M. and Stassen, H.G. (1989). A Computer aided system for the training of Aphasia patients, STAP. In: *Proc. First European Conference on the Advancement of Rehabilitation Technology,* Maastricht (Neth.) (1990) pp. 210-212.

Goezinne, J., Stassen, H.G. and Visse, B. (1990). LUCY, a Universal Communication Aid. In: *Proc. 1st European Conf. on the Advancement of Rehabilitation Technology,* Maastricht (Neth.), 30-32.

Visse, J.B., Goezinne, J. and Stassen, H.G. (1990). LUCY, a Universal Communication Aid. In: *North Sea Conf. on Biomedical Engineering,* Antwerpen, pp. 416-422.

Papenhuijzen, R. and Stassen, H.G. (1993). Fuzzy Set Theory for Modelling the Navigator's Behaviour. In: *Proc. 5th IFAC/IFIP/IFORS/IEA Symp. on MMS. Analysis, Design, and Evaluation,* The Hague (NL), ISBN 0-08-041900-3, pp.53-59.

Johannsen, G., Levis, A.H. and Stassen, H.G. (1993). Theoretical Problems in MMS and their Experimental Validation. In: *Proc. 5th IFAC/IFIP/IFORS/IEA Symp. on MMS. Analysis, Design, and Evaluation,* The Hague (NL), ISBN 0-08-041900-3, pp. 19-29.

Bos, J.F.T., Stassen, H.G. and Lunteren, A. van. (1993). Aiding the operator in the Manual Control of a Space Manipulator. In: *Proc. 5th IFAC/IFIP/IFORS/IEA Symp. on MMS. Analysis, Design and Evaluation,* The Hague (NL), ISBN 0-08-041900-3, pp. 215-220.

Stassen, H.G. and Wieringa, P.A. On the human perception of complex industrial processes. In: *Proc. 12th IFAC Congress* (1993), Sydney, ISBN 0-08-042215-2, pp. 441-446.

Stassen, H.G. (1994). Measurement and Modelling, an Iterative Approach. In: *Proc. Symp. on Measuring in an Interdisciplinary Research Environment,* Delft, Delftse Universitaire Pers. ISBN 90-407-1085-6, pp. 83-109.

Smets, G.J.F. and Stassen, H.G. (1995). Telepresence and telemanipulation, a survey of current literature. In: Prepr. 6th IFAC/IFIP/IFORS/IEA Symp. on Analysis, Design and Evaluation of Man-Machine Systems. Boston (USA), pp.13-23.

Other major publications, including major reports and popular articles

Stassen, H.G. and Bommel, P. van (1966). Détermination de quelques caracteristiques stochastiques des déviations des files de rails, en vue de l'étude du mouvement de lacet, ORE-report DTJ, Utrecht, 117 p.

Stassen, H.G. (1967). Random lateral motions of railway vehicles. Dr. thesis, Delft, 150 p.

Stassen, H.G. (1969). Mens en Machine (Man and Machine). Inaugural Lecture, DUT, Delft, 19 p. Also published in: Alcuinus, TH-Aken, (1969), pp. 309-314.

Lunteren, A. van and Stassen, H.G. (1969). *Annual Report 1969 of the Man-Machine Systems Group.* Report WTHD-21, DUT, 102 p.

Brands, J.Th., Ellfers, G. and Stassen, H.G.; In 't Veld, J. (1971). Project education: Opinion survey in the Netherlands. NUFFIC Bulletin 15 (2), 3-16.

Stassen, H.G. et al. (1973). Progress Report Jan. 1970 – Jan. 1973 of the Man-Machine Systems Group. Report WTHD-55, DUT, 320 p.

Stassen, H.G., Veldhuyzen, W. and Glansdorp, C.C. (1973). The helmsman's behavior in the control of slowly responding ships (In Dutch). *De Ingenieur* 85 (43), 850-852.

Stassen, H.G., Soede, M. and Luitse, W.J. (1974). The Lightspot Operated Typewriter: The valuation of a prototype. Report WTHD-65, DUT, 19 p.

Soede, M., Dieten, J.S.M.J. van and Stassen, H.G. (1974). On the acceptance, functional gain and mental load in arm prosthesis and orthosis control. Report WTHD-66, DUT, 17 p.

Luitse, W.J. and Stassen, H.G. (1974). A multi-disciplinary team for the treatment of patients with defect upper extremities: a need for optimal rehabilitation. Report WTHD-67, DUT, 17 p.

Stassen, H.G. and Lunteren, A. van (1974). Automatic data processing (In Dutch). *Polytechnisch Tijdschrift 29* (7), 205-218.

Stassen, H.G. (1976). Training programs in Ergonomics at the Universities of Technology in The Netherlands. Preprints NATO-Symposium on University Curricula for Ergonomics/Human Factors Engineering, Berchtesgaden, 9 p.

Stassen, H.G. (1977). *Progress Report January 1973 until July 1976 of the Man-Machine Systems Group.* Report WTHD-95, DUT, 263 p.

Stassen, H.G., Bekey, G.A. and Lyman, J. (1977). Human modelling in man-machine systems. Preprints Conf. on Cybernetic Models of the Human Neuromuscular System, Henniker (USA), 22 p.

Luitse, W.J. and Stassen, H.G. (1977). The treatment of patients with a lesion of the Plexus Cervico Brachialis (In Dutch). *Ned. Tijdschrift voor Arbeids-Ergotherapie* 5 (5), 142-148.

Stassen, H.G. and Eland, A.P. (1977). The Role of the proprioception in manual control. *Annual Report of the ZWO Foundation for Biophysics*, Den Haag, pp. 66-70.

Stassen, H.G. (1979). Assistive devices: A contribution to the rehabilitation of physically disabled persons? (In Dutch). Royal Netherlands Industrial Fair, Utrecht, 12 p.

Stassen, H.G. (1979). Man-machine aspects in the rehabilitation of unilateral arm amputees (In Dutch). *Proc. KIVI-Workshop on Biomedical Engineering,* Enschede, THT, 4 p.

Stassen, H.G. (1981). Man Machine Systems: The future problem in the design of automated systems (In Dutch). *De Veiligheid* 57(9), 391-396.

Stassen, H.G. (1981). Ergonomic disciplines do not play a role in the design process! (In Dutch). *Economisch Dagblad,* Den Haag, 1 oktober 1981, Vol.42, No.10.512, p. 17.

Stassen, H.G. and Lunteren, A. van (1981). System approach of the functional defect (In Dutch). *Boerhaave Course Book on Upper Limb Defects,* Leiden, pp. 263-280.

Lunteren, A. van and Stassen, H.G. (1981). System approach analysis of the upper extremity prosthesis (In Dutch). *Boerhaave Course Book on Upper Limb Defects,* Leiden, pp. 299-311.

Stassen, H.G. and Lunteren, A. van (1981). Evaluation methods (In Dutch). *Boerhaave Course Book on Upper Limb Defects,* Leiden, pp. 379-393.

Bakker, H., Feer, M. van de, Zuithoff, M.J. and Stassen, H.G. (1981). Evaluation of communication aids for severely handicapped persons (In Dutch). *Tijdschrift voor Sociale Geneeskunde*, A. en S.V. 59 (7), 207-212.

Eland, A.P. and Stassen, H.G. (1981). Identification of the adaptive feedback of the human motor system using the response difference method. *Annual Report of the ZWO Foundation for Biophysics* 1980-1981, Den Haag, pp. 162-167.

Lunteren-Gerritsen, G.H.M. van, Zuithoff, M.J., Stassen, H.G. and Lunteren, A. van (1981). Acceptance of arm prostheses: A field study (In Dutch). Report WBMT-MR-N-188, DUT, 293 p.

Soede, M., Hooykamp, P.C. and Stassen, H.G. (1981). Survey on in commercially available communication aids in the Netherlands (In Dutch). *Proc. Conf. on Handicap en Techniek,* Rotterdam, SWOG/EUR, pp. 175-181.

Stassen, H.G. and Soede, M. (1981). The importance of communication for the physically disabled (In Dutch). *Proc. Conf. on Handicap en Techniek,* Rotterdam, SWOG/EUR (1981), pp. 160-165.

Jaspers, R.B.M., Louw, C.J.M. de, Lunteren, A. van and Stassen, H.G. (1982). An evaluation of the rehabilitation process of patients with a lesion of the Plexus Brachialis (In Dutch). *Boerhaave Course Book on Plexus Brachialis Lesions,* Leiden, RUL, pp. 65-89.

Bakker, H. and Stassen, H.G.: Measurement and control in rehabilitation medicine (In Dutch). Proc. NVA, Utrecht (1982), pp. 52-61.

Lunteren, A. van and Stassen, H.G. (1982). *4th Progress report of the Man-Machine Systems Group over the period July 1976 through December 1982.* Report WTHD-161, DUT, 286 p.

Stassen, H.G. (1985). What may be the real contribution of the engineer to the treatment of afasia patients? (In Dutch). In: *Attention for the Disabled Human Being* (In Dutch). Lisse, Swets en Zeitliner, The Netherlands (1985), pp. 137-145. ISBN 90-265-0619-8, pp. 137-145.

Stassen, H.G. and Jaspers, R.B.M. (1985). Modelling and simulation in rehabilitation medicine (In Dutch). *Boerhaave Course on Rehabilitation Medicine,* Rotterdam, pp. 47-51.

Stassen, H.G. and Cool, J.C. (1986). External prostheses and orthoses for the upper extremities (In Dutch). *KIVI-Symposium on the Artificial Human,* Utrecht, pp. 39-49.

Stassen, H.G. (1986). Review on research of communication studies for the disabled (In Dutch). In: J.W. Smeets and J.J. Alexander. *Handicap and Communication.* Delft, DUP, ISBN 90-6275-276-4, pp.19- 26.

Ravenzwaay, E.T. van and Stassen, H.G. (1986). Man-Machine Interface Guidelines: Level 3. Delft, DUT, Fac. ME, EWICS-TC6, 72 p.

Stassen, H.G. (1987). Man as supervisor of complex industrial processes: Positive and negative developments (In Dutch). In: *Proc. KNAW Dept. of Physics* 96 (3), 9 p.

Stassen, H.G. and Papenhuijzen, R. (1990). Comments on the introduction of Symptom Based Procedures: An Ergonomic Investigation. DUT/Min. of Social Affairs, Delft/The Hague, (1990), 20 p.

Stassen, H.G. (1991). Onderzoek voor gehandicapten is een persoonlijke keus. *Assortiments Informatie,* 8 (2), 28-30

Stassen, H.G. (1991). Tom Sheridan Honoured. *IEEE-SMC Newsletter,* pp. 10-11.

Stassen, H.G. (1992). Post-Symposium Report on MMS '92. *Robotica* Vol. 10, Part 6, pp. 577-578.

Stassen, H.G. (1992). Summary on the MMS '92 *IEEE-SMC Newsletter/IFAC Newsletter,* 2 p.

Stassen, H.G., Goezinne, J. and Visse, B. (1993). Communication: A Basic Human Need. Communication Aids for Children and Adults. *Communication Outlook* 15 (1), 5-10.

Stassen, H.G. (1994). Supervision of Industrial and Rehabilitation Processes. Analogies and Differences. *In Proc. KNAW, Dept. of Physics,* 9 p.

Stassen HG, Grimbergen CA (1995). Technische ontwikkelingen in de geneeskunde. Kennis van informatiesystemen voor elke arts noodzakelijk. *Medisch Contact* 50(51/52):1649-1651.

DUT theses supervised by Henk G. Stassen

Master's Theses (most written in Dutch)

Design of a bicycle simulator. A.J.J. Timmermans, 1963

Construction and tune up of the bicycle simulator. F.J. Vergouwen, 1964.

Measurements with a bicycle simulator. J.T.H. Koelink, 1965

Identification of a mathematical model of the rider of a bicycle simulator. A. van Lunteren, 1967.

Human behavior during tactile manual control. A.W.A. Meyer, 1970.

A communication aid for severely physically disabled persons. M. Soede, 1970.

Modeling the rider of a bicycle simulator extended with a course following task. A.R. van Heusden, 1970

The use of deterministic test signals in the identification of human operator describing functions. W.P. de Kraker, 1970.

Proprioceptive feedback. H.J. de Groot, 1970.

A model to describe human decision making during the control of variant systems. P. Lemaire, 1971.

Modeling the control behavior of the helmsman of a supertanker. W. Veldhuyzen, 1971.

Simulation of a state space human controller model. L.M.M. Meyers, 1971.

The control of the grasping function of the WILMER hand prosthesis. G.J. Blaauw, 1971.

Arm motion analysis. H.J. ter Maat, 1971.

The relation between visual input and EEG during manual control tasks. W.P.J. van Berkel, 1971.

Feasibility study of an active thumb pronation in externally powered hand prostheses. G.A.J.M. van Ditzhuizen, 1972.

Hand function analyses, differences between dominant and non-dominant hand. J.J.C. Holthaus, 1972.

Human control behavior with deterministic forcing functions. H.J. Vink, 1972.

Transfer function estimation of the bicycle simulator. J.H.M. van Eijndhoven, 1972.

Identification in a closed loop without external forcing functions. J.N.M. de Jong, 1972.

Proprioceptive control of arm prostheses. J.Th. Mooij, 1972.

Human operator decision models. H.L.C.M. Stapper, 1972.

Human operator control behavior with a three-state display. E. Barents, 1973.

Design of a stimulator for electro-retino-graphy. W.M.C. Aarts, 1973.

Manual control in a dual control task with separate loops. R. van der Ven, 1973.

Manual control in a dual control task with coupled loops. A.E.J. Doyer, 1973.

The optimal control model. R.A. van Wijk, 1973.

The critical instability task. P.M.F. Oomen, 1974.

Design of a muscle stiffness transducer for prosthesis control. T. Gerbranda, 1974.

A helmsman control behavior model. H. van Rooyen, 1974.

On the mental load in arm prosthesis control. J. Vermeulen, 1974.

The optimal control model; the inverse control problem. P.A.H.T. Jongman, 1974.

The use of minicomputers in electro-retino-graphy. C.J. van Niekerk, 1974.

Criteria for the judgement of functional gain of hand prostheses for unilateral amputees. R.A. Pimontel, 1975.

Simulation of the helmsman's control behavior. P.C. van Holten, 1975.

Mental load studies in prothesis control by task analysis methods. M.E. Elias, 1975.

Advanced displays for maneuvring of large ships. H.W.J.M. van Gendt, 1975.

Model studies of the motor unit potential. S.J. Schepel, 1975.

The influence of preview displays on manual control behavior. T. Heyer, 1975.

A mathematical model of the treatment of quadriplegics. R. Hoogendoorn, 1975.

The bias in the identification of the optimal control model. J.H.A. Borgman, 1976.

Transfer functions of the smooth eyeball motions. L. Kretzschmar, 1976.

A prediction model for the treatment of quadriplegics. H.W. Croon, 1976.

Identification of the state variables describing the treatment of quadriplegics. J.A. van den Berg, 1976.

A linear model of the helmsman's control behavior. R.E. Schermerhorn, 1976.

The dynamics of a muscle stiffness transducer placed in the lower arm fitting. E.J. van Dijk, 1976.

Biofeedback for the training of muscle function in rehabilitation. M.L. Kooyman, 1976.

On the modeling of human control and decision behavior in partially automated systems. W. Hupkes, 1976.

An iterative method for parameter estimation in a closed loop system. M. Mellaard 1976.

The realization of a test set-up for research on human control and decision behavior in partially automated systems; verification of the model. W.I.H. Belien, 1977.

The realization of an estimation method for the parameters of the human operator optimal control model. R.A. Viergever, 1977.

Motion accuracy. D.W. Aalbers, 1977.

Analysis of the model outputs of the rehabilitation process of quadriplegics. J. van der Marel, 1977.

Identification of the parameters of the human operator optimal control model. H.W. Schneider, 1977.

Technical evaluation of hand prostheses. W.J. Tebeest, 1977.

The SCIT as a method to estimate human rest capacity during mentally loading tasks. R. Volmar, 1977.

Modeling human control behavior during navigation of a 200.000 TDW tanker. J. Scheepbouwer, 1977.

Modeling and parameter estimation of the rehabilitation process of quadriplegics. T. Verbruggen, 1977.

The stability of the blood flow control system in heart tissue. T. Gerretsen, 1978.

Theory and verification of a human supervisory control model. B. Sastra, 1978.

Software design for a set up to study human supervisory behavior. W.B. Verbeek, 1979.

Digital simulation of a distillation column for human supervisory behavior studies. J.B. de Wit, 1979.

Measurement of mental load during arm prosthesis control, based on task analysis and on information theory. W.A. Goudriaan, 1979.

Application of Fitts law to joystick control. F.G.M. Kooyman, 1979.

The design of human supervisory control behavior experiments. E.P. Tamminga, 1979.

Man-machine communication for a rehabilitation information processing system. J.C. Schuurman, 1980.

Fault detection, diagnosis and correction by human operators in the engine room of a supertanker. J.M.J.M. Eekhout, 1980.

Parameter estimation methods applied in human supervisory control. K. Frieling, 1980.

The distribution of flow through a linear network of channels: An analog of the capillary bed of the heart muscle. P.A. Wieringa, 1980.

A utility plant simulation for human supervisory behavior research. P.P. Gommers, 1980.

A micro processor based trend recording display system. P.A. van Dorp, 1981.

The control of an arm orthosis locking mechanism. P.A. Bootsma 1981.

Motion analysis of the shoulder mechanism. G.M. Pronk, 1981.

Task evaluation of a three-grip hand prosthesis. J. Dekker, 1981.

Software developments for the observer/controller/decision model. R.J. van der Veldt, 1981.

Ergonomical aspects of an alarm system in the process industry. B.J.A. van Rixel, 1981.

Parameter estimation in a model describing human supervisory behavior. G.A. van Oostveen, 1981.

Modeling of the oxygen distribution in a three-dimensional capillary bed. H.J. Jansma, 1982.

Effects of trend information on human supervisory control behavior in slowly responding systems. P.J. Lute, 1982.

Identification of the vestibulo-ocular system. P.G.M. Conijn, 1982.

A validation of prediction methods for duration of the freshman study in the mechanical engineering department. A.A.M. Kop, 1982.

Computer aided training of aphasia patients. J.A.M. Vriend, 1982.

Dual-mode models. R.J. de Groot, 1983.

Simplified training procedures for complex systems. S. Dorresteijn, 1983.

Mobility constraints of the shoulder girdle: A finite element model. P. Gelderblom, 1983.

Design of predictive displays. N. Gitz, 1983.

The risk of shutting down the Dodewaard nuclear power plant. G. Heslinga, 1983.

Analyses of the rehabilitation process of patients with a lesion of the plexus brachialis. R.B.M. Jaspers, 1983.

The effect of heterogeneous oxygen distribution on the arterial blood flow. A.W.H. Lambregts, 1983.

Fault management and decision theory. M.B. Mendel, 1983.

Experimental investigation of methods for body-powered arm prostheses. J. Postmes, 1983.

Modeling the ligaments of the shoulder girdle. A. van der Padt, 1983.

The control of bicycles under wind and road surface disturbances. P. Zuyderwijk, 1983.

Oxygen shunting in capillary bed model of the heart muscle. P.P. Bolier, 1984.

The benefits of prediction in supervisory control. R.W. Jansen, 1984.

Partial identification of the control of saccadic eye motions. D. Pobuda, 1984.

Design and evaluation of the man-machine interface of a co-generation power plant simulation. W.W. Snoeck, 1984.

Open loop control in goal directed arm movements. A.C.J. Taal , 1984.

Describing functions of the interaction between car driver and steering wheel. W.J. van Dort, 1984.

Adaptive behavior in goal directed arm movements. N.P.M. Conijn, 1985.

Control aspects of a new hand prosthesis prototype. M.H. Danz, 1985.

Control room ergonomics of a high pressure power plant. A.A. de Muynck, 1985.

Speech generation for the system for training of aphasia patients, STAP. A. Hummel, 1985.

Prediction error based parameter estimation of the respiratory control system. E.J. Meerwaldt, 1985.

Simulation of the behavior of a navigator. R. Papenhuijzen, 1985.

Nonparametric identification of the coronary flow dynamics. A.A.J. van der Vegt, 1985.

State prediction of plexus brachialis lesions. R. Happee, 1986.

A navigator model based on fuzzy set theory. H. Noback, 1986.

A state oriented predictive controller. W.A. van den Boomgaard, 1986.

Application of two parameter estimation methods for identification of the respiratory system. T.A.E. van der Leij, 1986.

Supervision of a bottle filling station. W.A. Blaauwboer, 1987.

Human reliability assessment at the educative nuclear reactor. A.D.L. de Vos, 1987.

Design of a supervisory system for the Dutch Steel Company. F.A. Fokker, 1987.

Model estimation of the navigator's behavior. L.G.M. Metzemaekers, 1987.

Design and application of an external fixator for a shoulder arthrodesis. F.J.M. Nieuwenhuis, 1987.

The navigator's internal representation of the ship dynamics. M.J. Pfeiffer, 1987.

Analysis of convergent and divergent eyeball motions. M. Pobuda, 1987.

Shoulder blade mobility of patients with a shoulder arthrodesis represented in helical axis coordinates. P. Schultz, 1987.

Modelling the alertness state of newborns. E.J. Timmer, 1987.

Expert system for the treatment of patients with a leasion of the plexus brachialis. C. van Daalen, 1987.

ARMA model description of the rehabilitation process of quadriplegics. J.W. van Hoek, 1987.

The use of an electrode array for the data selection in EMG analysis. J. Blok, 1988.

Standardization of graphics for the "system for training aphasia patients." E. Kouwe, 1988.

System analysis of the coronary circulation: The influence of the left ventricular pressure in the coronary blood flow. R. Moers, 1988.

Design and evaluation of a classical human-machine interface for supervisory control. G.A.M. Ruegg, 1988.

Fault diagnosis based on the multilevel flow modeling concept. A. Sissingh, 1988.

Extension of the navigator model. P.A. van der Meulen, 1988.

Design of a manual control set-up for exercises in the Man-machine Systems Course. F.C. van Reijn, 1988.

Evaluation of a fuzzy set navigator model. I.M. Wuerzner, 1988.

Application of the multilevel flow modeling concept on an experimental process. L.J. Wytzes, 1988.

Design of a human-machine interface based on the multilevel flow modeling concept. C.A. Zemering, 1988.

Study of the number of operators needed in an experimental industrial process. P. de Koning, 1989.

An insert task with a space manipulator. H. Doorenbos, 1989.

The human ability to control forces. N.M.C. Drost, 1989.

Application of a tree search algorithm in a diagnostic expert system for plexus brachialis lesions. L.W. Meinders, 1989.

Static force relations in the shoulder girdle. J.J. Oranje, 1989.

Factors for 3D-perception in graphic displays. F. van de Klashorst, 1989.

EMG-measurements for the validation of the shoulder model. M.J. van der Hoeven, 1989.

A fuzzy-set model to describe the trajectory planning of a navigator. B.O. van der Meulen, 1989.

The application of micro-air bubbles in order to measure a transmural flow distribution in the myocard. G.J. van der Ven, 1989.

Modeling broad skeletal muscles. R. Veenbaas, 1989.

Measurements on two different control concepts for an upper arm prosthesis. R.A. Claassen, 1990.

Automatic detection of NMR- images of shoulder muscle contours. J.P. Kuntz, 1990.

Fault management in supervisory control supported by a human-machine interface based on the multilevel flow modelling concept. A.M. Smit, 1990.

Supervisory control with a human-machine interface based on the multilevel flow modelling concept. B.R.M. Terpstra, 1990.

The influence of xenon during load following maneuvers with a boiling water nuclear reactor. A.C. Bosma, 1991.

The control of a space manipulator from Earth. P. Breedveld, 1991.

Human limb control. The use of an internal representation in fast goal directed movements. V.C.J. Gerdes, 1991.

The influence of the heterogeneity of volume flows on the estimation of the metabolic capacity of the capillary bed. B.M. Krijgsman, 1991.

Multilevel flow models for real-time fault diagnosis of industrial processes. P.C. Riedijk, 1991

The design of an alarm system. E.J.H. Rijs, 1991.

On the modeling of the oxygen supply dynamics in the heart muscles. L.A. Rozendaal, 1991.

Calculation of muscle forces in the shoulder girdle. R.P.T. Sansaar, 1991.

Analysis and recognition of myoelectric activity during fast arm movement as part of the control of prostheses. M.P. Smits, 1991.

A perfusion system for in vivo research. M. van Delden, 1991.

A predictive controller as a human control behavior model. J.J. van den Bosch, 1991.

LORETREAT, an object oriented representation of the knowledge on plexus brachialis lesions. W.P.M. van Heerebeek, 1991.

Measurements on fracture healing in applications of an external fixator. T.M. Dubelaar, 1992.

Application of 2D-echocardiography in order to estimate local flow values. A.C. Kessler, 1992.

Simulation of the NADH fluorescence at the surface of a rat heart. M.L. Leune, 1992.

Design of a task analysis method to describe surgery processes. J.J. van der Weijden, 1992.

Measurement of the volume of the heart muscle by means of fluorescence microscopy. B. Visscher, 1992.

The shoulder during wheelchair propulsion, an inverse dynamic approach. M. Baan Hofman, 1993.

Design of a decision support system in a chemical plant. M. Bakker, 1993.

Noise analysis in the use of surface electrodes. K. Brinkman, 1993.

Laboratory evaluation of an operator support system. E.F.T. Buiel, 1993.

The function of force feedback in fast goal directed arm movements. E. de Beer, 1993.

The influence of geometrical changes on the exerted muscle force due to the use of a shoulder prosthesis. O. de Leest, 1993.

Operator support during fault management. J.H. Hoegee, 1993.

Model based image analysis of NMR images of the shoulder. B.L. Kaptein, 1993.

Determination of the 3D shoulder position by means of X-rays and palpation. S. Le Poole, 1993.

The development of an experimental setup to measure shoulder reflexes. T.A. Mattaar, 1993.

Identification of the control of the heart muscle blood supply. E.C.J. Nijdam, 1993.

Internal representation for goal-directed movements modeled by multilayer perceptrons. G. Schram, 1993.

Exploratory experiments with spastic patients. J.S. Sipkema, 1993.

Determination of dynamic effects on scapulo-humeral rhythm by means of X-ray video. E.R. Valstar, 1993.

The influence of surrounding oxygen on high flow NADH fluorescence. C.A.D. van den Brande, 1993.

Flow estimation in the heart muscle by means of ultrasound. J.C. van der Grinten, 1993.

Mobility constraints of the shoulder mechanism. T. van Dijk, 1993.

Design issues of a planning system to support human operators. M. van Hattem, 1993.

Membrane potential measurement of the vascular smooth muscle cell. A.F. van Houten, 1993.

Diagnosis support for the supervision task in the control room of an M-frigate. W.W. Verduyn, 1993.

Validation of the PLEXUS expert system. F.A. Voorhorst, 1993.

Sensitivity analysis of the shoulder model. P. Westen, 1993.

Analysis and improvement of laparoscopic surgery. G. Claus, 1994.

Modeling the dynamics of an autoclave. E.J. Roovers, 1994.

The indicator dilution technique used for the identification of the coronary bed. B.N. de Jongh, 1994.

On the modeling of the oxygen transport in the capillary bed. J.M.C. van der Kuil, 1994.

Determination of morphological parameters from MRI images. B.P.F. Lelieveld, 1994.

Modelling muscle dynamics. P.M. Overschie, 1994.

The ERA simulation model. A.Y. Diepenbroek, 1994.

A model for simulation of realistic tugboat assistance. N.H. Bakker, 1995.

Anticollision displays. M. de Beurs, 1995.

Dynamics of the coronary circulation. J.M. Cornelissen, 1995.

Modeling of the internal representation of fast goal directed arm movements using dynamic neural networks. J.D.F. van Erven Dorens, 1995.

CAB-SIM, a cause-based operator model for use in dynamic probabilistic safety assessments for nuclear power plants. F.J. Groen, 1995.

Design of a low friction mechanism for laparoscopic instruments. M.J. Horward, 1995.

3D-display for dredging activities. M.N. Jonkhof, 1995.

Search algorithm for the location of the origin of heart rhythm disorders. A. Latour, 1995.

Design and evaluation of an improved density observer for environmental dredging. B.H.A.L. Leenders, 1995.

Suppression of false alarms in the intensive care unit. W.G.A. Pauwels, 1995.

Development and implementation of CAB-Flight. F.M. Peek, 1995.

Continuous state estimation of patients in the intensive care unit by means of the SAPS II rating scale. M.P. van Rooij, 1995.

Market analysis for EEG video monitoring systems. J.H. Sprakel, 1995.

The role of calcium in vasomotion. E.E. Szekely, 1995.

Doctor's theses

Ship maneuvering under human control, W.Veldhuyzen, 1976, 104 p.

Evaluation of models describing human operator control of slowly responding complex systems, J.J. Kok, 1978, 236 p.

Evaluation of models describing human operator control of slowly responding complex systems, R.A. van Wijk, 1978, 236 p.

Identification of human operator describing function models with one or two inputs in closed loop systems, A. van Lunteren, 1979, 157 p.

On the mental load in arm prosthesis control, M. Soede, 1980, 237 p.

Identification of the adaptive feedback of the human motor system using the response difference method. A.P. Eland, 1981, 131 p.

Human supervisory control behavior: verification of a cybernetic model. T.N. White, 1983, 111 p.

Studies on human vehicle control. H. Godthelp, 1984, 160 p.

Car driving as a supervisory control task. G.J. Blauw, 1984, 127 p.

Visual and proprioceptive information in goal directed movements: a system theoretical approach. J.C. Ruitenbeek, 1985, 124 p.

The Influence of the coronary capillary network on the distribution and control of local blood flow. Delft, P.A. Wieringa, 1985, 219 p.

Fault management. W.L.Th. Thijs, 1987, 243 p.

Technique for human error sequence identification and signification. G. Heslinga, 1988, 167 p.

On the dynamics of the coronary circulation. J. Dankelman, 1989, 157 p.

Medical decision support. an approach in domain of brachial plexus injuries. R.B.M. Jaspers, 1990, 273p.

Supervisory control, telerobotics and society. T.B. Sheridan, 1991 (Honory degree).

The shoulder girdle. analysed and modelled kinematically. G.M. Pronk, 1991, 244 p.

The shoulder mechanism. a dynamic approach. F.C.T. van der Helm, 1991, 285 p

Man machine aspects of remotely controlled space manipulators. J.F.T. Bos, 1991, 177 p.

Biomechanical aspects of manual wheelchair propulsion. H.E.J. Veeger, 1992, 133 p.

The implementation of the ship manoeuvring model in an integrated navigation system. J.H. Wulder, 1992, 171p.

The control of shoulder muscles during goal directed movements, R. Happee, 1992, 185 p.

Modelling of perception and action in compensatory manual control tasks. J.C. van der Vaart, 1992, 248 p.

Validating medical knowledge based systems. C. van Daalen, 1993, 292 p.

Design issues of human operator support systems. J.M.A. Sassen, 1993, 223 p.

Towards a human operator model of the navigator. R. Papenhuijzen, 1994, 179 p.

Biophysics in Aircraft Control. A model of the neuromuscular system of the pilot's arm. M.M. van Paassen, 1994, 206 p.

Intramyocardial blood volume and myocardial oxygen exchange. C.P.B. van der Ploeg, 1994, 130 p.

Distribution planning support system. M. Mourits, 1995, 214 p.

Determinants of monitoring behaviour in a four-instrument monitoring task. an experimental study. M.A.M. Leermakers, 1995, 143 p.

Section A

CONTROL OF BODY MECHANISMS, REHABILITATION, AND DESIGN OF AIDS FOR THE DISABLED

Chapter 4

Musculoskeletal Systems: The Human Shoulder

Frans C.T. van der Helm
Gijs M. Pronk

On November 30, 1992, Gijs Pronk was killed in a train accident. He started the shoulder research project in the Man-Machine Systems Group, and expanded the project to the large research group which is now focusing on the shoulder. He was a great source of inspiration, a dedicated scientist, and, above all, a dear friend.

1   Introduction

Sometimes, due to a brachial plexus nerve lesion, muscles controlling the motions of the upper extremity are paralyzed. The muscles affected depend on the location of the lesion. If the patient lacks control of one or more joints, it is impossible to position the hand. The rehabilitation process attempts to provide the patient some sort of functional control of the joint motions (Stassen, 1989) (Fig. 1). If after two years of rehabilitation and, when applicable, attempts at neurosurgical repair, muscles crossing the shoulder joint are still paralyzed, it may be advisable to arthrodize the joint. In a shoulder arthrodesis, the humerus (upper arm bone) is fused to the scapula (shoulder blade) by surgically removing the articular cartilage and fixating the humerus with respect to the scapula. Subsequently, the combination of both bones will move as one. The humerus can again be positioned by action of the thoracoscapular muscles between the thorax and scapula. These muscles are in most cases unaffected by the brachial plexus lesion. In addition, the arthrodesis will prevent subluxation of the shoulder joint. Elbow control can be established by an elbow orthosis (Cool, 1989), allowing either an unconstrained hanging of the forearm or a fixed position of 90 degrees elbow flexion. Wrist control can be established by an arthrodesis or an orthosis. The ultimate goal is to recover the ability to position the hand and some force exertion of the hand. In addition, the arm is again under control during walking and running, which yields better cosmesis of wearing and protects the arm against damage. In other words, in the patient's perception the arm has again become a part of the body.

The fusion angles between the scapula and humerus are decisive for the functional results of arthrodeses. Uncertainty about the outcome of the surgery impedes the patient and the surgeon in making the decision to operate. In 1979 a rehabilitation center asked the Man-Machine Systems Group whether the optimal fusion angles could be predicted from the scapular motions and desired working range of the hand. However, no measurement methods were available, and hence no data, for the three-dimensional motions of the scapula (Pronk, 1981). In addition, biomechanical analysis of the function of the morphological structures of the shoulder (e.g. muscles, ligaments, and joints) was only qualitative and limited in scope. This provided the impetus for shoulder research at the Man-Machine Systems Group.

At the start of the research project, very little quantitative and qualitative information appeared to be available about three-dimensional motions of the shoulder, the force and moment contribution of the muscles, and intermuscular coordination. Two basic items were necessary to undertake this research project: the development of a biomechanical model describing the kinematic and dynamic behavior and the development of methods for recording the input and output of the model: the motions of the bones and muscle forces.

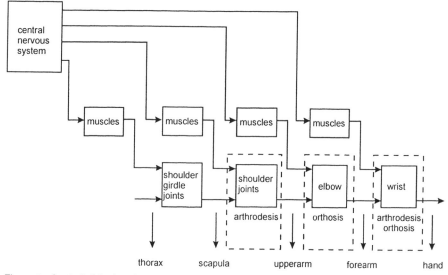

Figure 1. Control of the hand.

In this chapter a three-dimensional biomechanical model of the shoulder is outlined. Currently, to our knowledge, this is one of two shoulder models published (Karlsson, 1992; Van der Helm, 1991-1994[ab]). The model detail allows for application in a variety of circumstances, since it can offer predictions for most positions and force directions. The model is used to calculate muscle forces from recorded motions of the shoulder in an inverse dynamic analysis. It can also be used in forward dynamic simulations to study neuromuscular strategy. Since motions are input variables to the model, a new recording method is presented for the three-dimensional positions of the shoulder. Using the model, the control of shoulder stability and shoulder motions can be studied and validated in all of its complexity. During the project the initial goal of prediction of optimal fusion angles for shoulder arthrodesis has been replaced by a wider goal: to gain insight into the function of the shoulder in order to improve the diagnosis, treatment, and prevention of shoulder complaints.

2   The human shoulder

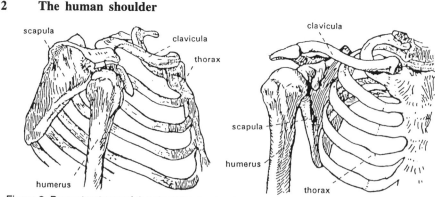

Figure 2. Bony structures of the shoulder mechanism.

The shoulder is one of the most complex joints in the human body. It consists of four bony parts, i.e., the thorax, clavicle (collar bone), scapula, and humerus (Fig. 2). In between are three joints, i.e., the sternoclavicular (SC) joint, the acromioclavicular (AC) joint, and the glenohumeral (GH) joint. Each of the joints has three rotational degrees of freedom, wherein the small translations can be neglected. The medial border of the scapula slides over the thorax, which results in a closed chain of thorax-clavicle-scapula. This thoracoscapular connection not only results in forced motions, but also couples the motion equations of the SC and AC joints.

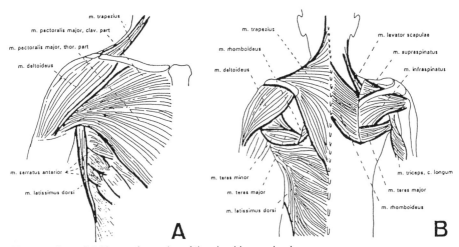

Figure 3. Superficial layer of muscles of the shoulder mechanism.
A: Frontal view B: Dorsal view.

Muscles are the actuators of the skeletal system. They convert chemical energy into mechanical energy and, furthermore, convert a neural input signal into a controlled force output. Muscles consist of a contractile part, the muscle fibers, and two tendinous parts on either side. Muscles have some important nonlinear dynamic properties. The fibers show a parabolic force length relation and nonlinear (active) damping characteristics. These contraction dynamics are combined with a nonlinear elastic tendon and nonlinear activation dynamics. One of the major problems in modeling muscles is getting a realistic set of parameters.

Seventeen muscles and three extra capsular ligaments control the motions of the shoulder (Fig. 3). These muscles have wide attachments and often span more than one joint. Electromyographic studies recording the electrical activity of the muscles have shown that parts of the muscles can contract independently.

3   Biomechanical model of the shoulder

The structure of the biomechanical model is based on the finite-element method (Van der Werff, 1977; Schwab, 1983; Jonker, 1988; Van der Helm, 1994[a]). Each relevant gross morphological structure is represented by an appropriate element, the mechanical behavior of which is known. By connecting the elements, the mechanical behavior of the complete mechanism can be calculated.

All of the joints are represented by three orthogonal hinge joints, constituting a ball-and-socket joint. The bones are represented by rigid bodies, defining the origins and insertions of the muscles in their local coordinate systems. The ligaments are represented by truss elements, which can shorten and elongate. Ligament forces are in the direction of the element and depend upon its length. The muscles are represented by truss or curved-truss elements. Curved-truss elements describe muscle lines of action curved around underlying bony contours. A specially developed *surface element* represents the connection between the medial border of the scapula and the thorax. Hence, the model incorporates the closed-chain nature of the shoulder.

A very precise parameter set is necessary to obtain valuable predictions. The data were obtained from an extensive cadaver experiment on both shoulders of seven human cadavers. As the morphology of bones, ligaments, muscles, etc. are very precisely balanced with each other, it was chosen not to average the data but rather to use complete sets of data from each cadaver. Data on the inertia of the segments, the geometry of the mechanism, and some of the muscle properties have been recorded (Veeger et al., 1991; Van der Helm et al., 1992). Inertia data were calculated from anthropometric measurements of the segments. The recorded geometry consists of articular surfaces (from which the joint rotation centers were estimated), ligament attachments, muscle attachments, the shape and position of bony contours, bony landmarks (for connection with motion recording experiments), and the scapulothoracic gliding plane (the surface of the thorax to support the scapula). Some of the muscles have wide attachment sides and a fanning course of muscle fibers (see Fig. 3).

A new theoretical approach has been used to derive the minimal number of muscle lines of action for each muscle to represent the mechanical effect. Each muscle affects the force and moment equilibrium of the joint(s) that it crosses. Six degrees of freedom is the maximum number of degrees of freedom the muscle can affect. Hence, six muscle lines of action is the maximum number of muscle lines of action necessary to represent the mechanical effect of the muscle. However, the number of muscle lines of action can be smaller, depending on the shape of the attachment sites and the fiber architecture (Van der Helm & Veenbaas, 1991). Muscles are represented in the biomechanical model by one to six muscle lines of action. A total of 95 muscle lines of action represent the 17 muscles of the shoulder. Two parts of one muscle are shown in Figure 4.

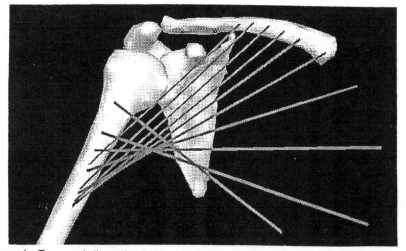

Figure 4.   Ten muscle lines of action are used to represent the mechanical effect of both parts of the m. pectoralis major, originating from the thorax and clavicle.

Input variables to the model are the motions of the bones. Output variables are the muscle forces, calculated by optimizing a performance criterion. Hence, the model uses an inverse-dynamic approach: The differential equations constituting the motion equations are transformed into algebraic equations by supplying the position, velocity, and acceleration of the generalized coordinates.

4   Kinematics of the shoulder

Three-dimensional motion recording of the shoulder is difficult, since the large excursions of the clavicula and scapula are hidden underneath the skin. Optical methods such as video or film cannot be used because markers glued to the skin do not represent the bony motions. For three-dimensional X-ray measurements, at least two cameras are necessary, and one must be able to distinguish the same marker point in both views. An approach using bony landmarks as markers failed because the accuracy of determining bony landmarks in X-rays was too low (Casolo et al., 1987). Högfors et al. (1991) used implanted tantalum balls as markers. The main disadvantage of the latter method is that it is invasive and is not allowed in the Netherlands as a research method.

In Delft a noninvasive method based on a palpation technique was developed (Pronk, 1987, 1991; Van der Helm & Pronk 995). The subject stands in a given position with the arms elevated symmetrically. Subsequently, 11 bony landmarks (Fig. 5) are retrieved underneath the skin by palpation (searching manually), and are digitized by a palpator (Pronk & Van der Helm, 1991). A *palpator* is a recording device, consisting of four links connected by four hinge joints. By recording the rotation of the hinge joints, the position of the endpoint can be calculated. The accuracy of the latest version of the palpator using optical rotation recording is about 0.1 mm. The position and orientation of the bones can be reconstructed using these 11 bony landmarks. The disadvantage of the palpation method is that it can only record static positions. However, recent experiments have shown that angular velocity during arm elevation has no significant effect on the scapular position with respect to the humerus (De Groot et al., 1995). Hence, static palpation recordings may be extrapolated to dynamic conditions. Thus, when the dynamic humerus position is recorded using, e.g., video cameras, the scapular position can be calculated using regression equations. To date, this is the only in-vivo, noninvasive, three-dimensional position recording method for the shoulder bones.

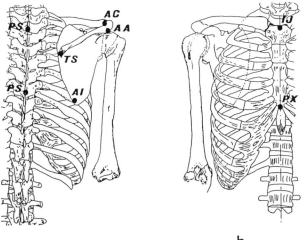

a.   b.

Figure 5:   In the palpation method, eleven bony landmarks are used to reconstruct the three-dimensional positions of the shoulder bones.

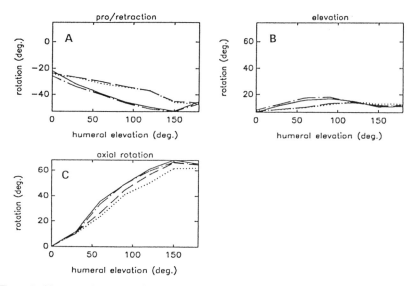

Figure 6. Mean rotation angles (N=10) for rotations of the clavicula with respect to its virtual reference position. A. rotation vertical axis, B. rotation forward/backward axis. C. rotation longitudinal axis. *ff* abduction unloaded, ------ anteflexion unloaded, -.-.-. abduction loaded, ... anteflexion loaded.

Medical terminology proved inadequate to describe the three-dimensional motions of the shoulder. In technical terms, the bony rotations were described using Euler angles. In order to interpret and compare Euler angles, a very strict definition of the initial position, rotation axes, order of rotation, and global or local rotation axes is necessary (Van der Helm & Pronk, 1995). Two definitions are used. Firstly, the orientation of the bones has been described with respect to the thorax, starting from a virtual reference orientation in which the bones were aligned with the global coordinate system (Figs. 6 to 8). In the virtual reference position, the clavicle is parallel to the left-right axis, the scapular plane is parallel to the frontal plane of the body, and the humerus is parallel to the vertical axis. The advantage of this definition is that the orientation for executing the task with respect to the gravitational field can be detected and compared between subjects and patients. The largest rotations of the scapula and clavicula (respectively, around an axis perpendicular to the scapular plane and around the longitudinal axis of the clavicula) appear to be about the same. Since the clavicula during shoulder motions remains almost perpendicular to the scapular plane, these rotation axes almost coincide.

Secondly, the orientation of the bone can be described with respect to the proximal bone, i.e., the rotations of the joint in between are described. Figures 9 to 11 show the results. Figiure 10 shows that the rotation in the acromioclavicular joint between clavicula and scapula is very small. As mentioned earlier, this is due to the fact that the clavicula is perpendicular to the scapular plane and the major rotations are around coinciding axes. Rotations of the scapula are due to rotations in the sternoclavicular joint between the thorax and clavicula (Fig. 9). Rotations in the glenohumeral joint between the scapula and humerus are up to 110 degrees (Fig. 11). This shows that about one third of the rotation of the humerus with respect to the thorax is caused by motions of the scapula.

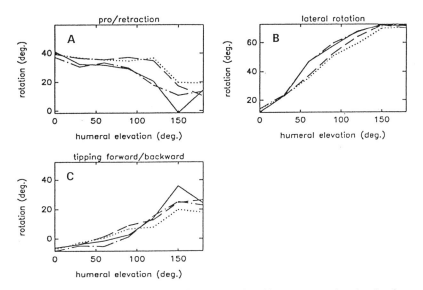

Figure 7. Mean rotation angles (N=10) of the scapula with respect to its virtual reference position. A. rotation vertical axis, B. rotation forward-backward axis, C. rotation longitudinal axis. *ff* abduction unloaded, ----- anteflexion unloaded, -.-.-.abduction loaded, ... anteflexion loaded.

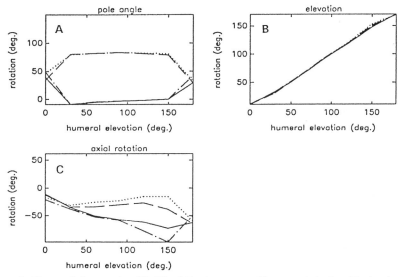

Figure 8: Mean rotation angles (N=10) of the humerus with respect to its virtual reference position. A. rotation vertical axis, B. rotation forward/backward axis, C. rotation longitudinal axis. *ff* Abduction unloaded, —— Anteflexion unloaded, -.-.-. Abduction loaded, ... Anteflexion loaded.

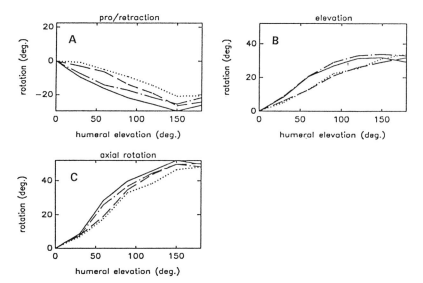

Figure 9: Mean rotation angles (N=10) for rotations of the sternoclavicular joint. A. rotation vertical axis, B. rotation forward/backward axis, C. rotation longitudinal axis. *ff* Abduction unloaded, ---- Anteflexion unloaded, -.-.-. Abduction loaded, ... Anteflexion loaded.

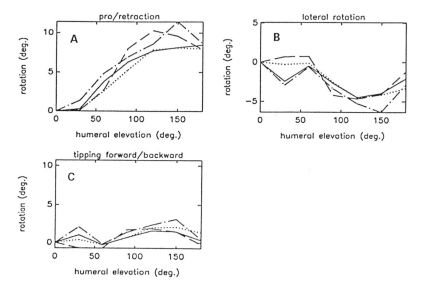

Figure 10: Mean rotation angles (N=10) for rotations of the acromioclavicular joint. A. rotation vertical axis, B. rotation forward-backward axis, C. rotation longitudinal axis. *ff* Abduction unloaded, ---- Anteflexion unloaded, -.-.-. Abduction loaded, ... Anteflexion loaded.

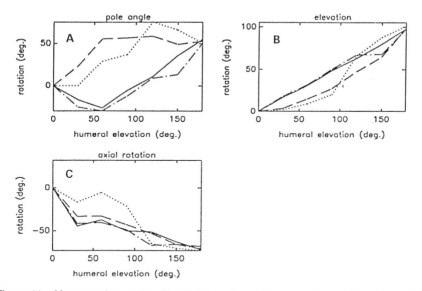

Figure 11: Mean rotation angles (N=10) for rotations of the glenohumeral joint. A. rotation vertical axis, B. rotation forward-backward axis, C. rotation longitudinal axis. ff Abduction unloaded, ---- Anteflexion unloaded, -.-.-. Abduction loaded. ..... Anteflexion loaded.

5   Inverse dynamics

It is impossible to directly measure the muscle forces in the human body. Therefore, the input and output variables of musculoskeletal models cannot be used for parameter optimization. Instead, using a recorded set of parameters for the model, the muscle forces are calculated. Using the finite element model of the shoulder, the motion equations are numerically derived. Palpation recordings of various types of motions have been used as input to the biomechanical model, sometimes in combination with video recordings (Veeger et al., 1993; Van der Helm & Veeger, 1995). The net moments can be calculated using the position, velocity, and acceleration of the generalized coordinates. The muscle forces can be calculated from these net moments. Since more muscles are generally present than motion equations, a unique solution is found in an optimization procedure. The optimization criterion must reflect some physiological quantity, such as minimization of energy consumption. A common choice is to replace minimization of energy consumption by minimization of the sum of squared muscle stresses (= muscle forces divided by cross-sectional area). Van der Helm (1994[a]) showed that if a muscle has a favorable moment arm for the required task, the muscle is active with all somewhat realistic, nonlinear physiological criteria. The recorded EMG cannot distinguish between force amplitudes resulting from different optimization criteria. Thus musculoskeletal models cannot be validated in a strict sense. In the inverse dynamic optimization, the position, velocity, acceleration, and external forces are input variables, while muscle forces are output variables of the biomechanical model.

In the shoulder model several other constraints are added in the optimization procedure. Muscle forces are positive (= traction forces) by definition and are limited to the maximal forces as determined by the cross-sectional area of the muscle. An inverse dynamic muscle model has been developed that takes into account muscle dynamics in the force calculation. This means that the increase or decrease of muscle force is limited in time with respect to previous calculated muscle forces (Happee, 1995). Forces in the scapulothoracic gliding plane are limited to compression forces, and forces in the ligaments are limited to traction forces. A special nonlinear constraint is included that prevents dislocation forces in the glenohumeral joint between the scapula and humerus. If the muscle forces are known, other dependent forces, such as ligament forces and joint reaction forces, can be calculated.

Useful information about muscle function can be derived using a biomechanical model. Previously, the function of muscles was only qualitatively known. If the muscle shortens during a motion, it was concluded that the muscle had a function for this motion. EMG recordings actually showed whether muscles were active during certain motions. However, the intermuscular coordination and quantitative muscle forces were unknown. In three-dimensional space, a muscle not only has a moment arm for the desired moment, but also shows some undesired moments about the other axes. These moments must be compensated by other muscles, which could in fact be a very important function for these muscles.

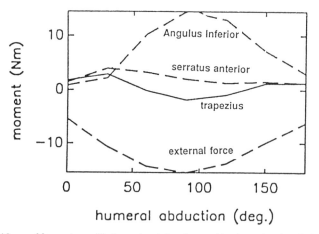

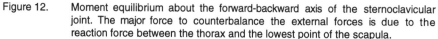

humeral abduction (deg.)

Figure 12.   Moment equilibrium about the forward-backward axis of the sternoclavicular joint. The major force to counterbalance the external forces is due to the reaction force between the thorax and the lowest point of the scapula.

For an example, Figure 12 shows the moment equilibrium around the forward-backward axis of the sternoclavicular joint during unloaded humeral abduction (sidewards elevation of the arm). Musculus serratus anterior is the most important muscle to counterbalance the moments due to the gravitational forces acting on the arm. However, it appeared that the major counterbalancing forces are due to the reaction forces between the thorax and scapula. In other words, the scapula moves to the side of the thorax during humeral abduction. This is a very efficient position energetically, since relatively small muscle forces are needed to maintain this position (Van der Helm, 1994[b]).

6   Direct dynamics

A forward simulation with the shoulder model is successful if the positions predicted from the muscle forces (or from the neural input to the muscles) agree with the recorded positions. Therefore, an inverse dynamic simulation is very efficient, since the desired output is fed to the (inverse) model, and the requested input (muscle forces) will result. The disadvantage of this approach is that it evades the problem with which the central nervous system is faced: how to determine the optimal neural input to the muscles for a desired trajectory of the hand, given the kinematic redundancy of the upper extremity (more degrees of freedom [DOF] in the joint than DOF in the endpoint) and actuator redundancy (more actuators than DOF). In addition, an inverse dynamic solution could represent an unstable position. In a direct dynamics approach (i.e., integrating the differential motion equations), an optimal solution will constitute the optimal joint trajectory as well as the optimal load sharing by the muscles. These types of simulations can reveal important aspects of the neuromuscular control of human motions.

For a proper assessment of ligament function, forward simulations are more appropriate (Pronk et al., 1994). In an inverse dynamic simulation, the stresses in the ligaments depend on their strain. The strain of the ligament is calculated from the noisy position recording. Hence, the stress in the ligaments will be noisy as well. In a forward simulation, a noisy neural input signal will be filtered by the inertia of the skeletal and the muscle dynamics; hence, a smooth pattern of ligament stresses will result.

To find the optimal solution to the direct dynamic approach is very costly computationally. The input signal must be parameterized as a function of time. Subsequently, numerous forward simulations must be performed in order to obtain the value of the objective criterion and its numerical derivatives with respect to the parameters of the input signal. For a comparable large-scale musculoskeletal system (the lower extremity during walking), computing time can take as long as 77 hours CPU on a Cray (Anderson et al., 1993). Since the shoulder model is of comparable size and has more degrees of freedom, one can expect similar computational efforts to obtain an optimal solution for the forward dynamic problem. For the time being, it has been concluded that forward dynamic optimization is not feasible for the shoulder.

A forward dynamic version of the shoulder model has been developed (Van der Helm, 1994[a]). The muscle dynamics of each muscle has been modeled using a third-order Hill-type muscle model, with a nonlinear activation dynamics, nonlinear force-length relation, nonlinear force-velocity relation, and nonlinear elastic properties. Parameters have been roughly determined based on some gross morphological properties. There is definitely a need for more accurate muscle parameters, which are currently being recorded. Input variables have been taken from the output of inverse dynamic simulations. However, due to the lack of a proper optimization algorithm, no results have been published yet.

7   Stability

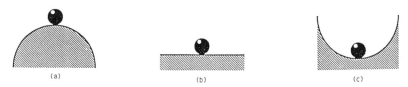

Figure 13.   Even though the ball is in equilibrium in a, b, and c, it is unstable in a and b and only stable in c. Quasistatic stability can be calculated using the second derivative of the potential energy.

Results of inverse dynamic simulations cannot be used without consideration of forward simulation. Muscle forces calculated in inverse simulations result in a force and moment equilibrium of these muscle forces, inertial forces, and external forces. However, it is not guaranteed that this equilibrium is stable (Fig. 13). In a stable equilibrium, perturbations will result in an increase in potential energy, which will drive the mechanism back to the equilibrium position. If in a forward simulation an unstable equilibrium position is encountered, small perturbations (even numerical errors) can result in a deviating motion pattern.

Therefore, a different approach is used for the inverse dynamic simulations. Constraints are added so that the equilibrium is stable: The second derivative of the potential energy with respect to the generalized coordinates must be positive. This is identical to the constraint that the eigenvalues of the joint stiffness matrix all be positive

(Van der Helm & Winters, 1995). A major factor in the joint stiffness matrix is the stiffness of the muscles. Simulations revealed that for unstable musculoskeletal systems (e.g., the head balancing on the neck or the arm raised above shoulder level), intrinsic muscle stiffness was not sufficient to achieve a stable position. In simulations of quasistatic positions, it was shown that inclusion of proprioceptive reflex activity is necessary for stability (Van der Helm & Winters, 1995). Results from stable inverse dynamic simulations will be used for forward simulations. Currently only quasistatic positions (posture maintenance) have been simulated. In forward simulations the inevitable time delays due to the reflex loop will introduce dynamic effects, which could result in instabilities due to the interaction between muscles.

8   Control mechanisms

For motion control and posture maintenance, feedback information is necessary for performing a task. Visual feedback, tactile feedback from the skin, position feedback from the joints, and length, velocity, and force feedback from the muscles are available. Of this information, only the muscle feedback is directly coupled to the motor neurons activating the muscles. It is likely that the other feedback sources are not used for direct motion control, since they require more processing and hence add additional time delays in the feedback loop.

Proprioceptive sensors are present in the muscles that provide information about the fiber length and contraction velocity, and about the force in the tendons. The transport speed of sensory signals along the nerves is limited to 40 to 100 m/sec, substantially below the speed of sound. This causes time delays in the reflex loop of between 30 and 70 msec, depending on the location of the muscle with respect to the spinal cord. These time delays result in a phase-shift of the closed-loop motion control. Since the inertial system functions as a double integrator, resulting in a 180 degree phase shift, closed-loop position control is impossible with only sensory information about position (Rozendaal, 1995; Fig. 14). Velocity information, preferably force information, is necessary to obtain a sufficient phase lead to compensate for the time delays.

However, due to the time delays, the gain in the closed-loop cannot be very high.

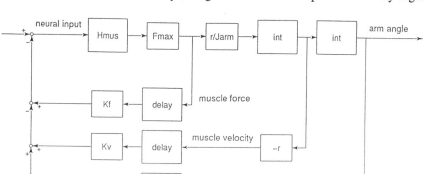

Figure 14:   Scheme of the musculoskeletal model under posture maintenance control. The muscle model consists of the first-order filters $H_{mus,1}$ and $H_{mus,2}$ and the constant F_{max} The arm is modeled by the double integrator including pre- and post-multiplicative constants. The sensory feedback control by the gains k_l, k_v and k_f and the time-delays (Rozendaal, 1995, with permission).

Therefore, closed-loop motion control would have a very limited bandwidth. Many fast motions must be open-loop (or feed-forward, in case of anticipation of perturbations) controlled. Perturbations are likely to be adjusted by an internal model of the limb and environment (Gerdes & Happee, 1994), such as in model-reference adaptive control. The internal model predicts the states of the system (in this case, the human arm), which are used to generate adequate neural input signals to the muscle. Delayed feedback information is used to adapt the internal model, and hence the effect on the controller output can be much more efficient than with conventional control schemes. Simulations showed a good congruence with experiments in which the external load of the arm was suddenly changed (Fig. 15).

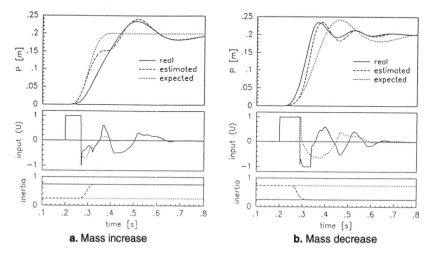

a. Mass increase   **b. Mass decrease**

Figure 15:   Simulated load perturbation. The step size is 20 cm. At the onset of movement, inertia is increased from 0.245 to 0.735 (kgm²), a, or decreased from 0.735 to 0.245 (kgm²), b. The "real" traces show the perturbed response, the "estimated" traces show the position and inertia estimated in the internal representation, and the "expected" traces describe the unperturbed response (From Gerdes & Happee, 1994, with permission).

As mentioned in the previous section, posture can be maintained by adapting the joint stiffness matrix. This concept is extended to the joint impedance matrix (including the velocity-dependent and inertia terms). Basically, there are two strategies to increase the impedance of the arm: either by co-contraction of antagonistic muscles at either side of the joint or by proprioceptive control (Brouwn, 1995). Proprioceptive control is more energy efficient, since muscles are only active when perturbations are actually perceived. The disadvantage of proprioceptive control is the time delays. Hence, only low-frequency perturbations can be rejected. Co-contraction stiffens the joint at higher energy costs. Since muscles have an intrinsic stiffness, perturbations are regulated without time delays, and this strategy also functions for higher frequencies. Future experiments will show at which frequencies there is a transition between these strategies.

9   Applications

The shoulder model has only been applied in its inverse dynamic version. The initial goal, prediction of the optimal fusion angles of the shoulder arthrodesis, has been superseded by a wider goal: to get insight in the function of the shoulder in order to improve diagnosis, treatment, and prevention of shoulder complaints. The model has been applied and validated for a number of standardized arm positions (forward and sideward elevation of the arm) (Van der Helm, 1994[b]) and for more complex tasks as fast, goal-directed arm motions (Happee & Van der Helm, 1995) and manual wheelchair propulsion (Van der Helm & Veeger, 1995).

One of the big advantages of model simulations is that parameters can be changed as they cannot be in reality. Van der Helm & Pronk (1994) studied the problem of the optimal fusion angles of the shoulder arthrodesis. They successively changed the fusion angles and evaluated the resulting workspace of the hand (assuming an elbow fixed in the 90 deg flexion position by an elbow orthosis). In addition, they were able to calculate the maximal force exertion in various directions.

A few hundred artificial shoulders are implanted each year in the Netherlands. One of the problems with a shoulder endoprosthesis (artificial shoulder joint) is that the functional result is not very good. The average maximal arm elevation after surgery is about 90 degrees. The major part of this elevation angle results from scapular motion. In other words, hardly any active motion is left in the shoulder joint after surgery. Passively, a full motion range can be obtained. This problem was studied using sensitivity analysis of the parameters affected by the prosthesis design (thickness, curvature) and surgical prosthesis (orientation, (re)location of the rotation center). A thorough analysis in which the effects on joint stability and moment arms were taken into account resulted in recommendations for the optimal values of these parameters (De Leest et al., 1995).

10   The robotic shoulder

For the development of the shoulder model and its control, many techniques and ideas from robotic control have been used. However, controlling a biological organ like the upper extremities, with its particular possibilities and impossibilities, can lead to new and inspiring notions about conventions in robotic control.

In robots, the configuration of human limbs is often imitated. However, the limitations of robots are in the area of the controllability of a complex mechanism. Controllability is enhanced by decoupling the DOF of the mechanism and/or the actuators controlling these DOFs. In the human body, most joints have multiple DOFs (i.e., ball-and-socket joints). In robots, joints are mostly restricted to one (rotational) DOF, i.e., a hinge.

Each muscle crossing a joint affects all DOFs, and hence the effect of muscle activation is strongly coupled. Muscles can only exert pulling forces, and therefore a double set of muscles is present to control each degree of freedom. Muscles are typical in-line actuators, i.e., they exert forces along a working line of action. Many muscles are polyarticular, i.e., they cross more than one joint. This configuration has some big advantages for controlling the stability of the joints. The stiffness of each joint can be controlled separately from the exerted moment by co-contracting antagonistic pairs of muscles. In addition, polyarticular muscles can couple the stiffness of more than one joint, i.e., perturbations at one joint will result in force changes at another joint. This is a direct and very efficient way to control the direction and magnitude of the stiffness field of the endpoint (Hogan, 1985).

Robots most often have moment actuators at their joints. A trajectory for each joint is calculated using inverse kinematics, and this trajectory is realized using feedback position control for each joint separately. The improvements using more advanced control schemes, i.e., using inverse dynamics, multivariable control, etc., are only marginal. Stiffness control of the endeffector is highly desirable in contact tasks; position control in contact tasks would result in very high forces. Stiffness control is most often realized by a force sensor in the tip or in the actuator. An inverse dynamic model of the robot is necessary for the actual control scheme.

One of the limits of (digital) feedback control is the sampling rate achieved, limiting the bandwidth. In humans there are significant time delays (30 to 70 msec) due to signal transportation along the nerves, and processing in the central nervous system. If feedback control were combined with these large time delays, only very low-frequency motions would be possible. Therefore, one may assume that humans use some kind of feedforward (or open-loop control) to overcome the time delays. An internal model of the extremity and the environment enables humans to incorporate feedback signals very effectively and to minimize the negatives of time delays: This is model-reference control.

11 Concluding remarks

The shoulder research in the Man-Machine Systems Group has resulted in a unique, detailed, three-dimensional model of the shoulder. Currently, to our knowledge, this is one of only two shoulder models published (Karlsson, 1992; Van der Helm, 1991 1994[ab]). The detail in the model allows for application in a variety of circumstances, since it can predict for most positions and force directions. In addition, the control of shoulder motions can be studied and validated in all of its complexity. It is interesting to note that the new research goals are closely connected to the research in the field of manual control previously pursued in our group. The major difference though is that the black-box models have been replaced by detailed, nonlinear white-box models.

References

Anderson, F.C., Ziegler, J.M. and Pandy, M.G. (1993). Dynamic optimization of large-scale musculoskeletal systems. Abstract 14th Meeting Int. Soc. Biomechanics, Paris.

Brouwn, G.G. (1995). Impedance control of simple arm models during posture by co-contraction. *Biological Cybernetics* (submitted).

Casolo, F., Colnago, M. and Cosco, P.P. (1987). A technique for dynamic analysis of joint motion using bi-planar cineradiography. Preliminary investigations on the shoulder complex. In: *Biomechanics* XI-B (Groot, G. de, Hollander, A.P., Huijing, P. and van Ingen Schenau, G.J., Eds.) Amsterdam, Free University Press pp. #1062-1069.

Cool, J.C. (1989). Biomechanics of orthoses for the subluxed shoulder. *Prosth. Orth. Int.* 13: 90-96.

De Groot, J.H., Valstar, E.R. and Arwert, H.J. (1995). Scapulo-humeral rhythm: A dynamic study. *Journal Biomechanics* (submitted)

De Leest, deRozing, P.M., Rozendaal, L.A. and Van der Helm, F.C.T. (1995). The influence of glenohumeral prosthesis geometry and placement on shoulder muscle forces. *Clinical Orthopaedics* (submitted).

Gerdes, V.C.J., and Happee R. (1994). The use of an internal representation in fast goal directed movements, a modelling approach. *Biological Cybernetics* (accepted).

Happee, R. (1995). Inverse dynamic optimization including muscular dynamics, a new simulation method applied to goal directed movements. *Journal of Biomechanics* (accepted).

Happee, R., and Van der Helm, F.C.T. (1995). The control of shoulder muscles during goal directed movements, an inverse dynamic analysis. *Journal of Biomechanics* (accepted).

Hogan, N. (1985). The mechanics of multi-joint posture and movement control. *Biological Cybernetics* 52: 315-331.

Högfors, C., Peterson, B., Sigholm, G. and Herberts P. (1991). Biomechanical model of the shoulder II. The shoulder rhythm. *Journals of Biomechanics* 24: 699-709.

Jonker, B. (1988). A finite element dynamic analysis of flexible spatial mechanisms and manipulators. Doctoral thesis, Delft University of Technology, The Netherlands.

Karlsson, D. (1992). Force distributions in the human shoulder. Doctoral thesis, Chalmers University of Technology, Göteborg, Sweden.

Pronk, G.M. (1981). Analysis of the motions of the bones of the shoulder girdle. MSc thesis A-309, Lab. for Measurement and Control, Dept. of Mech. Eng., Delft University of Technology, The Netherlands.

Pronk, G.M. (1987). Three-dimensional determination of the position of the shoulder girdle during huemrus elevation. In *Biomechanics* XI-B Groot, G de, Hollander AP, Huijing, P. and van Ingen Schenau, G.J. (Eds.), Amsterdam, Free University Press.

Pronk, G.M. (1991). The shoulder girdle: Analysed and modelled kinematically. Doctoral thesis, Delft University of Technology, The Netherlands.

Pronk, G.M. and Van der Helm, F.C.T. (1991). The palpator, an instrument developed for measuring the 3D-positions of bony landmarks in a fast and easy way. *Journal of Medical Engineering Technology* 15: 15-20.

Pronk, G.M., Van der Helm, F.C.T. and Rozendaal, L.A. (1994). Interaction between the joints in the shoulder mechanism: The function of the costoclavicular, conoid and trapezoid ligaments. *Journal of Engineering in Medicine* (Proc. Instn. Mech. Engrs) 207, 219-229.

Rozendaal, L.A. (1995). The necessity of velocity and force feedback for position control of the upper limb. *Biological Cybernetics* (submitted).

Schwab, A.L. (1983). Dynamics of mechanisms with flexible links. Rep. A-758, Lab. for Engineering Mechanics, Dept. of Mechanical Engineering, Delft University of Technology, The Netherlands.

Stassen, H.G. (1989). The Rehabilitation of Severely Disabled Persons. A Man-Machine System Approach. In: W.B. Rouse (Ed.). *Advances in Man-Machine Systems Research,* Vol.5, JAI Press Inc., ISBN 1-55938-011, 153-227.

Van der Helm, F.C.T. (1991). The shoulder mechanism: A dynamical approach. Doctoral thesis, Delft University of Technology, The Netherlands.

Van der Helm, F.C.T. and Veenbaas, R. (1991). Modelling the mechanical effect of muscles with large attachment sites: Application to the shoulder mechanism. *Journal of Biomechanics* 24: 1151-1163.

Van der Helm, F.C.T., Pronk, G.M., Veeger, H.E.J., Van der Woude, L.H.V. and Rozendal R.H. (1992). Geometry parameters for musculoskeletal modelling of the shoulder mechanism. *Journal of Biomechanics* 25: 129-144.

Van der Helm, F.C.T. (1994[a]). A finite element musculoskeletal model of the shoulder mechanism. *Journal of Biomechanics* 27: 551-569.

Van der Helm, F.C.T. (1994[b]). Analysis of the kinematic and dynamic behavior of the shoulder mechanism. *Journal of Biomechanics* 27:, 527-550.

Van der Helm, F.C.T. and Pronk G.M. (1994). The loading of shoulder girdle muscles in consequence of a glenohumeral arthrodesis. *Clinical Biomechanics* 9: 139-148.

Van der Helm, F.C.T. and Pronk G.M. (1995). Three-dimensional recording and description of motions of the shoulder mechanism. *Journal of Biomechanical Engineering* 117: 27-40.

Van der Helm, F.C.T. and Veeger, H.E.J. (1995). The use of a musculoskeletal model of the shoulder mechanism in wheelchair propulsion. Accepted by *Journal of Biomechanics.*

Van der Helm, F.C.T. and Winters, J.M. (1995). Stability and optimization for large-scale 3-D postural neuromusculoskeletal systems: Theory and implications. *Journal of Biomechanics* (submitted).

Van der Werff, K. (1977). Kinematic and dynamic analysis of mechanisms: A finite element approach. Doctoral thesis, Delft University of Technology, The Netherlands.

Veeger, H.E.J., Van der Helm, F.C.T., Van der Woude, L.H.V., Pronk, G.M. and Rozendal, R.H. (1991). Inertia and muscle contraction parameters for musculoskeletal modelling of the shoulder mechanism. *Journal of Biomechanics* 24: 615-629.

Veeger, H.E.J., Van der Helm, F.C.T. and Rozendal, R.H. (1993). Orientation of the scapula in a simulated wheelchair push. *Clinical Biomechanics* 8: 81-90.

Chapter 5

PLEXUS: A Medical Knowledge Based System for Brachial Plexus Injuries

Els van Daalen
Rob B.M. Jaspers

1 Introduction

Today, computers are part of all areas of health care. Hospital information systems, medical instrumentation, and assistive devices cannot be envisaged without computer systems. The potential of computer assistance in medical decision making has been explored extensively for the past three decades. The introduction of such decision support systems, however, was not undisputed. The appreciation of their ability to improve health care often conflicted with their potential effects on medical practice. This involved such issues as the depersonalization of medical care and the loss of jobs. However, due to the rapid growth of medical knowledge and the increasing specialization of medicine that accompanies it, together with the introduction of computers in many areas of daily life, the medical community has become more amenable to computerized decision support. Increasingly, information processing has become an important task for primary care physicians, as well as for experts in specialized fields of medicine.

In this context, research was conducted to assess the potential of medical decision support systems (Jaspers et al., 1982; van Daalen et al., 1993). As part of this research, a knowledge based system has been developed for brachial plexus injuries, i.e., nerve injury in the area between the neck and the arm (Jaspers, 1990; van Daalen, 1993). This system, PLEXUS, is the subject of this chapter.

2 Background

The project originates from a research program started in 1968 at the Laboratory for Measurement and Control at Delft University of Technology on the development and evaluation of externally powered prostheses and orthoses for the arm (Stassen, 1989). The project was carried out in cooperation with the De Hoogstraat Rehabilitation Center in Utrecht. Among other things, this research program gave rise to laboratory and field evaluation studies of prostheses (van Lunteren et al., 1983) and the development of decision support systems for the treatment of patients with spinal cord injury (Stassen et al., 1980). As a result of the latter project, the total rehabilitation time for patients with spinal cord injuries decreased without diminishing the quality of the treatment.

A similar system theoretic approach for brachial plexus injuries was studied (Jaspers et al., 1982). A major impediment to the application of methods based on data in the domain of brachial plexus injuries was the lack of a large reliable set of patient data (Jaspers, 1990). However, there are a number of experts in the field of brachial plexus injuries. Therefore, it was decided to represent knowledge about brachial plexus injuries in a knowledge based system. The aim of the PLEXUS system is to assist neurologists, neurosurgeons, orthopedic surgeons, rehabilitation physicians, and traumatologists in the diagnosis of and treatment planning for brachial plexus injuries. The system has been developed in cooperation with the departments of neurosurgery at Leiden University Hospital and De Wever & Gregorius Hospital in Heerlen.

3   Brachial plexus injuries

3.1 ANATOMY OF THE BRACHIAL PLEXUS

The brachial plexus is a network of nerves that is situated in the area between the neck and the arm, and that innervates the muscles of the shoulder, arm, and hand. In addition, the nerves of the brachial plexus provide sensory function for the arm and hand, and also carry autonomic fibers, which stimulate the sweat glands and constrict blood vessels. In order to provide the background that is necessary for understanding the rest of this chapter, the anatomy of the brachial plexus is briefly described. Detailed discussion may be found in Kerr (1918), Sunderland (1968), and Leffert (1985).

The (general) anatomy of the brachial plexus is shown in Figure 1. The brachial plexus generally originates at the five spinal nerves C5, C6, C7, C8, and T1. The spinal nerves are formed by the union of motoric nerve rootlets and sensory nerve rootlets that arise from the spinal cord. The spinal nerves join and divide to form a network of nerves.

Part of the brachial plexus is situated above the clavicle. This section is called the supraclavicular part of the brachial plexus, and part of the network is situated below the clavicle. This section is called the infraclavicular part of the brachial plexus. Supraclavicularly, C5 and C6 usually join to form the truncus superior (upper trunk), C7 forms the truncus medius (middle trunk), and C8 and T1 make up the truncus inferior (lower trunk), as seen in Figure 1. A number of nerves leave the plexus supraclavicularly.

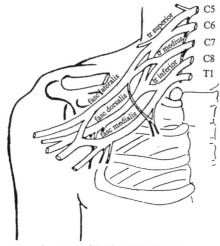

Figure 1.   Anatomy of the brachial plexus

The trunci each divide into an anterior (front) part and a posterior (back) part. The anterior parts of the truncus superior and the truncus medius join and form the fasciculus lateralis (lateral cord). The posterior parts of all three trunci join to form the fasciculus dorsalis (posterior cord), and the anterior part of the truncus inferior forms the fasciculus medialis (medial cord). The fasciculi are situated infraclavicularly, i.e., below the clavicle. Finally, the fasciculi divide into the peripheral nerves that supply the muscles in the arm.

Individual variations of the general anatomy may occur. The brachial plexus may, for example, be formed by the roots C4 to C8 or T1. This is called a prefixed plexus. Another possibility is a postfixed plexus, which is formed by roots C5 or C6 to T2. Such variations hamper the diagnosis of brachial plexus injuries.

3.2 PATHOLOGY AND ETIOLOGY OF BRACHIAL PLEXUS INJURIES

Brachial plexus injuries may be characterized according to the locations injured and the severity of the injury. The locations that may be injured consist of the anatomic structures already discussed, e.g., the spinal nerve C5, truncus superior, or fasciculus lateralis. The severity of a brachial plexus injury may be classified according to the structures in the nerve that are affected. A peripheral nerve consists of nerve fibers that contain the axons, surrounded by supportive tissues.

The classification of Seddon (1943) is well known and is defined as follows:

Neurapraxia: A lesion in which there is no axonal degeneration.
Axonotmesis: A lesion characterized by complete interruption of axons, but with preservation of the supporting structures of the nerve.
Neurotmesis: A lesion of such severity that all essential parts of the nerve are destroyed.

Another type of injury is one in which the nerve roots are avulsed (torn away) from the spinal cord; this may affect motoric or sensory rootlets alone, or both. Avulsions are the most serious type of plexus injury.

The brachial plexus may be damaged by traction, compression, penetration, or nontraumatic causes. Traction is the most frequent type of injury in brachial plexus lesions. These injuries are usually due to a forceful widening of the angle between the shoulder and the neck, or between the upper arm and the trunk. They most frequently occur during traffic accidents. Approximately 70% of traumatic brachial plexus injuries are caused by such accidents, and approximately 70% of the lesions in accidents involve the use of a bicycle or motorcycle (Narakas, 1985).

3.3 DIAGNOSIS

In order to determine an appropriate therapy, it is necessary to obtain a precise diagnosis, including the exact locations within the brachial plexus that are injured and the severity of the injury. Since it is not possible to directly measure the state of the nerves, indirect measurements must be used. This requires extensive neurological, neurophysiological, and radiological examinations, the results of which have to be interpreted and combined with patient history information. The most important data that are needed are discussed briefly here. For a more complete discussion, see Jaspers (1990).

Patient history. The exact cause of the injury and additional traumata may provide clues regarding the extent and severity of the injury. For instance, high velocity injuries are often more severe than injuries that take place at a low velocity.

Neurological examination. The motor function (muscle strength) examination is of utmost importance for determining the locations within the brachial plexus that are injured. Sensory examination of the segmental or peripheral nerve innervation areas provides further information. Other important clinical signs are Tinel-Hoffman's sign, which indicates the presence of a lesion in which a connection with the spinal cord has been preserved, and the presence of Horner's syndrome, which is indicative of root avulsion of at least T1.

Neurophysiological examination. Electromyography (EMG) provides additional information about the state of the muscles and allows the testing of muscles that are otherwise inaccessible. The EMG may reveal findings, such as early reinnervation, that may not yet be detectable by physical examination. Other neurophysiological examinations are measurement of somatosensory evoked potentials (SEP) and sensory nerve action potentials (SNAP).

Radiological examination. A plain X-ray of the clavicle, cervical spine, scapula and humerus is required. Any damage to these bone structures is also indicative of the severity of the lesion.

Furthermore, the area where the spinal nerves leave the spinal cord is investigated. A cervical myelography combined with a CT-scan can provide information as to whether nerve roots are avulsed.

3.4 THERAPY

Sharp lesions, such as stab wounds, have to be treated surgically immediately. In all other cases, associated injuries, such as fractures, vascular injuries, and head trauma, are treated in the early stages when an accurate diagnosis still has to be established. Conservative treatment is started to prevent contractures and control pain, and orthoses are provided.

When an accurate diagnosis is available, a decision is made as to whether the patient should be considered for neurosurgical repair. Generally, only more serious injuries that will not recover spontaneously are treated surgically. Depending on the nature of the injury, different neurosurgical procedures may be performed. One such procedure is nerve grafting. A nerve graft is a length of donor nerve, which may be taken from the leg, in order to replace the injured part of a nerve. In plexal root avulsions (nerve roots that have been torn away from the spinal cord), the only possible means to restore continuity is by coaptation with neighboring nerves, either from within the brachial plexus (intraplexal nerve transfer) or from outside the plexus (extraplexal nerve transfer). It may be necessary to use both nerve grafts and nerve transfers in order to reconstruct a brachial plexus.

When nerves are expected to be continuous but no recovery occurs, a possible operative procedure is the removal of scar tissue. This procedure is called neurolysis. It is indicated only in late cases and is a potentially hazardous procedure. After neurosurgery, conservative treatment is again necessary to mobilize joints, to re-educate reinnervating muscles, and for psychosocial support.

When the final prognosis is definite, secondary surgery may be considered. This may, for instance, entail transferring certain muscles or tendons, or fixation of certain joints. A comprehensive discussion of the treatment of brachial plexus injuries may be found in Alnot and Narakas (1989).

4   Need for assistance

To investigate whether there is a need for assistance in the domain of brachial plexus injuries, patient files were studied to determine whether patient management could possibly improve if physicians were to use a computer advisory system. Secondly, a study was carried out to determine whether there is a recognized need for assistance on the part of the physicians who are the potential users of such a system.

4.1 OBJECTIVE NEED FOR ASSISTANCE

In order to investigate the difficulties associated with the diagnosis and management of patients with a brachial plexus injury, a retrospective study was performed on 136 patients who had been referred to the De Hoogstraat Rehabilitation Center from different hospitals across The Netherlands. Of these 136 patients, 93 patients had been admitted to the rehabilitation center before 1981 and 43 patients were admitted between 1981 and 1985. A number of problems were identified, including the following:

- Localization of brachial plexus injuries is a very complex process, due to the complex anatomy of the brachial plexus and to possible anatomic variations.
- In the early stages of the injury, there are often associated injuries that require immediate attention, such that the brachial plexus injury is left unattended.
- Diagnosis is often neglected by the referring clinic because neurosurgical possibilities may not be known, or because physicians may have a pessimistic view of the results of reconstructive neurosurgical procedures. Of the 43 patients who were admitted to De Hoogstraat between 1981 and 1985, and who were involved in the investigation, 21% had not received any additional diagnostic tests in the referring clinic other than motor and sensory examinations. In 47% of the 136 patients who were studied, the referring physician had not recorded a diagnosis in the patient file, indicating the site, extent, or severity of the injury.
- Patients are often referred to a rehabilitation center at a late stage. For the patients who were admitted to De Hoogstraat after 1981 and who were involved in the investigation, the average time to admission was 12.5 months and the median time to admission was 3 months.
- Only few patients are treated neurosurgically. In the group of patients admitted to De Hoogstraat between 1981 and 1985, the percentage receiving nerve repairs was 25%.

Due to these problems, on average treatment results for the patients in this survey seem to be worse than the results found in the literature (Narakas, 1981; Dolenc, 1982; Gusso et al., 1985). The full results of the study have been described by Jaspers (1990). It is clear that there is a need to improve the diagnosis of brachial plexus injuries and also to increase the awareness of the possible benefits of neurosurgical treatment and the necessity to refer brachial plexus patients to a specialist center at an early stage.

4.2 SUBJECTIVE NEED FOR ASSISTANCE

In addition, the need for assistance for potential users of a computer advisory system was investigated. Grolman (1989) performed a preliminary study of 67 neurologists in The Netherlands. For this investigation a questionnaire was developed and distributed among the neurologists. Since only 19 of the 67 questionnaires were both returned and at least partly completed, careful interpretation of the results is required. Some results of the study are shown in Table 1.

Table 1.   Results of a questionnaire distributed among 67 neurologists in The Netherlands

Number sent out	67			
Number returned and completed	19 (28%)			
Question	very/yes	fair/some	poor/no	no answer
System will be used	7 (37%)	4 (21%)	5 (26%)	3 (16%)
Would use system personally	8 (42%)		5 (26%)	6 (32%)
Expect problems for introduction	2 (10%)	7 (37%)	6 (32%)	4 (21%)

It can be seen from Table 1 that 11 of the 19 neurologists thought a computer program in the domain of brachial plexus injuries would be used in practice if made available, and 8 of 19 physicians said they would personally use the decision support system.

It can thus be seen that these physicians are positively inclined toward using a decision support system for the diagnosis and treatment planning of brachial plexus injuries, although there are some physicians who do not think that such a system would be used. The neurologists also indicated a number of requirements that will have to be met by such a system. An advisory system must be of impeccable medical quality, and it must not be time consuming to use. Furthermore, such a system must be user friendly and well validated.

5   The knowledge-based system PLEXUS

In the previous section a need for assistance and a mildly positive attitude among neurologists towards a decision support system in the domain of brachial plexus injuries have been demonstrated. A decision support system has been developed in cooperation with two Dutch brachial plexus experts, Prof. Dr. R.T.W.M. Thomeer of the Academic Hospital in Leiden and Dr. A.C.J. Slooff of the De Wever & Gregorius Hospital in Heerlen. The system, PLEXUS, is meant for physicians who are not specialists in brachial plexus injuries.

In order to request advice from PLEXUS, the physician enters patient data into the computer, the computer then reasons with the patient specific data and uses the general knowledge concerning brachial plexus injuries that is stored in the system to generate patient specific advice regarding

- the locations that are injured,
- the severity of the injured locations, and
- the preferred treatment.

PLEXUS uses patient history information and the results of radiological examinations, in addition to neurological and neurophysiological data. The PLEXUS advice is shown to the physician on the computer screen. The most important aspects of the system will be discussed in the following. A detailed description of the architecture of the knowledge based system PLEXUS may be found in Jaspers (1990).

5.1 KNOWLEDGE BASE

PLEXUS comprises two knowledge bases that are implemented in the Delfi 2+ expert system shell (de Swaan Arons, 1991). The first knowledge base, PLEXAKT, contains the knowledge necessary for localization of brachial plexus injuries and the second knowledge base, TREAT, contains knowledge about the severity of injuries and treatment planning. The architecture of both knowledge bases is now discussed.

The diagnostic knowledge base PLEXAK.   The aim of the diagnostic module is to determine the exact location of the structures within the brachial plexus that are injured. The solution strategy has been implemented according to the following general method. The data that are entered into the system by the physician are first abstracted into meaningful intermediate concepts. Following data abstraction, the concepts are checked for possible inconsistencies. Significant inconsistencies are reported to the physician. Using the intermediate concepts, a rough localization of the injury is then performed using production rules (if-then rules). Based on the rough localization, the exact injured locations are found by means of a hypothesize-and-test algorithm that hypothesizes possible injury combinations and tries to find the combination that best explains the motoric deficit in the arm.

As already explained, three different tasks may be distinguished: data abstraction, heuristic match, and refinement. The representation of these tasks is now explained.

Data abstraction. The data abstraction knowledge converts the data into intermediate concepts to be used for further reasoning, and then detects possible inconsistencies and incompleteness in the data that have been entered. The data abstraction knowledge is represented in the form of production rules.

A number of production rules test whether the data that is entered into the computer is consistent. There are, for instance, various tests that provide similar information. By checking whether the results of these examinations contain the same information, the consistency of the data can be checked.

When inconsistencies are detected, certain rules containing messages are presented that indicate the inconsistencies that have been found. The system shows these messages on the computer screen and requests the user to perform the tests again, but also goes on reasoning with the evidence it has, using the results of the tests that are most reliable. The production rule formalism is well suited to this kind of knowledge, since a certain action has to be taken when certain evidence is found. An example of a consistency checking rule can be seen in Figure 2.

```
CONSISTENCY CHECKING RULE


IF
[ Tinel.location = "supraclavicular" ]
AND
[ c5.Tinel_radiating ]
AND
NOT [ c5.lesion ]
AND
NOT [ c5.sensibility = "anaesthetic" ]
AND
NOT [ c5.sensibility = "hypaesthetic" ]
THEN
CONCLUDE patient.required_examination := "check_sign_of_Tinel"   CF (1.000)
CONCLUDE patient.required_examination := "sensibility"   CF (1.000)
EXECUTE plexusremark
WRITE ****************************************************************
WRITE    The Tinel-Hoffman sign is radiating towards the c5-dermatome, indicating
WRITE    a lesion of spinal nerve c5. But there is no sign of motoric or sensory
WRITE    disability of this spinal nerve. Therefore I advise you to check again
WRITE    the sensibility of the c5-dermatome as well as towards which dermatome the
WRITE    Tinel-Hoffman sign is radiating.
WRITE ****************************************************************
```

Figure 2   Consistency checking rule

It is clear that inconsistencies can only be found when redundant information is present. The detection of inconsistencies allows the system to deal with uncertain information (measurement noise) to a certain extent, and it also makes the system more robust for uncertain knowledge, such as individual variations, because it can detect whether this is present. In addition, there are rules that detect whether insufficient evidence is available to find the injured locations. Messages will be shown to the user when this is the case. When additional information would be required to perform a better localization, the system will provide a diagnosis based upon the information that is available but also indicates that it would be able to provide improved localization if more information were present.

Heuristic match.. The heuristic match task provides a rough localization of the injury. It contains empirical associations between the intermediate concepts and the conclusions that may be drawn. Production rules have been used to represent the surface knowledge used for the heuristic match task.

Evidence from patient history, and neurological, neurophysiological, and radiological examinations is used to draw possible conclusions. An important drawback of rule based systems is that when a large collection of rules is used, the structure will usually not be transparent. To overcome this problem, all the evidence has been classified into five different categories. The categories of evidence are used in different kinds of production rules as follows:

Triggering facts:	Facts that immediately lead to a certain conclusion, regardless of other facts that may be present.
Necessary facts:	Facts that have to be present for a certain conclusion to be true.
Exclusionary facts:	Facts that immediately lead to exclusion of a certain conclusion.
Corresponding facts:	Facts which will increase the certainty of a conclusion that has already been established; however, the presence of such a fact alone will not lead to the conclusion being true.
Irrelevant facts:	Facts that are not relevant for a certain conclusion.

This classification of evidence makes the uncertainty in the suggestive strength of each piece of evidence for each hypothesis explicit, rather than quantifying it. The transparency of the system also improves by using the classification of evidence, since it shows the relationship of each piece of evidence to each hypothesis.

The strategy is to use the evidence in such a way that, first as many hypotheses as possible are confirmed and dismissed, and then to prove any additional hypotheses by weighing positive evidence against negative evidence. For that purpose the various kinds of evidence are used in four different kinds of production rules:

Triggering rules:	Rules that use triggering facts to immediately establish a hypothesis.
Pruning rules:	Rules that use exclusionary facts to immediately exclude a hypothesis.
Evaluation rules:	If no sufficient exclusionary evidence is available to rule out a hypothesis, and sufficient necessary facts are present, the hypothesis is postulated to be processed further.
Confirmation rules:	Corresponding facts are used in order to become more certain about a hypothesis that has already been established.

A sample pruning rule is shown in Figure 3.

```
PRUNING RULE

IF
[Tinel.location = "supraclavicular"]
AND
[c5.Tinel_radiating]
THEN
CONCLUDE c5.exclude_avulsion := TRUE   CF (1.000)
FI
ENDRULE
```

Figure 3   An example of a pruning rule from the heuristic match task

The heuristic match task provides a rough localization of the injury. After the heuristic match task, it may be certain for some locations whether or not they are injured. For other locations, the empirical associations that are used are not deep enough to be able to exactly determine whether they are injured. In general, a number of competing hypotheses will be present that constitute a set of mutually exclusive diagnoses that can each account for the symptoms considered by the heuristic match task. Therefore, this task is complemented by a deep refinement task to find the exact injured locations.

Refinement. The refinement task, for exact localization of the injury, is based on deep knowledge of the structure and function of the nerves of the brachial plexus. In the brachial plexus, 41 different possible injury locations have been distinguished. Usually, one brachial plexus lesion consists of more than one injured location. Therefore, in theory, there are 2^{41} different injury combinations.

It is not feasible to enumerate all possible combinations and to test which combination best explains the data. Thus, it is not possible to use a hypothesize and test approach for every combination that may occur. Therefore, during the heuristic match task, all evidence about certain hypotheses is gathered and as many hypotheses as possible are confirmed or excluded based on shallow knowledge and on knowledge that builds up the combinations from individual locations. In this way, the search tree consisting of all possible injury combinations is pruned.

Following this pruning, the possible combinations that remain can be hypothesized and tested to see which combination best explains the motoric deficit in the patient's arm. For efficiency reasons, a heuristic search algorithm is applied that searches for the optimal solution without requiring every possible combination to be tested.

Representation of brachial plexus for refinement.. For PLEXUS the abstraction level of the deep representation is at the level of the nerves. The structure and function of the nerves are represented in the computer. The structure of the brachial plexus is the way in which the nerves are connected to each other and finally to the muscles of the shoulder, arm, and hand. The function is represented as the conduction of signals directed through the network towards the muscles.

If there is no injury, then the signals from the central nervous system can be conducted through the network of nerves and the muscles have full function. If there is an injury somewhere in the pathway between the spinal cord and a muscle, the signal will not be conducted completely and the muscle will only function partially. If all pathways to a muscle are blocked, then the muscle will not function at all.

Thus when a certain injury combination is hypothesized, from the structure of the pathways and the function of passing on the signal, a prediction can be made as to whether a muscle will function fully, function partially, or not function at all. This can then be compared with the actual muscle strengths that were measured by the physician during the neurological examination. Part of the structure of the nerves is shown graphically in Figure 4.

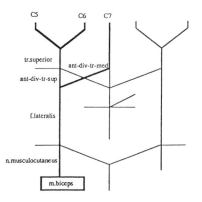

Figure 4   Graphical representation of part of the structure of the brachial plexus

In Figure 4, the innervation of the biceps muscle has been highlighted. The structure of the highlighted part of the brachial plexus may be represented in the computer in the following manner:

biceps = (((C5 + C6) * truncus-superior * anterior-division-truncus-superior + C7 * anterior-division-truncus-medius) * fasciculus-lateralis * n-musculocutaneous)

The function of the pathways of nerves that lead to the musculus biceps can also be represented. Assume that C5 and C6 have a more prominent part in the innervation of the musculus biceps than the spinal nerve C7, and assume that an intact nerve has an innervation value of 1,   and a defect nerve has an innervation value of 0, then the function can be represented in binary relations. An example of a binary relation for the musculus biceps is the following:

innervation-biceps = (((2/5*C5 + 2/5*C6) * truncus-superior * anterior-division-truncus-superior + 1/5*C7 * anterior-division-truncus-medius) * fasciculus-lateralis * n-musculocutaneous)

Depending on the parts of the plexus that show a defect, the innervation of the biceps can be calculated. It may range from 0 to 1. All muscles can be represented in this way. When a certain combination of injured nerves is hypothesized, the innervation of the muscles can be predicted by these binary relations. The predicted innervation of the muscles can then be compared with the real innervation that was found by the physician during the neurological examination.

Heuristic tree search for refinement.   As was stated earlier, it would still be very inefficient to hypothesize all possibilities, i.e., all combinations of injured locations that remain after the heuristic matching task. Therefore, a heuristic tree-search algorithm is used to find the best possible combination of injured locations without hypothesizing every possible combination. The A* algorithm, adapted for trees instead of graphs, (Rich, 1983) has been implemented in PLEXUS (de Lind van Wijngaarden & Furth, 1987; Meinders, 1989).

From the initial state determined by the heuristic match task, this Algorithm for Knowledgeable Trees (A^{kt}) generates candidate solutions for the locations of the injury in a breadth first manner. Subsequently it employs an evaluation function to assess the quality of the generated candidates and a heuristic function to estimate the improvement that can be gained by further expanding the tree and evaluating the children of a candidate solution. To guarantee that the optimal solution will be found, the heuristic function is implemented in such a way that it overestimates this potential improvement in the diagnosis, or at best is a precise estimate.

For PLEXUS, the algorithm incorporating a heuristic search consists of 41 different locations that may be injured. The locations that are known to be intact or defective after processing the production rules are set at their final value. This means that the search tree is pruned, since certain branches of the search tree do not have to be expanded. The locations that are unknown after processing the production rules are evaluated by the A^{kt} algorithm.

The principle of parsimony has also been incorporated. This principle implies that when different solutions present the same value of the evaluation function, the combination of the least number of injured locations that explains the symptoms is selected as the correct diagnosis. Since the brachial plexus is a redundant network of nerves, there may be several explanations for a certain motoric deficit. For instance, an injury of two nerves more distal (lower) in the plexus may cause the same motoric deficit as one injury more proximal (higher) in the plexus. This is, for instance, the case in a possible injury of both n.axillaris and n.radialis as opposed to an injury of the fasciculus dorsalis. This principle is used in most neurological localization programs (First et al., 1982; Hertzberg et al., 1987; Fisher, 1990). In PLEXUS, the principle of parsimony is applied after the production rules have been processed and is used only for the remaining essentially similar hypotheses.

Thus, the localizing strategy incorporated in PLEXUS consists of three phases:

- Data abstraction takes place with consistency checking and transformation of the data into meaningful concepts,
- Production rules are used for a rough localization of the injury and pruning of the search tree,
- The A^{kt} algorithm is applied for exact localization of the brachial plexus injury.

After localization of the injury, the treatment module is applied.

Knowledge base for providing treatment advice TREAT. In order to suggest a therapy, it is necessary to determine the severity of the injury and to distinguish between injuries that show spontaneous recovery and those that will not recover spontaneously. When the severity of the injury has been investigated, the treatment plan can be determined. The main aim is to differentiate among those patients who should be treated only conservatively from those who should be treated surgically, so that the patients who should be treated surgically may be referred to a specialist center for surgery.

The treatment planning module is completely rule based. First, production rules are applied to assess the severity of the injury. These rules use information about the localization of the injury and additional data obtained from the patient history, radiological examination, and more recent physical and neurophysiological examinations, so it can be decided whether any improvement has taken place.

Three different groups of injuries may be distinguished:

- Injuries that will recover spontaneously
- Injuries that will not recover spontaneously
- Injuries for which it is not yet known whether they will recover spontaneously

After the severity has been determined, the treatment planning knowledge is applied.

Conservative treatment is advised for injuries that will recover spontaneously. For nerve injuries that will not recover spontaneously, surgical treatment may be advised (although this will not always be the case).

The system distinguishes between three general surgical procedures. These are not specified in detail; that is the task of the physician in the specialist center who will perform the operation. When it is not yet possible to decide whether spontaneous recovery will take place, the system advises further diagnostic testing and another consultation with the system to be performed after a certain period of time.

5.2 USER INTERFACE

The PLEXUS user interface is meant for the input of data and the output of advice. Users may enter all relevant data into the computer by means of the user interface. When data entry has been completed, consultation with the knowledge based part of the system may be requested. The recommendations provided by PLEXUS are then shown on the computer screen.

The user interface has been designed according to the results of an investigation of the present practice of neurologists, and of neurologists' requirements regarding computer advice and the presentation of that advice (Grolman, 1989). Presently, the user interface runs on an Apple Macintosh computer and is implemented using the software package Hypercard. Interaction with the system requires no previous typing and computing experience.

The user interface is based on a well-known scheme devised by Merle d'Aubigné and Deburge (1967). This scheme is seen in Figure 5. The way in which this scheme has

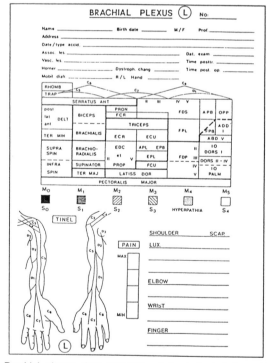

Figure 5. Brachial plexus data recording form (adapted from Merle d'Aubigné & Deburge, 1967, with permission)

been represented on the computer screen is seen in Figure 6. This scheme is the first page of the user interface; it shows a summary of the data that have been entered. The actual data entry is carried out on subsequent pages of the interface. An example of such a screen is seen in Figure 7.

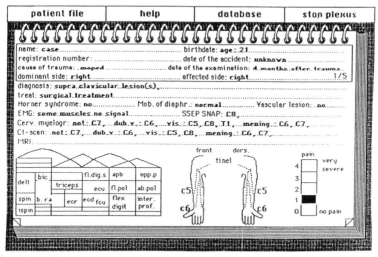

Figure 6.   Summary information represented on the computer screen.

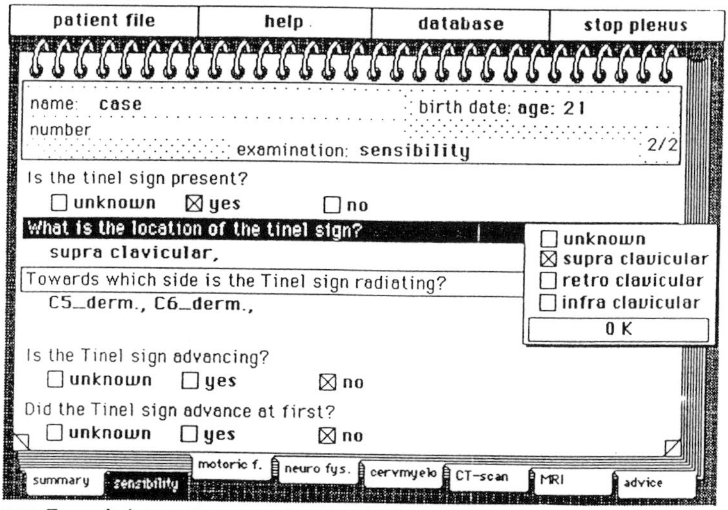

Figure 7.   A data entry screen for neurological examinations.

Most of the data entry is carried out using the computer mouse by clicking on the relevant answer possibility. The data entry pages have been divided into five different sections:

- Patient history
- Neurological examinations

- Neurophysiological examinations
- Radiological examinations
- Advice.

It is possible to quickly skip to the next section by clicking on the relevant section name at the bottom of the screen. Each of these sections contains various screens on which data may be entered. One may proceed to the next page by clicking on the dog-ear in the bottom right-hand corner of each screen.

When all relevant data have been entered into the computer, the physician can request advice from the knowledge based system. It is not necessary to answer all questions in order to perform a consultation. It is up to physicians to gather the data that they think are relevant for a specific patient. A consultation can be requested by clicking on the consultation option in the advice section. The system will then reason with the data and the knowledge represented in the knowledge based system, and the diagnosis and the treatment plan are shown to the user in textual form. The injured locations are also shown in a graphical representation of the anatomy of the brachial plexus. An example of possible graphical output is shown in Figure 8.

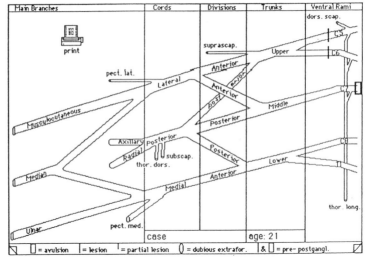

Figure 8.   Injured locations (, I) shown in a schematic graphical representation of the brachial plexus.

6   Evaluation

The final objective of any expert system is to improve patient care. Before this can be studied clinically, laboratory testing should indicate that the system is reliable, does not produce potentially hazardous advice, and is potentially useful. During its development, PLEXUS has been tested by introducing both retrospective cases and by using a large number of computer generated cases. A number of studies involving the cooperating brachial plexus specialists were also performed.

In addition, two more formal empirical studies were carried out (van Daalen, 1993; Van Daalen & Malessy, 1995). The first study was a comparison of PLEXUS to international brachial plexus experts; the second study was a clinical evaluation of the problem-solving performance, usability, and acceptability of the system. The preliminary validation studies will first be discussed briefly, after which the two formal studies will be dealt with more extensively.

6.1 PRELIMINARY VALIDATION

During the development of PLEXUS, the performance of the system was tested using about 100 test cases consisting of retrospective patient data provided by the cooperating experts. The results of these cases were reviewed and the system was updated until it was felt that the performance on these training cases was at an expert level. At this point, preliminary evaluation studies of the system's output were performed.

The test cases in a first test of system performance consisted of 15 retrospective cases that originated from one of the cooperating experts. The data were entered into the computer by the researchers. The diagnostic and treatment advice provided by PLEXUS was compared with the diagnosis and treatment that was determined by the expert who actually saw the patients. Percentages of poor, fair, and good cases were calculated. The system did not give any poor recommendations. 80% of the diagnoses were judged to be good, and 87% of the treatment plans were judged to be good.

To gain a better insight into the level of expertise reached by the system in comparison with human experts, a further preliminary performance evaluation entailed a double-blind evaluation involving both cooperating experts. Both cooperating experts provided data on the 10 latest patients on whom they had operated. The test cases were entered into the computer by the experts themselves. The system was tested against the two cooperating experts, who each diagnosed the 10 cases that did not originate from them. The origin of the diagnoses was blinded, and the treating experts were asked to rate both the system's diagnoses and the other expert's diagnoses of the treating expert's patients on a five point scale. The number of cases in which the system had the same score as the nontreating expert, the number of cases in which the system had a better score than the human expert, and the number of cases in which the human expert scored higher were determined. In 7 out of 20 cases, the knowledge based system's advice was judged to be better than that of the human expert, and in 5 out of 20 cases the expert's advice was judged to be better.

During this evaluation, aspects of user interaction were addressed implicitly, since the physicians themselves entered the data. At that time, the user interface was a textual interface, and the interaction between the user and the knowledge based system consisted of the physician typing the answers to the questions that were posed by the knowledge based system. This user interface proved to be inadequate for physician use of the system. In order to solve this problem, the present graphical interface was developed (Grolman, 1989). The graphical interface was informally evaluated by videotaping sessions in which the experts interacted with the knowledge based system. The user interface was updated on the basis of information obtained during these sessions.

A knowledge based system should be tested on a range of test cases. However, knowledge based systems are often developed for domains for which large numbers of test cases are not available. This means that there is usually a lack of cases for testing the system. Furthermore, since solving the problem often requires expertise, it is usually not possible for the developer to directly draw up suitable test cases, and it is too time consuming for an expert when large numbers of test cases are needed. In order to adequately validate a knowledge based system, some kind of method for test case generation is often needed. One method of test case generation is to use an inverse knowledge based system, which upon being given a diagnosis generates the input data that correspond to the diagnosis. Although this method has a number of limitations and is not a substitute for empirical studies, validation on a large range of test cases, either using real or generated test cases, is necessary for all systems that are meant for actual use since it allows many more cases to be tested than in an empirical evaluation study.

Shwe et al. (1989) described a system that generates scripts of test data. A similar approach has been applied to PLEXUS (Voorhorst, 1993). Upon entry of a diagnosis, a test case generator provides input data for the knowledge based system. The test case generator uses diagnoses as its input. It reads these diagnoses from a file. This file consists of combinations of injured brachial plexus locations. Because in theory there are 2^{41} different injury combinations it is not possible to test all of them. Therefore, the diagnosis file is filled up by running a computer program that chooses different injury combinations based on certain criteria that depend on the goal of the study. Possible choices could be to test extreme diagnoses or to test diagnoses that are clinically relevant. For instance, the program could determine all injury combinations, existing of less than six injured locations in which the injury locations are situated close to one another. In this way choices can be made as to the diagnoses that are used for producing test cases.

When this test case generator was used for PLEXUS, various errors and omissions in the knowledge base were detected. One conclusion reached was the fact that although PLEXUS establishes a diagnosis on the basis of various data, the final diagnosis is not tested for consistency. In a few test cases this led to an inconsistency. Furthermore, the anatomic model used by PLEXUS shows a few slight differences from the literature. One conclusion of a sensitivity analysis carried out with the test case generator shows that the system is quite sensitive to variations in the muscle strength measured. Therefore, correct measurement of muscle strength by the physician is important.

6.2 FORMAL EVALUATION

Two formal empirical studies were carried out. The first study was a comparison of PLEXUS to international brachial plexus experts; the second study was a clinical evaluation of the problem-solving performance, usability, and acceptability of the system. The expert system was not altered during the course of the formal studies.

Comparing PLEXUS with brachial plexus experts. PLEXUS was compared with four internationally acknowledged and independent brachial plexus experts from Switzerland, France, Spain, and Great Britain. This investigation of the system's problem solving performance consisted of three different rounds. In the first round, the four international experts were asked to submit retrospective data on 10 patients, and to add their own diagnoses and treatment plans. The diagnoses and treatment plans were removed from the data for the next round. The first round only consisted of data gathering. The data were not presented to the computer at this stage.

In the second round, the data from these 40 patients were entered into the computer. In addition, each expert received the data of 15 of the patients he had not submitted and was asked to provide a diagnosis and treatment plan for these cases. This resulted in three opinions per case: one provided by the attending expert who had actually treated the patient, one by PLEXUS, and one by an expert who had not treated the patient. In order to perform a direct comparison, a standard of performance is necessary. The clinical diagnosis established by the expert who actually saw the patient was used as the standard for patients who were treated conservatively, and the per-operative diagnosis was used as the standard for patients who were treated surgically. Using this standard, the opinions provided by the computer and by the nontreating experts were analyzed.

Since a brachial plexus traction injury is frequently not confined to one lesion site within the brachial plexus, each location was diagnosed, including its severity. The diagnoses were analyzed by calculating the mean number of true-positive, false-positive, and false-negative answers per patient as given by the computer and by the nontreating experts when compared with the treating expert. For example, if PLEXUS concluded root avulsion C5, C6, and C7, whereas the treating expert diagnosed avulsion C5 and C6, the result of the comparison would be that PLEXUS produced two true-positive answers and one false-positive answer for that individual patient.

With respect to treatment, the choice was between three decisions: to refer the patient for neurosurgery, to provide conservative treatment only, or to wait and postpone the final treatment decision. The agreement between PLEXUS and the treating experts, on the one hand, and between the nontreating experts and the treating experts, on the other hand, was calculated.

As mentioned earlier, three opinions per case became available during the second round. In the third round, rather than using the treating experts' opinion as the standard, nontreating experts were asked to judge the opinions. Each expert was asked to judge all three opinions for each of the 15 patients he had not submitted and had not seen in the second round of the evaluation. This was done blindly, i.e., the origin of the advice was removed and for each patient the three opinions were stated in random order. The judgement categories ranged from very good to very poor on a five point scale. The Wilcoxon-Mann-Whitney test was used, and significance was concluded at the 5% level.

The second round of this study revealed that the mean number of true-positive answers per case given by PLEXUS could not be shown to differ significantly from the number of true-positive answers given by the nontreating experts (Table 2). This was also true for the number of false-negative answers. The number of false-positive answers provided by the system was significantly higher. For one case, no diagnosis was given by the treating expert. Therefore, a total of 39 diagnoses could be used in the comparison of PLEXUS and the treating experts. The total number of cases diagnosed by the nontreating experts is more than 39 (Table 2). Removal of the cases that were used twice yielded the same results.

Table 2   PLEXUS diagnostic performance in the international evaluation

comparison	number of cases	TP mean ± SD	FP mean ± SD	FN mean ± SD
PLEXUS vs. treating experts	39	2.2 ± 1.7	2.7 ± 1.9	1.8 ± 1.4
non-treating vs. treating experts	47	2.5 ± 1.8	1.5 ± 1.3	1.5 ± 1.3

Abbreviations: *TP* mean number of true positive answers per patient; *FP* mean number of false positive answers per patient; *FN* mean number of false negative answers per patient.

The results of the comparison of treatment plans (Table 3) show that the proportion of agreement between PLEXUS and the treating experts is somewhat higher than between nontreating and treating experts, since nontreating experts more often decided to postpone the final decision. The Kappa coefficient of agreement (Cohen, 1960; Fleiss, 1981) was calculated and showed no significant difference between the score obtained by PLEXUS and the score obtained by the nontreating experts. Cases evaluated by more than one nontreating expert were omitted from the sample, i.e., each patient was only evaluated once.

Table 3.   PLEXUS treatment performance in the international evaluation

comparison	number of cases	proportion of agreement	Kappa mean ± SE
PLEXUS vs. treating experts	40	0.88	0.61 ± 0.14
non-treating vs. treating experts	31	0.74	0.40 ± 0.14

Els van Daalen and Rob Jaspers

In the third round, the experts blindly judged the opinions provided by PLEXUS and the treating and nontreating experts on a five point scale. Only the cases that were rated on the five point scale have been taken into account. With regard to the diagnosis, PLEXUS received a somewhat lower score than the experts (Table 4), although the differences were not significant.

Table 4. Scores received by PLEXUS, the treating experts and the nontreating experts in a blind judgement of diagnoses and treatment plans

category	number of cases	PLEXUS mean* ± SD	treating experts mean ± SD	non-treating experts mean ± SD
diagnosis	37	0.84 ± 1.01	0.95 ± 0.94	1.07 ± 0.97
treatment plan	29	1.38 ± 0.78	1.14 ± 0.69	0.75 ± 1.35

*The judgement scale ranged from -2 to + 2.

Analysis of the diagnostic performance in comparison to internationally acknowledged brachial plexus experts shows that the accuracy of the recommendations provided by PLEXUS is good. However, the system does produce a somewhat higher proportion of false-positive answers (see Table 2). With respect to treatment (see Table 3), the nontreating experts more often advised the patient to wait for a period of time than PLEXUS and the treating physicians did. This may be due to the fact that the treating experts did actually see the patients. The full results of the formal evaluations of PLEXUS have been described by van Daalen (1993).

Clinical evaluation. Before a formal clinical evaluation can be performed, a laboratory evaluation must have demonstrated adequate performance, safety, reliability, potential usefulness, and satisfactory human-machine interaction. Only few medical knowledge based systems, including PLEXUS, have been tested clinically (Adams et al., 1986; Bankowitz et al., 1989; Murray, 1990; Wyatt, 1989). PLEXUS was evaluated clinically in four large hospitals in The Netherlands. The neurologists involved were asked to use the system prospectively for all patients presenting with a brachial plexus injury. Before receiving the advice, the physicians were asked to enter their own diagnosis and treatment plan.

At the end of the 1.5 year evaluation period, two brachial plexus experts were asked to judge the diagnoses and treatment plans provided by the physicians and the advice given by PLEXUS. The origin of the opinions was removed and opinions were placed in random order. The experts judged the opinions on a five point scale. Results were analyzed by investigating the number of times the judging experts rated PLEXUS as being equal, better, or worse than the physicians. When the knowledge based system answer was judged worse, the reasons for this were analyzed in detail.

At the end of the evaluation period in the Dutch hospitals, the data on 19 patients had been entered into the system. Seven of these were prospective cases. Concerning diagnosis, PLEXUS scored equal to or better than the physicians in at least 84% of the cases. Regarding treatment, PLEXUS scored equal to or better than the physicians in at least 68% of the cases (Table 5).

Table 5. Number of times PLEXUS was rated to be equal, better, and worse than the physicians in the clinical evaluation

Category	Judge*	% equal	% better	% worse
Diagnosis	e1	53	42	5
	e2	31	53	16
Treatment	e1	37	31.5	31.5
	e2	42	37	21

*Two experts (e1 and e2) were involved in the clinical evaluation to judge the answers.

The objective of the clinical evaluation of PLEXUS was to investigate whether the system really has the capacity to assist physicians. Ideally, a large group of physicians assisted by the system should be compared with another large group not using the system. The assisted and unassisted situations should be compared by using the actual patient outcome. In this evaluation study, due to the relatively small number of physicians, the difference between the physicians would become too large when comparing two groups. The variability of treatment results, depending on, among other things, the variability of the neural tissue and the surgical techniques applied, in combination with a three year follow-up period needed for final assessment, also hamper such a study. Thus, the purpose of the study was to investigate physicians' unassisted and assisted opinions. However, only few physicians entered their final opinion after receiving the advice. Therefore, the setup was restricted to investigating diagnoses and treatment plans provided by the physicians and by PLEXUS.

Experience during the clinical evaluation showed that before a clinical evaluation is carried out, potential users should also be involved in a laboratory evaluation to study potential usefulness and usability. This would allow a number of the problems and suggestions that result from a clinical evaluation to be identified and corrected prior to a clinical study.

The results of the clinical evaluation of PLEXUS show some discrepancies between the two judging experts. In some cases this was due to a lack of information in the patient file. In others, neither of the diagnoses was completely correct and it was a matter of choice for the judging experts as to how heavily certain errors or omissions were penalised. With regard to the treatments, in a number of cases one of the judges differentiated more between the treatments, whereas the other stressed the importance of setting the indication for surgery per se.

During the clinical evaluation, data were not entered by the physicians as readily as had been expected. A reason for this may be that computer handling does not routinely belong to their daily work.

The usability of the computer system was investigated by analysing videotapes of the physicians working with the system. The videotapes were analyzed using a number of usability criteria (Ravden and Johnson, 1989). Ravden and Johnson (1989) describe an evaluation checklist for assessing usability of computer based application systems. Each of the first nine sections of the checklist is based on criteria that a well-designed user interface should aim to meet. The criteria are the following: visual clarity, consistency, compatibility with user conventions and expectations, informative feedback, explicitness, appropriate functionality, flexibility and control, error prevention and correction, and user guidance and support. Five hour-long videotapes were made during the clinical evaluation. The videotapes were played back and the time spent by the physicians on each interface screen was recorded. In addition, any important remarks made by the physicians and any observations that could be of interest to the usability of the interface were written down for each screen. The transcripts were then analyzed using the nine usability criteria mentioned earlier. This method used to evaluate the interface proved to be a very useful method from which a great deal of information was obtained.

Results of the usability study show that to improve the usability of the system, the consistency of some of the methods for data entry has to be improved, some of the items have to become more compatible with user conventions, and the system requires some additional functionality. With regard to additional functionality, PLEXUS could offer more assistance in entering the data into the computer. Furthermore, at present PLEXUS is aimed at providing advice after all data considered to be relevant by the physician are entered into the computer. The system is not especially geared at giving intermediate advice concerning further tests to be conducted after only the clinical data have been entered. The physicians indicated that they would appreciate advice about further examinations to be conducted after this point.

The conclusion that the possibilities for assisting the physicians in data gathering and data entry should be improved also arose from the performance study discussed earlier.

Acceptability was investigated by means of a short questionnaire distributed among the physicians. This included requesting their opinion as to whether the system would be used if generally available, whether they would use it personally, and asking them to mention positive and negative aspects of the system. The acceptability questionnaire revealed that, on the one hand, some time and effort are needed to enter patient data into the computer and , on the other hand, the systematic overview of the patient data is appreciated.

7   Conclusions

The evaluation of PLEXUS has shown that the system performs well, although certain areas still have to be improved. The answer to the question concerning the applicability of PLEXUS, and indeed of medical knowledge based systems in general, remains largely unanswered. A number of physicians indicated that they would use the system if it were generally available, whereas during the clinical evaluation the system was not used as readily as might have been expected. The acceptability of the system will require further investigation, since incorporation of some of the suggestions resulting from the evaluation studies will require significant alterations to the system.

Quality of care is becoming an increasingly important issue in the area of medicine. Clinicians will have to explicitly record and be able to justify the complete process of patient analysis and treatment. Furthermore, medical knowledge is advancing rapidly, and it is difficult to keep up to date with the most recent medical research. Therefore, it is expected that physician use of computers will become increasingly important. These may not be computer systems that perform complete patient analyses, but, for instance, programs that combine and show patient information in such a way that it helps the physician to correctly analyse the patient and decide on the appropriate therapy.

Acknowledgments

The development and validation of PLEXUS would not have been possible without the cooperation and expert advice provided by Dr. A.C.J. Slooff, Prof. Dr. R.T.W.M. Thomeer and Dr. M.J.A. Malessy. Prof. dr. J-Y. Alnot, Dr. R. Birch, prof. Dr. A.O. Narakas and Dr. A. Santos Palazzi are kindly acknowledged for taking part in the international evaluation of PLEXUS. In addition, we thank Dr. C.W.G.M. Frenken, Dr. P.J. de Jong, Dr. V. van Kasteel, Dr. J.F. Ploegmakers, Dr. M. Prick, Dr. R. Schellens, and Dr. T.W. van Weerden for their cooperation in the clinical evaluation of PLEXUS. The hardware for the clinical evaluation of PLEXUS was provided by Apple Nederland B.V.

References

Adams, I.D. *et al.* (1986). Computer aided diagnosis of acute abdominal pain: a multicentre study. *British Medical Journal*, 293: 800-804.

Alnot, J.Y. and Narakas, A. (eds.) (1989). *Les Paralysies du Plexus Brachial.* Expansion Scientifique Française, Paris.

Bankowitz, R.A., McNeil, M.A., Challinor, S.M., Parker, R.C., Kapoor, W.N. and Miller, R.A. (1989). A computer-assisted medical diagnostic consultation service: Implementation and prospective evaluation of a prototype. *Annals of Internal Medicine*, Vol. 110, pp. 824-832.

Cohen, J. (1960). A coefficient of agreement for nominal scales. *Educational and Psychological Measurement*, Vol. 20, No. 1, pp. 37-46.

Daalen, C. van, Stassen, H.G., Thomeer, R.T.W.M. and Slooff, A.C.J. (1993). Computer assisted diagnosis and treatment planning of brachial plexus injuries. *Clinical Neurology and Neurosurgery*, Vol. 95, pp. S50-S55.

Daalen, C. van (1993). Validating medical knowledge based systems. Ph.D. Thesis, Delft University of Technology, ISBN 90-370-0089-4.

Daalen, C. van and Malessy, M.J.A. (1995). Brachial plexus lesions: validating the expert system PLEXUS. *Neuro-Orthopedics*, Vol. 1g, No. 2, pp. 68-78.

Dolenc, V. (1982). Diagnostic et traitement des lésions du plexus brachial: A propos de 100 cas. *Neurochirurgie*, Vol. 28, pp. 101-105.

First, M.B., Weimer, B.J., McLinden, S. and Miller, R.A. (1982). LOCALIZE: Computer-assisted localization of peripheral nervous system lesions. *Computers and Biomedical Research*, Vol. 15, pp. 525-543.

Fisher, W.S. (1990). Computer-aided intelligence: Application of an expert system to brachial plexus injuries. *Neurosurgery,* Vol. 27, No. 5, pp. 837-843.

Fleiss, J.L. (1981). *Statistical Methods for Rates and Proportions.* NY: John Wiley and Sons.

Grolman, J.R.D. (1989). A user interface for the medical expert system PLEXUS (in Dutch). M.Sc. Thesis, Faculty of Industrial Design, Delft University of Technology.

Gusso, M.I. *et al.* (1985). Il trattament fisiokinesiterapico delle lesioni traumatiche del plesso brachiale. *Minerva Medica*, Vol. 76, pp. 635-640.

Hertzberg, T.M., Tremblay, G.F. and Lam, C.F. (1987). Computer-assisted localization of nervous system injuries. *Computers and Biomedical Research*, Vol. 20, pp. 489-496.

Jaspers, R.B.M., Louw, C.J.M. de, Lunteren, A. van and Stassen, H.G. (1982). Evaluation of the rehabilitation process of patients with a brachial plexus lesion (in Dutch). *Proceedings of the Boerhaave Course on Traumatic Brachial Plexus Injuries*, Leiden, pp. 65-89.

Jaspers, R.B.M. (1990). Medical Decision Support: An approach in the domain of brachial plexus injuries. Ph.D. Thesis, Delft University of Technology, ISBN 90-370-0028-2.

Kerr, A.T. (1918). The brachial plexus of nerves in man, the variations in its formation and branches. *American Journal of Anatomy*, Vol. 23, pp. 285-395.

Leffert, R.D. (1985). *Brachial Plexus Injuries.* Churchill Livingstone, New York.

Lind van Wijngaarden, D.G. de and Furth, V.H. (1987). *Diagnosis of Brachial Plexus Injuries.* Delft University of Technology.

Lunteren, A. van, Lunteren-Gerritsen, G.H.M. van, Stassen, H.G. and Zuithoff, M.J. (1983). A field evaluation of arm prostheses for unilateral amputees. *Prosthetics and Orthotics International*, Vol. 7, pp. 141-151.

Meinders, L.W. (1989). Application of a tree-search algorithm in a diagnostic expert system for brachial plexus injuries (in Dutch). M.Sc. Thesis, Report A-431, Laboratory for Measurement and Control, Delft University of Technology.

Merle d'Aubigné, R. and Deburge, A. (1967). Etiologie, évolution et pronostic des paralysies traumatiques du plexus brachial. *Revue de Chirurgie Orthopédique et Réparatrice de l'Appareil Moteur*, Vol. 53, No. 1, pp. 23-42.

Murray, G.D. (1990). Assessing the clinical impact of a predictive system in severe head injury. *Medical Informatics*, Vol. 15, No. 3, pp. 269-273.

Narakas, A.O. (1981). Brachial plexus surgery. *Orth. Clin. North Amer.*, Vol. 12, pp. 303-323.

Narakas, A.O. (1985). The treatment of brachial plexus injuries. *International Orthopaedics*, Vol. 9, pp. 29-36.

Ravden, S.J. and Johnson, G.I. (1989). *Evaluating Usability of Human-Computer Interfaces: A Practical Method.* Ellis Horwood Limited, Chichester.

Rich, E. (1983). *Artificial Intelligence.* McGraw-Hill, New York.

Seddon, H.J. (1943). Three types of nerve injury. *Brain*, Vol. 66, pp. 237-288.

Shwe, M.A., Tu, S.W. and Fagan, L.M. (1989). Validating the knowledge base of a therapy planning system. *Methods of Information in Medicine*, Vol. 28, pp. 36-50.

Stassen, H.G, Lunteren, A. van; Hoogendoorn, R., Kolk, G.J. van der, Balk, P., Morsink, G. and Schuurman, J.C. (1980). A computer model as an aid in the treatment of patients with injuries of the spinal cord. In: *Proceedings of the ICCS*, Cambridge, Massachusetts, IEEE, pp. 385-390.

Stassen, H.G. (1989). The rehabilitation of severely disabled persons. A man-machine system approach. In: W.B. Rouse (ed.), *Advances in Man-Machine Systems Research*, Vol. 5, JAI Press Inc., pp. 153-227.

Sunderland, S. (1968). *Nerves and Nerve Injuries*. Churchill Livingstone, Edinburgh.

Swaan Arons, H. de (1991). Delfi: design, development and applicability of expert system shells. Ph.D. Thesis, Delft University of Technology, ISBN 90-6275-734-0.

Voorhorst, F.A. (1993). Validating PLEXUS using generated test cases (in Dutch). M.Sc. Thesis, Report A-628, Laboratory for Measurement and Control, Delft University of Technology.

Wyatt, J. (1989). Lessons learnt from the field trial of ACORN, an expert system to advise on chest pain. In: B. Barber et al. (eds.), *Proceedings MEDINFO'89*, North-Holland, pp. 111-115.

Chapter 6

Identification of the Coronary Circulation

Jenny Dankelman
Jos A. E. Spaan

In this chapter we describe the most important results of our research concerning the identification of the regulation and mechanics of the coronary circulation. This research has been performed in close cooperation between the Laboratory of Measurement and Control of the Delft University of Technology and the Cardiovascular Physiology Group, now working at the Department of Medical Physics of the University of Amsterdam. This cooperative effort is more than ten years old and has resulted in an integration between system and control theory, modeling and simulation (TUD), on the one hand, and physiology, animal experiments, and validation (UvA), on the other hand. The coronary circulation supplies the heart muscle with blood. In a healthy heart, flow through the coronary vascular system is adapted to the metabolic needs of the heart muscle. However, in disease, flow may fall short in relation to needs. The purpose of the identification process is not purely descriptive, but is also to understand the determinants of coronary flow control and flow impediment.

Identification of the control and the mechanics of circulation is complicated for several reasons. Firstly, it is unknown how the different mechanisms that are involved in the flow control work. Therefore, it is difficult to give a full description of the system by physical laws. Secondly, the vessels are distensible and compressed by heart contraction. Since resistances are not constant, the system behaves nonlinearly. The use of standard identification techniques requires linearization of the system around one or more working points. However, in the beating heart it is often impossible to analyze the system in a particular working point since variables or parameters have ranges that are too wide. A third problem is that most identification techniques result in a description based on input-output behavior, whereas the main purpose of identification of the coronary circulation is to get more insight in the internal mechanisms of the system observed. A final complication is the limited possibility of measurements due to the fact that the phenomena studied take place on a micro scale and inside the heart muscle. This means that most direct measurements of significant variables are impossible. These four problems make the use of standard identification techniques based on linearization impossible.

Since we are not only dealing with unknown parameters, but also with uncertainty of the physical model structure, simulation models are developed on the basis of the limited knowledge of the system and on hypothesized (control) mechanisms. Simulated pressure -flow relations are compared with experimental data and fits are optimized by parameter adjustment. Critical experiments are designed to further test these models. In case the model structure itself allows accurate prediction of signals that can be measured, the parameter values needed for this have to be realistic. When necessary, the model structure should be changed guided by new hypothesized mechanisms. Specific examples of model development and evaluation are given. The results of other studies on the coronary circulation hinging on the cooperation between the departments are described.

1 Introduction

The heart is the pump that supplies all organs with blood, and hence must work continuously to sustain life. It is aided by the coronary circulation, which ensures that the heart muscle, or myocardium, is perfused with blood. The vascular bed distributes the blood flow and, therefore, functions as a network to supply nutrients (e.g., oxygen) and to remove waste products (e.g., adenosine).

The heart contains four chambers (Fig. 1A). During ventricular contraction (systole) blood inside the left ventricle is squeezed into the aorta, while blood inside the right ventricle is squeezed into the pulmonary artery. During relaxation (diastole) the right and left ventricles are refilled with blood from the right and left atria, respectively. The atria contract slightly earlier than the ventricles facilitating the filling of the ventricles.

The two feeding arteries for the coronary circulation originate from the aorta just above the aortic valve. These arterial vessels lie on the epicardium of the heart and branch into smaller arteries penetrating the myocardium (Fig. 1). These arteries branch into arterioles, which in turn supply the capillary bed, where material exchange with the surrounding tissue takes place. The capillaries drain into the venules, which join into veins running over the epicardium, mostly paralleling the coronary arteries. Finally, the blood is drained into the right atrium.

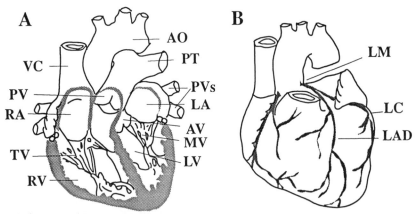

Figure 1. Anatomy of the heart. Panel A shows a schematic cross section of the frontal view of the heart. For the left part of the heart, blood from the pulmonary veins (PVs) flows into the left atrium (LA). During diastole and facilitated by atrial contraction, it flows into the left ventricle (LV) via the mitral valve (MV). When the left ventricle contracts, the blood is squeezed via the aortic valve (AV) into the aorta (AO). For the right part of the heart there is a similar circuit from the vena cava (VC) to the right atrium (RA) and via the tricuspid valve (TV) into the right ventricle (RV). After ventricular contraction it flows through the pulmonary valve (PV) into the pulmonary trunk (PT). Panel B shows the position of the epicardial coronary arteries on the epicardium of the heart. RC, right coronary artery; LM, left main coronary artery; LC, circumflex artery; LAD, left anterior descending branch.

A healthy coronary circulation system is a prerequisite for the heart to work properly. The amount of blood flowing through the myocardium depends on (1) mechanical factors and (2) the activation of smooth muscle tone in the arterioles (regulation, coronary flow control). Identification of the subsystems within the coronary circulation would be enhanced if quantities such as pressure in the myocardium and flow in the capillaries could be measured. Also, an accurate description of the structure of the vascular network would be helpful in predicting input-output relations. However, lack of information on all these points has resulted in the development of various hypotheses for mechanical interaction between heart contraction and coronary flow, on the one hand, and the control of flow, on the other hand. Some phenomena related to mechanical factors and smooth muscle activation are described in the following sections.

In this chapter coronary flow control and coronary flow mechanics are first described in detail (Sections 2 and 3)(Dankelman, 1989). Then a model for coronary regulation is presented (Section 4). The predictions of this control model are evaluated with experiments that are described in Section 5. The experimental responses show a mechanical effect. The mechanical factor is analyzed with four different mechanical models in Section 6. In the discussion in Section 8 some additional completed studies are described.

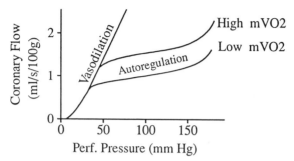

Figure 2. Schematic representation of the pressure-flow relation in the vasodilated coronary bed and in the autoregulated bed. In the autoregulated bed this relation is given at two levels of metabolism.

The effect of autoregulation and metabolic adaptation on coronary flow is illustrated in Figure 2. It depicts the coronary flow as a function of perfusion pressure in the vasodilated bed (vessels are maximally dilated), and at two levels of metabolism when autoregulation is intact. At equal levels of perfusion pressure the coronary flow in the vasodilated bed is higher than in the autoregulated bed. In the flat region of the autoregulated pressure-flow curves, the flow is relatively independent of perfusion pressure. Furthermore, the flow in the autoregulated bed is higher at higher oxygen consumption.

2   Coronary flow control

Contraction and dilation of smooth muscle cells in the arteriolar vessel wall change the arteriolar resistance and regulate coronary flow. The control of coronary flow demonstrates itself by two well-known phenomena: autoregulation and metabolic adaptation. Autoregulation is the intrinsic ability of an organ to maintain a relatively constant blood supply following changes in perfusion pressure (Feigl, 1983). This phenomenon is responsible for keeping the blood flow constant at constant metabolism despite changes in arterial pressure. Metabolic adaptation is the phenomenon of the adjustment of blood flow to the level of the metabolism of the heart. The response to increased heart rate is an increase in metabolic activity, leading to an increase in coronary blood flow and oxygen supply to the heart. Metabolic adaptation increases blood flow with increasing metabolic demand to supply more oxygen. This adaptation is illustrated by the shift of the autoregulation curve to a higher value of flow at a higher metabolism (Mosher et al., 1964).

The static and dynamic behavior of coronary flow control is consistent with that of a control system based on feedback. From control engineering theory it is known that the existence of an error signal is essential for a feedback system. Hence, in order to understand the control loop regulating coronary flow we have to identify an endogenous factor that fulfills the role of error signal. The requirement for such an error signal is that it has to increase coronary resistance when flow is above its working point and to decrease coronary resistance when flow is below it. We have previously shown that on the basis of the assumption that tissue oxygen pressure is the controlled variable, the steady-state behavior of coronary flow control can be described very well (Drake-Holland et al., 1984, Vergroesen et al., 1987). We assessed whether the same assumption would result in an adequate description of the dynamic response of the coronary circulation to a sudden change in metabolism, as occurs with a sudden change in heart rate (Dankelman et al., 1989a). The previous oxygen model was extended to describe the dynamics of coronary control (Section 4). The theoretical analysis resulted in some predictions that were suitable for experimental verification (Section 5).

3   Coronary flow mechanics

The coronary arterial blood flow and coronary venous blood flow in the beating heart are pulsatile. This is illustrated in Figure 3 by the coronary flow signals measured in an anesthetized goat. The left main coronary artery is artificially perfused with constant pressure independent of the aortic blood pressure. The great cardiac vein is drained with a constant pressure. During systole the coronary arterial blood flow is significantly lower than during diastole (Spaan et al., 1981b; Spaan, 1991). The coronary venous flow, however, is higher during systole than during diastole. The opposite variation of coronary venous and arterial flow induced by heart contraction indicates that flow variations are not simply due to resistance variations but that intramyocardial blood volume varies periodically. One cannot conclude from the decreased systolic arterial flow that contraction impedes coronary flow. The same reasoning applied to the coronary venous flow would lead to the conclusion that cardiac contraction augments flow. It is obvious that the heart contraction has an important influence on the coronary blood flow. It appears that the flow is not linearly related to arterial-venous pressure difference, which indicates that the coronary system is a nonlinear system.

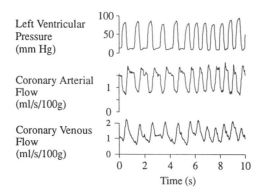

Figure 3. Simultaneously measured left ventricular pressure and arterial and venous flow at constant perfusion pressure. The phase opposition between arterial and venous flow, as well as their relation with cardiac contraction, can be seen.

A proper system analytical analysis combining the effects of control and mechanical interaction of heart contraction and the coronary circulation is a prerequisite for a quantitative estimate of the response of coronary arterial resistance to a metabolic stimulus (Dankelman et al., 1990). In Section 6 this interaction is evaluated.

4   Oxygen model for coronary flow control

4.1 DESCRIPTION

In the model for coronary flow control it is assumed that the coronary resistance controls flow and that coronary resistance is related to tissue oxygen pressure (pO_2). Furthermore, it is assumed that tissue pO_2 is equal to venous pO_2. The tissue pO_2 is then related to oxygen consumption, arterial pO_2, the oxygen binding capacity of the blood and blood flow.

The model is schematically drawn in Figure 4. To describe the dynamics of metabolic flow adaptation to a change in heart rate, two time-limiting processes have to be accounted for: the dynamic changes of tissue pO_2 induced by a change in oxygen consumption and/or flow and the dynamic change of arteriolar resistance caused by a change in tissue pO_2. The transient in tissue pO_2, namely $pO_{2,t}(t)$, depends on the supply of arterial oxygen to tissue, $o_s(t)$, (represented by the arrow F_1) and oxygen consumption, $mVO_2(t)$. F_2 represents the relation between the change of arteriolar resistance and the change in tissue pO_2. With constant pressure perfusion (Pp) the coronary blood flow ($q_a(t)$) is the dependent variable and varies with the arteriolar resistance ($r_a(t)$). This latter relation is indicated by F_3: $q_a(t)=Pp/r_a(t)$. In the case of constant flow perfusion, the perfusion pressure is the dependent variable and resistance does not affect flow, and hence F_3 is not active.

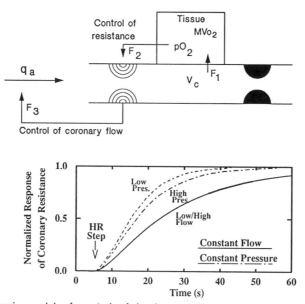

Fig. 4. Dynamic model of control of local coronary resistance. Panel A: Schematic representation of the model for regulation. mVO_2, oxygen consumption by tissue; V_c, capillary blood volume; q_a, coronary blood flow; F_1, oxygen extraction from the blood; F_2, adaptation of the arteriolar resistance to a change in tissue pO_2; F_3, change in $q_a(t)$ after a change in resistance. Panel B: Model simulation of the regulation of coronary resistance to a change in oxygen consumption. Dotted curves: normalized response of coronary resistance with constant pressure at high and low pressures, respectively. Solid curves: simulation with high and low flows. Constant pressure perfusion shows a faster response and is dependent on the level of pressure. Constant flow perfusion shows a slower response with a speed independent of the level of flow (F_3 is not active). Note that these dynamic differences are not the result of different parameter values since these were equal for both models. (After Dankelman et al. 1989a).

The oxygen content in blood and tissue depends on the oxygen saturation of hemoglobin and oxygen solubility in tissue. Hemoglobin is the protein dye that colors our blood red and is contained in red cells. It can reversibly bind with oxygen. Oxygen saturation depends on oxygen pressure in an S-shape manner (oxygen dissociation curve). Based on a mass balance for oxygen and assuming that tissue and capillary space behaves as one well- mixed compartment, one may write

$$(Hb1.36KV_c + s_tV_t)\frac{dpO_{2,t}(t)}{dt} = o_s(t) - mVO_2(t), \text{ where} \tag{1}$$

Hb = hemoglobin concentration in g/100 ml blood,
1.36 = oxygen binding capacity of 1 g hemoglobin in ml O_2/g Hb,
K = slope of the linearized oxygen dissociation curve in 1/mm Hg,
V_c = volume of the capillary space in ml/100 g,
s_t = solubility of oxygen in tissue in ml O_2/ml tissue/mm Hg,
V_t = volume of the interstitial and extracellular space in ml/100 g,
$pO_{2,t}(t)$ = oxygen pressure averaged over tissue and capillary blood in mm Hg,
$mVO_2(t)$ = oxygen consumption in tissue in ml O_2/s/100 g, and
$o_s(t)$ = oxygen flux out of the capillary space in ml O_2/s/100 g.

For $o_s(t)$ (F_1 Fig. 4A) one may write

$$\begin{aligned}o_s(t) &= Hb\ 1.36\ q_a(t)\ [S_a - s_v(t)] \\ &= Hb\ 1.36\ q_a(t)\ K\ [PO_{2,a} - pO_{2,v}(t)], \text{ where}\end{aligned} \tag{2}$$

$q_a(t)$ = coronary arterial blood flow in ml/s/100 g,
$S_a, s_v(t)$ = arterial and venous oxygen saturation, respectively, and
$PO_{2,a}, pO_{2,v}(t)$ = arterial and venous oxygen pressure, respectively.

In the steady state the left-hand side of Eq. (1) is zero. In this case Eqs. (1) and (2) combined result in the relation of Fick used in physiology to calculate oxygen uptake of an organ. In the non - steady state Eqs. (1) and (2) describe approximately a first - order process with a time constant:

$$t_1 = \frac{Hb\ 1.36\ K\ V_c + s_t\ Vt}{Hb\ 1.36\ K\ Qa_m} \tag{3}$$

where Qa_m is the mean coronary blood flow. An estimation of this time constant can be obtained from capillary volume, which is about 6% of total tissue volume. Hb = 8 g/100 ml blood; $s_t = 2.8.10^{-5}$ ml O_2/ml tissue/mm Hg; K = 0.024/mm Hg; and Qa_m = 1.5 ml/s/100g. This results in a time constant of approximately 4 s.

For the non-steady state the following relation was assumed between arteriolar coronary resistance, $r_a(t)$, and tissue pO_2, $pO_{2,t}(t)$, (F_2 in Fig. 4A):

$$t_2\ \frac{dr_a(t)}{dt} + r_a(t) = K_1\ pO_{2,t}(t). \tag{4}$$

In the steady state this equation relates resistance to tissue pO_2 as was the case in the original oxygen model. For the non-steady state t_2 is the characteristic time constant for the adaptation of resistance to a change in tissue pO_2. The physiological basis must be found in the rate of response of smooth muscle tone to a change in stimulus and all other factors as yet undefined. A discussion on the rate of coronary regulation is found in Section 8. In case of constant flow perfusion, the signal for control of resistance is not influenced by flow, as it is with constant pressure perfusion. This latter effect is indicated by F_3 in Figure 4A.

4.2 MODEL PREDICTIONS

Using the model, the resistance variation following the different simulated interventions can be calculated. The results are presented as normalized resistance variations to allow for a comparison of the course of the resistance variations obtained from different interventions. The normalized response is defined as

$$\text{normalized response} = \frac{r_{cor}(t) - R_0}{R_1 - R_0}, \text{ where} \tag{5}$$

$r_{cor}(t)$ = coronary resistance,
R_0 = resistance before the simulated change in oxygen consumption, and
R_1 = resistance after the new steady state has been reached.

Because of its definition, the normalized response equals zero at the start of the response and equals unity in the new steady state. This allows us to evaluate the course of the response independent of magnitude and direction of the oxygen consumption (e.g., heart rate) change.

From a steady state, mVO_2 in the model was suddenly changed by 20%. The resulting normalized resistance responses are shown in Figure 4B. The dotted curves in Figure 4B depict the responses when pressure was kept constant and flow was allowed to change. The response with constant flow perfusion is given by the solid curve in Figure 4B. The model predicts that coronary adjustment is slower with constant flow perfusion than with constant pressure perfusion. Moreover, the model shows that with constant flow perfusion the rate of the response is independent of the flow level, but that with constant pressure perfusion the rate of response is pressure dependent. The experiments described in the following section were designed to test these predictions experimentally.

5 Experiments

5.1 PREPARATION

The experiments, performed at the Department of Medical Physics and Informatics (Academic Medical Center, Amsterdam), were done on anesthetized goats. A left thoracotomy was performed and the third and fourth ribs were removed. The pericardium was opened and a cradle was formed. The His bundle was destroyed by local injection of formaldehyde, and the right ventricle was paced. A stainless-steel cannula was inserted into the left main coronary artery. Blood from the cannulated left carotid artery was circulated through the perfusion system (Spaan et al., 1981a), a flow probe, and back via the cannula (Fig. 5). Perfusion pressure (Pp) was measured at the cannula tip via a small-lumen catheter connected to a pressure transducer. A clamp between the flow probe and steel cannula could produce a large resistance on the perfusion line, giving the perfusion system the characteristics of a constant flow source. The great cardiac vein was cannulated as well. Blood from the cannulated great cardiac vein drained into a similar pressure-controlled reservoir via a flow probe. Venous blood was pumped back into the left jugular vein via a heat exchanger and filter. A catheter tip manometer was inserted through a purse-string suture in the left atrial appendage into the left ventricle.

5.2 EXPERIMENTAL RESULTS

The heart rate was altered stepwise among 60, 90, 120, and 150 beats/min at different levels of constant-flow or constant-pressure perfusion. Coronary arterial pressure and flow were averaged per beat. Then the coronary pressure/flow ratio was calculated. The coronary pressure/flow ratio only reflects resistance in the steady state or in conditions where flow and/or pressure vary slowly enough to neglect capacitance effects.

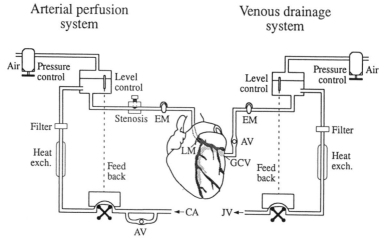

Figure 5. Arterial perfusion and venous drainage system of experimental preparation.
EM, electromagnetic flow probe; LM, left main coronary artery; CA, carotid artery; GCV, great cardiac vein; JV = jugular vein. With a high pressure proximal to a high stenosis, a constant flow was obtained. AV, cuvette for measurement of arteriovenous oxygen content difference. For a detailed description, see the text.

The experimental results with constant pressure and constant flow perfusion are shown in Figure 6. With constant perfusion pressure, coronary flow decreased with heart rate as a consequence of constriction due to metabolic adjustment. With constant flow perfusion, this constriction was measured as an increase in perfusion pressure. However, the normalized responses of the pressure/flow ratio increased only after an initial fall as shown in Figure 6. This dip in normalized response was consistently found after a change in heart rate, either up or down, and is analyzed in the next section.

It is clear that the normalized responses change more rapidly at constant-pressure perfusion than at constant-flow perfusion. The times in seconds after the heart rate step at which the pressure/flow ratio was changed 50% (t_{50} values) for the responses are 22.2 ± 0.5 s (n = 73) (mean $\pm$ S.E.) for constant-flow perfusion, $14.4 + 0.6$ s (n = 39) for low-pressure perfusion and 17.0 ± 0.6 s (n = 44) for high pressure perfusion. No difference is seen between the two levels in constant-flow perfusion. However, at constant-pressure perfusion the pressure/flow ratio responds faster at the lower perfusion pressure. The differences in these t_{50} values correspond with the predictions given in Figure 4.

6   Mechanical models for the coronary circulation

The responses of driving the pressure/flow ratio on an abrupt change in heart rate show a fast initial reversed phase followed by a slow phase caused by regulation. The interpretation of the initial rapid change of the coronary system will be made on the basis of models that are capable of taking the interaction between heart contraction and the coronary vasculature into account. At the moment there is no commonly accepted model for this interaction (Klocke et al., 1985, Spaan 1985; Krams et al., 1989). The usefulness of the application of the different models of coronary mechanics will be assessed for their usefulness in describing the regulatory response.

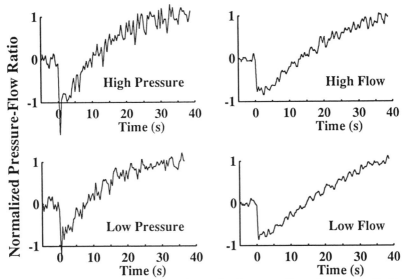

Figure 6: Summary of the dynamic change of normalized pressure-flow ratio as a result of heart rate changes. Left hand side panels are results obtained with constant perfusion pressure and right hand side panels with constant flow perfusion. Upper panels show the averaged results of all experiments with high pressure and high flow and bottom panels show averaged results obtained with low pressure and low flow. Redrawn from Dankelman et al. 1989a.

The basic model structure for regulation of coronary blood flow set up in the previous section will be extended by the addition of various different elements extracted from existing models for the passive coronary bed. These extended models will be tested on their ability to explain the occurrence of the fast initial phase followed by the slow-regulation phase under the different conditions of constant-pressure and constant-flow perfusion. We will then attempt to infer the rate of change of coronary resistance as an indication of the change in vascular smooth muscle tone from the coronary responses.

The predictive value for five different models will be tested. These extensions to the control model concern the mechanical interaction effects between cardiac contraction and coronary flow and represent different hypotheses. For the interpretation of this initial phase of the response, the following models presented in the literature and explained later were analyzed:

A   Linear intramyocardial pump model; tissue pressure or elastance driven
B   Intramyocardial pump model with variable venous resistance
C   Varying elastance model
D   Waterfall model
E   Model with waterfall and a small intramyocardial compliance

The five models are depicted schematically in Figure 7.

In the first model intramyocardial compliance is assumed distal to the arterial resistance, as proposed by Spaan et al. (1981a). The compliance (C) of the vascular bed is the change in blood volume (V) caused by a change in the pressure difference over the vascular wall (P): $C = dV/dP$. Since pressure outside the vessels is generated by contraction, this is denoted as tissue pressure driven. However, there is also the possibility that pressure is generated by altering stiffness of the wall. This is denoted as elastance driven. We will not elaborate further on these differences here. The second model is an extension of the first model. It is assumed that the venous resistance is affected by mechanical changes as follows: The venous resistance is related to the inverse

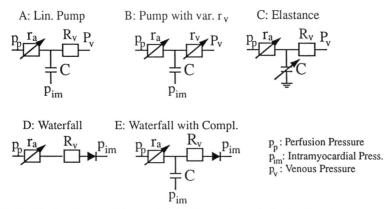

Figure 7. Electrical analogies of the models for the mechanical effects of contraction on flow. The upper three panels represent the two pump models and the varying elastance model. In the linear pump model, only r_a is variable and depends on regulation. In the second model, r_v is also variable and depends on transmural pressure. In the third model, C is variable and depends on the level of contraction. The bottom two models are based on the waterfall model. In the right model, a small compliance has been placed before the waterfall element.

of the squared volume according to the law of Poiseuille (for tubes of constant length) and the volume is given by the filling of the intramyocardial compliance. In both the pump models (A, B) and half of the left ventricular pressure is assumed intramyocardial compliance of 0.08 ml/mm Hg/100 g is used (Spaan et al. 1981a), and half of the left ventricular pressure is assumed to be the mean intramyocardial pressure.

The third model is also an extension of the first one. This model represents a simple interpretation of the elastance model (Krams et al., 1989). The varying elastance model is based on the changing elasticity of the heart wall during contraction (elastance = 1/C dP/dV). In our model the changing elastance is represented by the changing compliance of the vessels wall. In the two waterfall models (D, E) the outflow resistance is assumed to be constant and the venous outflow pressure equal to the intramyocardial pressure, $p_{im}(t)$ (Downey & Kirk 1975, Klocke et al., 1981, 1985). Model E is a combination of models A and D. The waterfall model is extended by intramyocardial compliance. The five mechanical models were combined with the model for regulation in order to analyze the effect of a step in intramyocardial pressure on the simulated pressure/flow ratio.

7   Simulation results

The responses generated by the five extended models are depicted in Figure 8. The parameters for the models with their values are as follows: C = 0.08 in the two pump models (Fig. 7A, B); C = 0.007 ml/mm Hg/100 g in the waterfall model with compliance (Fig. 7E), and $t_1 = 4$ s and $t_2 = 20$ s. In the elastance model, the dependency of C on heart rate is unknown. However, the effect of a change in $p_{im}(t)$ in the linear pump model on the responses is similar to a change in C in the elastance model. Therefore, if we assume that C changes with the heart rate step and that the value of C after the heart rate step in the elastance model is equal to the value of C in the linear pump model, then the simulation results of models A and C will be similar. Note again that the prediction of the differences between constant-flow and constant-pressure perfusion and the prediction of the models with regulation alone are the result of model behavior and not of altered values of parameters.

The response of the pure regulator model is illustrated by the solid lines in Figure 8. The normalized pure regulatory responses show a steady increase to the end value; it does not contain an initial dip as in the experimental responses. With constant-flow perfusion, it simply is the response of two first-order systems with time constants t_1 and t_2, respectively. With constant-pressure perfusion there is a feedback loop. This leads to a faster response compared with that obtained with constant-flow perfusion. The responses of the five extended models differ, but all exhibit the characteristic of the experimental response (Fig. 8). All five models produce a response with an initial dip in the normalized response of the pressure/flow ratio, although the dip is deeper with constant-pressure perfusion than with constant-flow perfusion and is followed by a slow increase to a steady level.

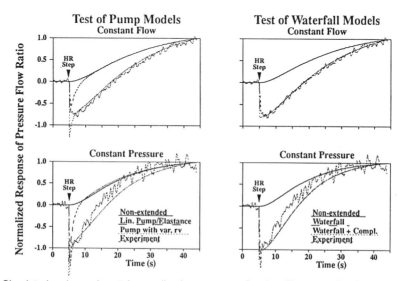

Fig. 8 Simulated and experimental normalized responses of p_p/q_a. The normalized responses are produced by: (—) the regulation model alone , (- - -) extended with the pump model (left panels) or waterfall model (right panels) and (......) extended pump model with variable venous resistance (left panels) or waterfall+compliance (right panels). The noisy signals are the experimental responses. Upper panel: simulation with constant flow perfusion. Lower panel: simulation with constant pressure perfusion.(Redrawn from Dankelman et al. 1990.)

8   Discussion

The distensibility of the coronary vessels has complicated the interpretation of experimentally measured pressure and flow signals in terms of arteriolar resistance. If the dynamics of the coronary bed in a constant contractile state of the arteriolar smooth muscle were known, more detailed information could be obtained on the mechanical dynamics in this contractile state. This would be helpful in understanding the mechanism of regulation. However, the dynamics in a constant contractile state are not known.

In the case of constant-flow perfusion, the half-time of the response of the linear pump model is equal to that of the regulatory model (Fig. 4). The t_{50} value produced by the linear pump model with constant-pressure perfusion is smaller than that produced by the regulator model alone, so this model predicts a half-time for regulation in the experiments larger than that in the experimental pressure/flow ratio. Table 1 gives the predicted half time for the response of arteriolar resistence to a step change in tissue oxygen concentration corrected for the different assumed mechanical influences. The differences between the t_{50} value of extended and non-extended models are used as estimate of the mechanical influence on the response.   The values in this table are obtained by extracting these differences from the experimental t_{50} value.

The t_{50} value of simulated normalized pressure/flow ratio is influenced by the variable venous resistance in both perfusion conditions. In both cases it results in an estimation of t_{50}, that is larger than the response for regulation should be. In the case of constant-flow perfusion, there is no interaction between the regulatory response and the passive response, however, it is not possible to extract the response time for regulation from the normalized response of the pressure/flow ratio. In the case of constant flow perfusion, the half-time for the response of regulation is about 7 s faster than those of the pressure/flow ratio. With constant-pressure perfusion, the difference is less. The responses with the pump model with variable resistance fit the experimental results better than those of the linear pump and elastance models. The elastance model used in the simulations is a very simple one. A better model would be the combination of the elastance model with a variable venous resistance. In this case the results would be comparable with the pump model with variable resistance (model B).

Table 1. Predicted half time for the response of arteriolar resistance

	t_{50} (s) Constant flow	t_{50} (s) Constant pressure
Linear pump /elastance model	22	16
Pump with var. r_v	15	9
Waterfall model	15	9
Waterfall with C = 0.007	15	9
Waterfall with C = 0.08	15	10

It is almost impossible to discriminate between the mechanical models on the basis of these experiments. The simulations in Figure 8 show that the pump model with variable venous resistance and the waterfall with a small compliance could produce responses that best agree with the experimental responses. Furthermore, the simulations show that the t_{50} value of the normalized response could not be translated in a simple way to the response time of regulation in either case. The response time is reduced due to the mechanical effects. The difference between these two models is too small to discriminate on the basis of these experiments. In all cases, however, the simulated response time with constant-flow perfusion is slower than those with constant-pressure perfusion. It can be concluded that the estimated half-times for regulation on a heart rate step are about 9 s for constant pressure and about 15 s for constant-flow perfusion, supporting the assumed value for the response time of smooth muscle.

Dynamic responses following changes in perfusion were analyzed as well. The control model presented earlier predicts that the response following step changes in flow with flow constant in the response will be independent of the direction of the perfusion step, whereas the responses following a step in pressure with pressure constant in the response will be faster and will depend on the direction of the step. These predictions were tested in anesthetized goats and dogs (Dankelman et al., 1989b, 1992). The responses following step changes in flow and following an increase in perfusion pressure were as predicted. However, the response following a decrease in perfusion pressure was much faster than predicted and showed an overshoot. This discrepancy between the model and experimental findings suggests that with the dynamic response to a perfusion perturbation an additional mechanism might be involved besides those active during response to a perturbation of metabolism. Moreover, it appears that the dynamic responses in the dog are much faster than in the goat, although earlier studies showed similar static behavior of coronary flow in the dog versus the goat (Vergroesen et al., 1987). The directional sensitivity in responses with perfusion changes, observed earlier in goats, is normally absent in dogs. The differences in the rate of regulation were sought in the response of the arterioles within the hearts of the two species. However, in the goat the response to a decrease in perfusion was as fast as in the dog, while the response to an increase in perfusion was as slow as the response to a change in heart rate in the goat. This specific direction sensitivity for a mechanical stimulus might still be present but obscured by the overall faster response of the coronary resistance vessels. It is therefore not clear whether the occurrence of direction sensitivity in the goat is related to an overall slower response of the coronary circulation.

In a theoretical study, the effects of pressure-dependent changes in vascular volume, resistance, and capacitance in the coronary circulation have been analyzed by a distributed mathematical model of the coronary circulation (Bruinsma et al., 1988). This model did not include regulation of coronary blood flow and is evaluated only for the fully dilated vasculature. In the nonlinear intramyocardial pump model, the coronary circulation was divided into arteriolar, capillary, and venular compartments. Resistance and compliance of all these compartments were pressure dependent. It was concluded that the resistance variations affected by heart contraction during the cardiac cycle were mainly in the venular compartment. Furthermore, interpretation of transients in coronary flow and/or pressure by models containing fixed resistance and capacitance may seriously underestimate intramyocardial capacitative effects and characteristic time constants for pressure-induced resistance changes.

The intramyocardial blood volume distribution between vessels involved in oxygen exchange and more distal vessels was estimated in anesthetized goats (Van der Ploeg et al., 1993). Measurements of arterio-venous oxygen difference (AVO_2) transients induced by a flow step could be characterized by two phases: delay time, and slow change to a new steady state. AVO_2 responses were fitted by a two-compartment model consisting of a well-mixed compartment from which oxygen is consumed with volume Vm and a distal unmixed compartment without O_2 exchange with volume Vunm. The rate of change of AVO_2 transient depends on Vm, while the delay time depends on Vunm. Measurements resulted in Vm = 9.9 ± 1.1 ml/100 g and Vunm = 3.8 ± 0.3 ml/100 g. Maximal vasodilation caused a significant increase in Vm, while Vunm did not change. Hence, the increase of intramyocardial blood volume induced by vasodilation must be looked for in the capillary bed and not in the coronary veins. These results of the volume estimations depend on the assumed oxygen exchange model. Therefore, errors were made in the volume estimations by not taking into account factors such as flow heterogeneity, different mixing sites, or Krogh-like oxygen exchange. These errors were estimated as well (Van der Ploeg et al., 1995a). These results indicate that the coronary volumes are well approximated by the estimations obtained with the two-compartment models used.

Interpretation of transients following changes in perfusion are complicated because a change in perfusion induces changes in myocardial oxygen consumption at a constant heart rate (Gregg, 1964). Determination of the dependency of oxygen consumption on perfusion pressure (Gregg's phenomenon) was also analyzed after decreasing the rate of change of regulation by a factor of four by administration of glibenclamide (Dankelman et al., 1994). The response of oxygen consumption (MVo_2) was calculated using the two-compartment model of Van der Ploeg et al., (1993). The rate of change of MVo_2 and of resistance were in the same range for each condition. The results indicate that transients in oxygen consumption depend on the rate of change of autoregulation. Furthermore, during transients large variations in oxygen consumption can occur when the rate of regulation is slow. It was concluded that normally vasomotor tone diminishes pressure induced-oxygen consumption (Gregg's phenomenon) (Dankelman et al., 1996). The dependency of myocardial oxygen consumption was also studied following a step change in flow perfusion at different heart rates (Van der Ploeg et al., 1995b). Heart rate affected the MVo_2 dependency on flow during control as well as during maximal vasodilation. The higher MVo_2 dependency on flow at a high heart rate could be explained by the capillary pressure dependency of Gregg's phenomenon.

9   Conclusions

Model analysis shows that for the analysis of the dynamic response of the coronary regulation, it is important to know the mechanical effects occurring at the same time. Model simulations show that, because of the interaction of regulation and mechanical changes in the coronary circulation, there is no direct relation between the rate of change of vascular tone or smooth muscle contractile state and the rate of change of the pressure/flow ratio. Furthermore, the half-time for regulation ranges from 9 to 16 s with constant-pressure perfusion and from 15 to 22 s with constant-flow perfusion, depending on the model used for the mechanical changes in the coronary bed. A model based on intramyocardial compliance or elastance in combination with variable venous resistance, and a model based on a waterfall and a small compliance, could explain the fast initial dip. However, we could not discriminate among the predictions of these three models.

The studies on the identification of the coronary circulation have shown that the cooperation between the Laboratory of Measurement and Control and the Department of Medical Physics and Informatics has been very fruitful. The combination of theoretical model studies with experimental studies has produced in new insights into several aspects of mechanisms that influence the control of coronary blood flow.

References

Bruinsma P., Arts T., Dankelman J. and Spaan J.A.E. (1988). Model of the coronary circulation based on pressure dependence of coronary resistance and compliance. *Basic Research in Cardiology* 83: 510-524.

Dankelman, J., Spaan, J.A.E., Stassen, H.G. and Vergroesen, I. (1989a). Dynamics of coronary adjustment to a change in heart rate in the anaesthetized goat. *Journal of Physiology* 408: 295-312.

Dankelman, J., Spaan, J.A.E., Van der Ploeg, C.P.B. and Vergroesen I. (1989b). Dynamic response of the coronary circulation to a rapid change in its perfusion in the anaesthetized goat. *Journal of Physiology* 419: 703-715.

Dankelman, J. (1989) On the dynamics of the coronary circulation. PhD thesis, Technische Universiteit Delft.

Dankelman, J, Stassen, H.G. and Spaan J.A.E. (1990). System analysis of the dynamic response of the coronary circulation to a sudden change in heart rate. *Medical & Biological Engineering & Computing* 28: 139-148.

Dankelman, J., Vergroesen, I., Han, Y. and Spaan J.A.E. (1992). Dynamic response of coronary regulation to heart rate and perfusion changes in the dog. *American Journal of Physiology* 263: H447-H452.

Dankelman, J., Van der Ploeg, C.P.B. and Spaan J.A.E. (1994). Glibenclamide decreased the response of coronary regulation in the goat. *American Journal of Physiology* 266 : H1715-H1721.

Dankelman J., Van der Ploeg, C.P.B. and Spaan, J.A.E. (1996). Transients in myocardial oxygen consumption following abrupt changes in perfusion pressure in the goat. *American Journal of Physiology* (In press)

Downey, J.M. and Kirk, E.S. (1975). Inhibition of coronary blood flow by a vascular waterfall mechanism. *Circulation Research* 36: 753-760.

Drake-Holland, A.J., Laird, J.D., Noble, M.I.M., Spaan, J.A.E. and Vergroesen I. (1984). Oxygen and coronary vascular resistance during autoregulation and metabolic vasodilation in the dog. *Journal of Physiology* 348: 285-299.

Feigl, E.O. (1983). Coronary physiology. *Physiology Review* 63: 1-205.

Gregg, D.E. (1964). Effect of coronary perfusion pressure or coronary flow on oxygen usage of the myocardium. *Circulation Research* 13(6):497-500.

Klocke, F.J., Weinstein, I.R., Klocke, J.F., Ellis, A.K., Kraus, D.R., Mates, R.E., Canty, J.M., Anbar, R.D., Romanowski, R.R., Wallmeyer, K.W. and Echt, M.P. (1981). Zero-flow pressure relationships during single long diastoles in the canine coronary bed before and during maximal vasodilation. *Journal of Clinical Investigation* 68: 970-980.

Klocke, F.J., Mates, R.E., Cantly, J.M. and Ellis, A.K. (1985). Coronary pressure-flow relationships. Controversial issues and probable implications. *Circulation Research.* 56: 310-323.

Krams, R., Sipkema, P. and Westerhof N. (1989). Varying elastance concept may explain coronary systolic flow impediment. *American Journal of Physiology* 257: H1471-H1479.

Mosher, P., Ross, J., Mcfate, P.A. and Shaw R.F. (1964). Control of coronary blood flow by an autoregulatory mechanism. *Circulation Res*earch 14: 250-259.

Spaan, J.A.E., Breuls, N.P.W. and Laird, J.D. (1981a). Diastolic-systolic coronary flow differences are caused by intramyocardial pump action in the anesthetized dog. *Circulation Res*earch 49: 584-593.

Spaan, J.A.E., Breuls, N.P.W. and Laird, J.D. (1981b). Forward coronary flow normally seen in systole is the result of both forward and concealed back flow. *Basic Res. Cardio*logy 76: 582-586.

Spaan, J.A.E. (1985). Coronary diastolic pressure-flow relation and zero flow pressure explained on the basis of intramyocardial compliance. *Circulation Res*earch 56: 293-309.

Spaan, J.A.E. (1991). Coronary blood flow, Mechanics, Distribution and Control. *Developments in Cardiovascular Medicine* 124, Kluwer Academic Publishers, Dordrecht, ISBN 0-7923-1210-4.

Van der Ploeg, C.P.B., Dankelman, J. and Spaan J.A.E. (1993). Functional distribution of coronary vascular volume in beating goats. *American Journal of Physiology* 264: H770-H776.

Van der Ploeg, C.P.B., Dankelman, J. and Spaan J.A.E. (1995a). Comparison of different oxygen exchange models. *Medical and Biological and Comp*uting 33: 660-661.

Van der Ploeg, C.P.B., Dankelman, J. and Spaan, J.A.E. (1995b). Heart rate affects the dependency of myocardial oxygen consumption on flow in goats. *Heart and Vessels* 10: 250-265.

Vergroesen, I., Noble, M.I.M., Wieringa, P.A. and Spaan, J.A.E. (1987) Quantification of O2 consumption and arterial pressure as independent determinants of coronary flow. *American Journal of Physiology* 252 (Heart Circ. Physiol. 21): H545-H553.

Chapter 7

In Search of Design Specifications for Arm Prostheses

Ton van Lunteren
Erna van Lunteren-Gerritsen

1   Introduction

The designer of arm prostheses for people with a missing arm or orthoses for people with a paralyzed arm is confronted with a number of questions when setting the goals for the product to be developed. What functions of the normal hand should be incorporated? What compromises are acceptable, taking into account the limitations imposed by the available means?

Over a period of about 25 years this question has been a major research theme of the Man-Machine Systems group of the Department of Mechanical Engineering and Marine Technology of the Delft University of Technology. The activities over this period can be characterized as a history of enlarging system boundaries. Roughly three overlapping stages can be distinguished:

- Investigation of basic principles (1967 to 1975)
- Laboratory experiments with technical prototypes and commercially available prostheses (1972 to 1982)
- Field studies with existing and newly designed clinical prototypes (1974 to 1990)

These stages clearly correspond to the state of affairs in the design group because there is a long way to go from a basic idea to a clinical prototype. Furthermore, increasing insight into the problem field due to monthly contact with persons with arm defects in rehabilitation centers involved has led to the earlier mentioned widening boundaries of the system to be considered. This work started from the control engineer's view of a prosthesis as a technical system to be controlled, but finally evolved into a system consisting of a wearer with his or her prosthesis in physical and social interaction with the environment. The three stages of development are considered in the following sections, with emphasis on field studies.

2   Investigation of basic principles

At this stage, the activities of the MMS group started more or less from ground zero. On the one hand, there were control engineers' technical questions about feedback paths and, on the other, questions about the normal values of forces and displacements of the human arm and hand. Almost all investigations were conducted by students as master's thesis projects, consisting of a literature study followed by experimentation. A brief summary is given.

2.1 TACTILE FEEDBACK VIA THE SKIN, EVALUATED BY MEANS OF A TRACKING TASK

When using an externally powered prosthesis, in those days mainly CO_2 powered because myoelectric prostheses were still in the experimental stage, the wearer of an arm prosthesis has no feedback about grasping force. An information channel that might be used is tactile stimulation of the skin. It had already been shown that electrical stimulation was not feasible, because the range between threshold level and pain level is only a factor three. For a CO_2 powered prosthesis, a small pressure chamber on the skin was thought to be a possible feedback device. After a literature study on the mechanical and sensory properties of the skin, and some introductory experiments with a number of stimulator prototypes (Verkaik, 1967), a number of tracking tasks were executed in which the number of display characteristics was varied. The choice of a tracking task was based on the availability of software for data processing in an earlier project on modeling human control behavior. The experiments were conducted with one or two pressure chambers with a diameter of 14 mm placed on the inner side of the thigh. Signal amplitudes were transformed to sinusoids with a constant velocity amplitude in the frequency range between 5 and 82 Hz, within which 20 different levels could be discriminated. Thus the frequency was an indication of the amplitude of the tracking error. Several different characteristics for the relation between error and stimulus frequency were applied.

It was found that the best tracking performance was obtained by presenting only the sign of the error, using an on-off stimulation with a frequency of 40 Hz. In this case the subjects introduced a dither to detect the zero value (Verkaik, 1968).

In a later stage, experiments used a mechanical vibrator with a 4 mm diameter spherical probe fitted in a ring of 25 mm interior diameter placed on the back of the subject. The input signal of the tracking task was normally distributed noise in a frequency range between 0.02 and 0.1 Hz. Also in this experiment, a number of display characteristics were applied, among which was the introduction of a dead zone. Both frequency modulation and amplitude modulation were applied. Besides tactile stimulation, auditory and visual displays were used in order to compare the performances with these different modalities. The best results in the tactile experiments were obtained with a three-state element with a 15 Hz frequency for negative values, a 50 Hz frequency for positive values, and a dead zone around the zero value of 2.5% of the input range. This dead zone was found to eliminate the need for a dither. It was found in experiments with two subjects that the ratio between the root mean square (RMS) values of error and input determined over a 3 minute interval was 0.15 in an experiment that lasted 5 minutes. This value was 0.11 in the optimal displays of the auditory and visual experiments. It is interesting to note that both subjects thought that the experiments with the tactile stimulation were less tiring (Meijer, 1970).

2.2 RANGE OF DISPLACEMENT, VELOCITY, AND ACCELERATION OF ELBOW MOTIONS

The plans in the design group to develop an externally powered elbow orthosis for persons with a paralyzed arm raised the question of desired motion characteristics. Such a design would also be valuable for a prosthesis for upper arm amputees. Taking into account the weight, power consumption, and volume available, it was doubtful whether the performance attained would be comparable with that of a normal arm. Nevertheless, knowing this performance would be useful to judge whether the specifications thus obtained could be considered acceptable. Therefore, a study was undertaken to gather data based on normal daily activities. In the literature (McWilliam, 1964), a maximum velocity of 2 rad/s was found for elbow motions in a repetitive task. However, it could not be considered to be representative of daily life activities. To obtain more realistic data for both arms, the elbow motions of a subject were measured over a 3 hour period of normal daily activity. When leaving out the periods in which no elbow motions were recorded, the distributions of velocity and acceleration could be described by a normal distribution with 3 s values below 1 rad/s for the velocity and below 10 rad/s^2 for the acceleration. No significant differences between dominant and nondominant elbow motion patterns were found (Ter Maat, 1971).

2.3 ANALYSIS OF SOME NORMAL HAND FUNCTION CHARACTERISTICS IN RELATION TO DOMINANCE

In this study (Holthaus, 1972) two aspects were considered, namely, the prehension force when lifting an object and finger dexterity in a peg in hole task. A device with a force transducer having a height of 30 cm and a width adjustable to 2 or 8 cm was made for the lifting task. It had a weight of 0.17 kg and could be loaded with an extra weight of

1 kg. The lifting tasks were executed by seven subjects. Each experiment was conducted 10 times. Although the dominant hands tended to slightly higher forces than nondominant hands in the four possible experimental conditions, these differences were not significant. For the light object the ratio between the grasping force and the gravity force lay between 1 and 4 for a 2 cm width and between 1 and 3 for the 8 cm width, dependent on the subject. For the heavy object, these values were 2 and 1.5. respectively. For the light object, the minimally necessary ratio was measured with one subject and was about 1. These values are based on the results of 5 out of the 7 subjects. For the other two subjects the forces were a factor of 6 or higher. So it can be stated that, especially in the case of light objects, the subjects tend to overdose their grasping force. This effect is somewhat less in the case of a large width of the object to be picked up. Based on these results it was concluded that there is no strong demand to provide a prosthesis with a proportional control for low prehension forces.

During the peg in hole task, the subjects were sitting at a table, with the forearm fixed by a V-shaped support. Subjects had to manipulate a cylinder in a vertical direction through a hole, without touching the edges. The gross movement was achieved by the wrist flexion-extension, while the precise motions were executed by finger manipulation. The tests were executed with seven subjects. The cylinder diameters were 10, 20, and 30 mm, while the holes were chosen in such a way that a play of 2, 4, 6, 8, 10 and 12 mm could be realized. During the experiments the error times, i.e., the time that the cylinder made contact with the edge of the hole, were measured. In all cases the error time was 1 s for a 2 mm play and decreased exponentially to 0.01 s for a 12 mm play. These results were independent of the cylinder diameter. On average, the error time for the nondominant hand was a factor 2 higher than for the dominant hand.

2.4 INFLUENCE OF PROPRIOCEPTION ON THE GENERATION OF CONTROL SIGNALS

An interesting contribution by Simpson in the field of prosthesis control is the concept of *extended proprioception* (Simpson, 1969). This means that if there is a one-to-one relation between the wearer's control action, e.g., the displacement of an input device, and the state of the prosthesis, e.g., the opening width, then the wearer has proprioceptive information of this quantity if he or she knows this relation. Therefore it is worthwhile to obtain quantitative information on this type of feedback.

Experiments were executed with forces and displacements of the first phalanx of the right-hand index finger. This control site was chosen because of a request from one of the rehabilitation centers for the development of a partial hand prosthesis. Two subjects participated. They were trained to generate a number of force and displacement values presented on a visual display by giving visual feedback of their outputs. During the experiments they had to generate their outputs without visual feedback according to four different stimulus patterns, one with two different levels executed 20 times, two with three levels, each executed 10 times, and one with five levels arranged as a staircase, executed 5 times. The switching time between succeeding levels was 5 s. The displacement tests were executed after completion of the force tests. Forces were in the range 0.5 to 5 N, and displacements were in the range 5 to 25 degrees. In general it can be stated that the average values over 5 s have a mean error of 10% to 20% of the required value both for displacement and force. However, the subjects had more difficulty in keeping the force level constant over a 5 s period than they had with a displacement level (Mooij, 1972).

2.5 FEASIBILITY OF MUSCLE STIFFNESS AS A CONTROL SIGNAL FOR PNEUMATICALLY POWERED PROSTHESES

Since about 1967 (Childress & Billock, 1970; Childress, 1985) myoelectric arm prostheses have become commercially available. The fact that the activity of the flexor and extensor muscles of the missing hand is used to open and close the prosthesis makes this type of control very attractive, in contrast to those controlled by shoulder motions. A vulnerable point, however, is susceptibility to electric disturbances and variations in skin resistance. In order to keep the advantages of control by the natural flexors and extensors, the possibility of using other physical quantities representing muscle contraction were investigated, e.g., muscle bulge (Simpson, 1966) or muscle stiffness, for which a sensor was announced by a commercial firm as an alternative for electrically powered arm prostheses (Näder,1970). However, 3 years later the system was still not available.

An investigation of the feasibility of such an attractive-looking possibility was started because this principle might be used to generate a pneumatic control signal for a pneumatically powered prosthesis. Therefore, a transducer was made based on the principle published. It consisted of a ring which was pressed against the skin above the muscle. Within this ring a small cylinder is pressed against the skin. The displacement of this cylinder for a given force is then a measure for the hardness of the underlying muscle. Based on a theoretical study making use of the results of an investigation of human skin properties (Vlasblom, 1967), a transducer was designed and tested experimentally (Gerbranda, 1974). Measurements were made with the stiffness transducer placed above the musculus extensor carpi radialis brevis, the force generated from which could be measured with an external force transducer, and then with another subject in the same manner with the musculus extensor digitorum communis. Sensor diameters of 4, 6, and 10 mm were tested, while the sensor depth was increased by steps of 0.2 mm. The transducer ring was pressed against the skin with six different loading forces in the range between 1 and 2 N.

The results roughly showed a cubic relation between the sensor output and the muscle force, with a parameter that depends on the transducer parameters. However, it should be mentioned that the output was very sensitive for the position above the muscle bulge. An accuracy of less then 2 mm was necessary. Based on the laboratory experiments, a miniature version of a pneumatic transducer with a 6 mm sensor diameter was designed and tested. However, when it was built into a prosthesis socket, it was found that in this case the sensor output was significantly lower. Moreover, the disturbances originating from changes in the elbow position and from small rotations of the socket made it impossible to generate useful control signals, and thus the project was abandoned.

2.6 INFLUENCE OF THE CONTROL SITE AND THE CONTROL COORDINATES ON THE POSITIONING OF A SIMULATED ARM

At the time of this investigation, at least all body-powered arm prostheses were controlled by shoulder motions, both for upper arm and forearm prostheses. In the first case more then one degree of freedom has to be controlled. Pneumatically powered upper arm prostheses had already been designed by Simpson (1969, Simpson and Kennworthy 1973, and Simpson and Smith, 1977), using Simpson's extended proprioception concept.

A number of experiments were conducted based on this idea to answer the question of whether the performance depended on the coordinate system in which the control task was executed (Solkesz, 1975). For these experiments a two degree of freedom task was chosen, i.e., positioning of the hand of a simulated arm on a visual display. Three coordinate systems were investigated for the control of the hand position, i.e., cartesian coordinates (x,y), polar coordinates (r,j), and joint angle coordinates (a,b). Control motions were generated by shoulder motions corresponding to horizontal and vertical angular rotations of the clavicle. Experiments were conducted with a two degree of freedom joystick as a reference. A test run consisted of the presentation of four arbitrarily chosen target positions indicated by a circle. Tests were executed with four different target diameters, 15, 30, 45, and 60 mm.

No significant differences in execution times were found between the coordinate systems. The task execution time for shoulder control was about 1.5 times that for hand control. These differences were mainly caused by the final phase of the positioning task. The same applies for the influence of the target diameter. The time decreased about exponentially as a function of the target diameter, such that the time for a diameter of 60 mm was 0.45 of that for a diameter of 15 mm, independent of the control site.

3   Laboratory experiments with technical prototypes and commercially available prostheses

The laboratory tasks described in this section mostly deal with prototypes built in the design group to see whether a certain idea was sufficiently promising for further development. A positive outcome implied that there still would be a long way to go. Therefore, in order to prevent false hopes, in this stage of development trials were not executed with amputees but with normal subjects. For this purpose the prosthesis was fitted over the left hand using a special construction. Clinical prototypes developed later were provided by the prosthesis team in the cooperating rehabilitation centers.

3.1 A FEASIBILITY STUDY OF AN ACTIVE THUMB PRO-/SUPINATION IN EXTERNALLY POWERED HAND PROSTHESES

Six different types of prehension can be distinguished for a human hand (Peizer et al., 1969). The available prosthesis type only had one degree of freedom, which clearly only provides a limited function. In the design group an adaptive hand had been developed (Cool & Hooreweder, 1971) consisting of five pneumatically powered fingers. The thumb had a fixed position opposed to the other fingers. The question arose as to whether a sideways movable thumb would improve the function of the hand without much cost in terms of extra attention. Therefore, a prototype was built with a movable thumb on a rotating base. The grasping function of the fingers and thumb was pneumatically driven. The thumb rotation was electrically driven and was mounted in such a way that the rotation axis fell approximately along the index finger. Effects of wrist rotation of the subject on the orientation of the hand were prevented in order to make the configuration compatible with that of an amputee. The prosthesis was controlled by transducers in the subject's hand. Four control modes were considered

A   One input signal and no sideways thumb rotation
B   One input signal with a coupling between finger flexion for grasping and thumb rotation
C   Two independent input signals
D   Two input signals, one for the finger flexion, while the thumb rotation resulted from a combination of the first and second control signal

Five subjects performed a number of tests. The first one was the so-called Carlson test (Carlson, 1970), which consisted of moving 12 cylindrical objects of different diameters, 0.5, 1.5, and 2.5 inches, and made of fabric, foam rubber, and hard PVC, from one place to another. The task completion time and number of errors were measured. The subjects also had to indicate their opinions on the controllability, attention required, and inconvenience of the control method on a five point scale. Finally, a number of daily living (ADL) tasks had to be executed, namely, lifting a marble, lifting a 3.5 by 2.5 inch box, lighting a match, grasping a plastic coffee cup, lifting a tea cup, holding a spoon for feeding, and handling a saucer. The activities were judged with reference to handling, natural appearance, and speed.

The results of the Carlson test did not show much difference between the error scores and execution times. However, the ratings of the subjects gave a clear ranking in the order of A, B, C, and D on all three scales. Mode A obtained an average score of nearly 5 and mode D had a score between 1 and 2. For the daily living tasks the same order could be found, although the differences were rather small. The final conclusion was that the addition of a thumb rotation is not likely to improve the function of the prosthesis (Van Ditzhuisen, 1972).

This investigation illustrates that the provision of a prosthesis with more functions does not necessarily lead to better performance, because it also poses a higher demand on the wearer's control actions, leading to a higher mental load. This leads to the following topic.

3.2 MENTAL LOAD IN ARM PROSTHESIS CONTROL

A prosthesis can be judged based on its functional possibilities, which depend on a number of parameters, such as opening width, grasping force, shape, and stiffness of the prehensor, energy source, and type of control. These possibilities can be investigated on a test bench. Such a study has indeed been executed in which experimental prototypes were compared with a number of commercially available hands and hooks (Tebeest, 1977; Zwart, 1979).

Whether a prosthesis can be used, however, will also depend on the ease of operation, which can only be investigated in a number of tasks that are thought to be representative for the use of a prosthesis in daily life. It can be hypothesized that whether or not an amputee will use his or her prosthesis for a certain activity will depend on not only the function it has to offer but also the mental load the use of this function imposes on the wearer. Therefore, an investigation was undertaken in which a number of tasks was executed by a group of subjects with different prosthesis types, and both task performance and mental load were measured (Soede, 1980).

In the literature (see, for instance, Moray, 1979) a number of methods have been described to measure mental load, none of which is ideal. Based on experience with mental load measurements in a number of other applications, the double-task method was chosen with a continuous secondary task. The method is based on the assumption that a certain mental capacity is available for task execution. A subject is asked to spend the mental capacity left over from the primary task on a secondary task. Thus his or her performance in the secondary task is an indirect measure for the mental load in the primary task. For the secondary task the subcritical instability task was chosen (Jex et al., 1966). This is a tracking task with an unstable controlled element, so that the subject is forced to keep busy. An increasing mental load in the primary task then leads to an increase in the tracking error. The instability parameter can be chosen in such a way that execution remains possible by first testing the subject's abilities with only the tracking task.

A problem, at least at the time this investigation was begun, was the choice of relevant prosthesis tasks, because little was known about the actual use of prostheses in practice. Therefore use was made of available ADL lists and rehabilitation centers training programs. Based on an analysis of these lists (Pimontel, 1975) which led to a classification into 14 groups (Huisman, 1976), six more or less abstracted tasks were chosen for the experiments:

- A switchboard task, in which one of two closely located switches had to be operated
- A fixation task, in which two pieces of board fixed to each other by two simple door magnets had to be pulled apart in such a way that one piece had to be fixed against the table by the prosthesis
- A circle tracking task, in which a joystick had to be moved by the prosthesis through a circular slot
- A pigeon hole task, in which a pin had to be put in a pigeon hole without touching the wall
- A manipulation task, in which the hand screwed a nut on a bolt held by the prosthesis

Five prosthesis types were used in the experiments: a cosmetic hand, a body powered hand, a body powered hook, a myoelectric hand, and a myoelectric hand with wrist rotation. The cosmetic hand was only used in the three tasks first mentioned. The myoelectric prosthesis was fitted in such a way that the wrist rotator had to be used, so the task was more difficult than when using other prostheses. All tasks were executed 6 times by 12 right-handed subjects.

The main results with reference to the prostheses with an active prehension were as follows: The body powered prosthesis performed better than the myoelectric one in adjusting a given prehension force, in the sense of resulting in a shorter execution time, a lower error score, and a lower mental load. In the switchboard task and the circle tracking task, more errors were made with the hook than with the other prostheses. Extension of the tasks with active use of the wrist rotator indeed yielded a larger mental load score.

About a year after completion of this investigation (Soede, 1980), a number of experiments were repeated on a smaller scale (Dekker, 1981; Houtman, 1982). In these experiments, an experimental prototype of the design group, the so-called three-grip hand (Zwart, 1979), was compared with a myoelectric hand as a reference. The experiments used 6 subjects. The prehension task and the manipulation task were omitted for technical reasons. In addition to the measurements of performance and mental load, a questionnaire was also used in which the subjects were asked to give a rating on a seven interval scale with reference to the feasibility for the task, attention required, controllability, and degree to which a prosthesis/task combination was fatiguing, physically as well as mentally. All tests were performed with different execution orders over 3 days.

It was found that the scores for the two prostheses were approximately equal, both for execution time and normalized tracking error. Moreover, the execution times were also comparable with those found by Soede for the myoelectric hand. The normalized tracking error was somewhat higher. This difference, however, could be attributed to a performance level in the tracking task in a single task condition that was lower than that of Soede's subjects. Furthermore, the correlation between the error scores in the secondary task and the judgements on both attention required and controllability was about 0.7, which was also the correlation between judgements on different days. Because the use of a questionnaire is much simpler than the execution of experiments with a double task, this method should be favored in this type of experiments.

4   Field evaluation of arm prostheses for unilateral amputees

4.1 BACKGROUND

This section deals with a field study among adult traumatic amputees with a very broad scope aimed at finding the optimal conditions to enable amputees to function again in their environments with the available technical means. It was executed in the period 1974 to 1981. This investigation was initiated by the multidisciplinary treatment team at De Hoogstraat, one of the cooperating rehabilitation centers. The goals were, on the one hand, to get optimal treatment of amputees and, on the other, to see what the positive and negative properties of existing prosthesis types were in order to design better prostheses. Therefore the investigation was directed at two main topics of interest, the amputee, the problems he or she has to cope with, the role the prosthesis plays in his or her life, and the prosthesis, its use, and its potential benefits and burdens (Van Lunteren et al., 1983).

4.2 METHODS AND DEFINITIONS

Five measurement methods were applied in the investigation. The following three were applied during a 1 day visit of the amputee to the rehabilitation center:

- Medical examination
- Two psychological tests
- Questionnaire consisting of 92 multiple choice questions, mainly about the prosthesis and its use

The following methods were applied during two 1 day visits of an occupational therapist at the amputee's home:

- Semi-structured interview based on 220 questions, covering psychological aspects and the prosthesis
- Observations of the actual use of the prosthesis for about 50 daily life activities

The group of participants satisfied the following selection criteria. They had a unilateral traumatic amputation that took place more than 1 year ago, had Dutch nationality, and were 18 years or older. They were obtained from a list of 101 amputees from two rehabilitation centers. Of this group 42 persons satisfied the selection criteria and were willing to participate. Before starting the investigation, the procedure was tested with five amputees from a third rehabilitation center. The list for the semi-structured interview initially consisted of a smaller number of items. As a result of this tryout, where the interview had a more open character, the interview was expanded to the present 220 questions.

The data obtained could be divided into two groups, quantifiable data and purely qualitative data. The first group consisted of all answers that can be classified into a limited number of categories. To this group belong the medical data, the results of the psychological tests, the multiple choice questionnaire, the observations of ADL, and also part of the interview data. These data could be reduced to histograms, frequency tables, and correlation coefficients.

The second group consisted of descriptions of events, thoughts and feelings, explanations of answers, etc. These data, in the form of literal citations, were arranged according to subject matter. For the analysis of the prosthesis actions, a number of mechanical functions were defined, which was found to be very useful to characterize the type of use of a prosthesis, i.e.,

- Fixation in the prosthesis
- Fixation between prosthesis and body
- Fixation between prosthesis and environment
- Bilateral fixation, i.e., between prosthesis and own hand
- Carrying
- Supporting
- Pushing or shoving

Two types of fixation in the prosthesis were distinguished: direct fixation, i.e. grasping an object directly from the environment, and indirect fixation, i.e., taking an object from the wearer's own hand.

Besides the mechanical function, a prosthesis also has a cosmetic function, i.e., a function in making the amputee look as natural as possible. In this context three types of cosmesis were defined:

- The passive cosmesis, i.e., the appearance of the prosthesis as such. This type of cosmesis is entirely prosthesis dependent.
- The cosmesis of wearing, i.e. the way the amputee moves in a natural way with or without a prosthesis, for instance, while walking on the street. This type of cosmesis is entirely wearer dependent.
- The cosmesis of using, i.e., the way in which the amputee uses the prosthesis in executing a task. This type of cosmesis depends both on the possibilities of the prosthesis and the proficiency of the wearer.

The group consisted of 39 men and 3 women; 40 persons were right-handed and 2 left-handed. In 26 cases the amputation was on the dominant side. The amputation levels were as follows: 12 above the elbow, 16 forearm, and 14 wrist. The time between the accident and amputation varied for 11 amputees between 1 and 5 years, 12 between 5 and 10 years, and 19 between 10 and 35 years. Four types of prostheses were involved: 5 persons used a cosmetic hand, 12 a body powered prosthesis (hand and/or hook), 19 a myoelectric hand, 2 a switch operated hand, and 4 used no prosthesis. For the majority of the group, medical factors did not obstruct the use of a prosthesis.

The psychological tests made it possible to compare the group on a number of psychological characteristics with the Dutch population in general. It was found that the group contained more extroverted people. On the test attitude scale, more self- defensive people were found. The other scales showed no significant differences.

4.3 THE PROSTHESIS AND ITS USE

One of the results of the observations was the fact that many activities were found to be executed in different ways, with or without the prosthesis. The ADL analysis resulted in a number of tables that give the relation between individual user, type of activity, and functional alternative. It was found that 26 activities out of the original list of 50 two-handed tasks were executed by at least four participants. Therefore this list was used for further processing. For each prosthesis type and amputation level, the average number of activities was determined for each of the earlier defined prosthesis functions, but also for other techniques, such as using one hand, some other body part, or with aid. For the body powered prosthesis a distinction was made between users who mainly use the hand and those who mainly use the hook. Only one person used the hand and hook about equally and, therefore, is counted as a half-person in both groups for calculation of the average. The results for a number of prosthesis functions are given in Table 1.

Here it should be mentioned that grasping (fixation with respect to the prosthesis) is mostly used indirectly. Direct grasping is an exception. Fixation with respect to the body or to the environment, for instance, a table, is an often used alternative to grasping. The bilateral fixation function does not occur in the table because it is mainly used in activities outside the list of 26 items. In the interviews it was mentioned as a frequently used function. When asked for what kind of activities they thought their prosthesis to be important, 28 people mentioned hobbies, 17 driving a car or riding a bicycle, 14 work, and 6 ADL. With regard to the question of what they liked and did not like about their prosthesis, a number of positive and negative items were mentioned. Positive comments referred to the cosmesis (35) the motor functions (16). Negative comments referred to the cosmesis (20), technical reliability (13), lack of touch (6), hindrance due to the harness (6), and hindrance due to the socket and perspiration (9).

Complaints with respect to the cosmesis mainly refer to the vulnerability of the cosmetic glove with respect to dirt, discoloration, and damage. It should be mentioned that criticism about the cosmesis comes from the same people who also mention its positive properties. In the same manner, comments about the technical reliability mainly come from the people who also appreciate its mechanical functions and who also have a user score above the group average.

With respect to cosmesis, it can be said that most amputees focus on passive cosmesis. Only a few people are aware of the cosmesis of wearing. One of the participants said that he had noticed how he walks when he saw himself mirrored in a shop window. After that he had trained himself to move in a more natural manner. Of the 42 participants, 22 still wear the same type of prosthesis that they first obtained, and the others changed over the course of time to another type or to no prosthesis at all.

Based on the individual scores for intensity of use of the motor functions of the prosthesis, the group could be divided into 21 users with a total score higher than average and 21 users with a lower than average score. For each of the groups the answers to a number of interview questions, the medical data, and the results of the psychological tests were analyzed. Besides the overall use of the prosthesis, the use of the grasping function alone was considered for those persons who had a prosthesis with this function. A chi-square test was used to determine the following relations and their significance levels:

Table 1.   Percentage of a list of 26 items where a given prosthesis function was used

Prosthesis and amputation level		Nr	Fixation with respect to			Pushing, shoving	Carrying
			prosthesis	environment	body		
Cosmetic hand	a	4	1	18	18	4	4
	b	1	-	-	4	4	-
Body powered hand	a	5.5	7	10	16	3	1
	b	1	-	-	8	-	-
Body powered hook	a	2.5	17	2	32	3	-
	b	3	3	9	5	3	-
Myoelectric hand	a	15	31	12	10	4	-
	b	4	12	10	7	3	3
Switch operated hand	a	2	4	2	-	-	-

    b = below elbow;   a = above elbow

- Lower arm amputees have a higher user score than upper arm amputees (0.05)
- Help from the environment is related to a lower user score (0.02)
- The latter statement also applies for the use of the grasping function only (0.05)
- Wearers of a myoelectric prosthesis use the grasping function more than wearers of a body powered one (0.02)
- Amputees who obtained their prosthesis within half a year after the amputation used their grasping function more than those who obtained it after a longer time interval (0.02)

4.4 COPING WITH THE AMPUTATION

In many cases the cause of the amputation, often an industrial accident, still keeps people occupied. The positive or negative reaction of the environment can also be important. The period between hospitalization and entering the rehabilitation center is a difficult one. People are no longer regarded as a patient and they discover the consequences of their handicap. Admittance to a rehabilitation center is often mentioned as a positive turning point. An important event is the provision of the first prosthesis. On the one hand, there is a feeling of being complete again, and, on the other, there is an alien object that has to be incorporated into their body scheme. Initially, attention is mainly focussed on the cosmetic function of the prosthesis. The motor function comes secondly. Thus people first discover what they cannot do with the prosthesis before they learn what they can do with it. The question of whether they can return to their old job or, if not, whether they can find another one is a matter of concern from the outset. The attitude of the amputee's family is very important. They can have a very positive influence by encouraging the amputee to do things on his or her own, which increases his self-confidence.

Of the 42 persons, 26 could be considered to have coped with their amputation, whereas 16 still had difficulties. It was found that extroverted people were mainly found in the group who had learned to cope with their amputation. Therefore, it could be deduced that the group as a whole represented a positive selection with respect to the way they had learned to cope with their handicap. Taking into account this effect, this means that, in general, the number of amputees who still have difficulties in coping with their handicap should amount to 60%, instead of the 40% found in this group.

4.5 SOME CONCLUSIONS

Besides the already mentioned outcomes, a number of additional statements can be made based on the outcomes of this study. For the majority of the group the cosmetic function of the prosthesis is very important. Many people who value the cosmesis highly pay much attention to the passive cosmesis but not to the cosmesis of wearing and using the prosthesis. As a result they often look conspicuous because of their tendency to hide the prosthesis.

Use of the direct grasping function is an exception. If an object is picked up directly by the prosthesis, in most cases the orientation differs from the desired one. Therefore, in two-handed tasks the object is given to the prosthesis by the user's own hand. The direct grasping function is only used for activities such as picking up a suitcase or grasping the handlebars of a bicycle.

In assessing the use of the prosthesis for task execution in general, it has to be borne in mind that the prosthesis has to compete with a number of alternatives, such as.

- Use of the stump
- Use of other body parts such as mouth or knees
- Asking for help from other people
- Avoidance of certain activities

Body powered prostheses often require compensatory motions in order to operate the grasping function in different locations. These compensatory motions may look unnatural, and they are avoided if there is a less obtrusive alternative. The body powered hook is especially advantageous for hobbies and jobs that require some manual skills because of the following properties:

- It provides good sight of the grasped object.
- It is not easily damaged.
- It is easy to clean.

The myoelectric prosthesis is valued because the wearers feel less hampered by it. In particular, below-elbow amputees do not need a harness. The latter point is also cosmetically attractive.

Whether an amputee will wear and use a prosthesis depends on its potential benefits and burdens. Benefits are the cosmetic and mechanical functions. Burdens can be divided into burdens inherent to wearing only and burdens related to the use of the prosthesis. Burdens inherent to wearing are the inconvenience caused by a harness if present, the fitting, and the burdens of a poor cosmesis. Burdens related to use are the physical and mental effort required to operate the prosthesis, the lack of touch, an obstructed view, the weight, and the susceptibility to dirt and damage. For many users the burdens of wearing are the most annoying properties, due to their constant presence. Unilateral amputees wearing a prosthesis do not consider the mental load as important. Some people with a long forearm stump who do not value the passive cosmesis prefer not to wear a prosthesis, thus avoiding its burdens. They often move very naturally and thus are less noticeable than some wearers of a prosthesis.

As a final conclusion it can be stated that no general rule can be given about the best solution, because how the benefits are weighed against the burdens always depends on individual preferences. For designers of arm prostheses, however, it is important to strive to reduce the burdens. Here the cosmesis and comfort of wearing are the dominant factors.

5   Field evaluation of arm prostheses for children born with one hand

The second field study was part of a design program for body powered hand prostheses for children. This investigation was undertaken between 1982 and 1990 (Van Lunteren & Van Lunteren-Gerritsen, 1989, 1992). The impetus was the development in the design group of a prototype of a body powered hand prosthesis without a harness for children with a unilateral forearm defect.   The prosthesis combined the advantages of a myoelectric prosthesis (no shoulder harness) with that of a body powered prosthesis (light weight) (Pistecky, 1983; Kruit & Cool, 1989). Instead of shoulder motions, elbow motions were chosen for control of the hand opening (Fig. 1). In the meantime this control principle had also been worked on independently by other designers (Richter, 1987; Meeks & LeBlanc, 1988). The designer of body powered prostheses for children has to take into account that the forces and displacements children can generate as a control input are much smaller than those generated by adults. This again has consequences for the opening width and pinch force of the prosthesis.

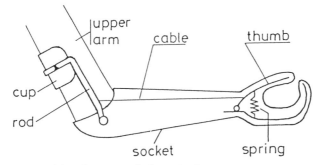

Figure 1.   Principle of the elbow controlled prosthesis.

In order to optimize the design, the following questions had to be answered:

* For what actions is a prosthesis important for this category of children?
* What are the minimal requirements for opening width and pinch force?
* For what actions is a wrist rotator important?

It was decided to look at the way existing prostheses are used, to make a prototype based on assumed specifications, and to try it out.   The first clinical prototype was available by the end of 1981.

5.1 METHODS AND DEFINITIONS

Based on experiences in the field study of adult traumatic amputees, an observation method was chosen in which a child with a prosthesis is followed over a normal school day from the moment he or she starts dressing until going to bed in the evening. If possible, a day is planned that includes both a physical exercise class and a class involving manual activities. During the day all actions of the child, with or without a prosthesis, are observed and the ways in which they are executed are noted. This means that the list of activities is not standardized, which makes a comparison more difficult. However, the only question being asked was for what activities do the children use their prosthesis. Moreover, it was felt that in a standardized test situation the child might try to do things differently from his or her normal pattern, i.e., in a way that was thought to please the observer.

Besides the functions that were used for a certain action, it was also observed whether the way an action was executed was unobtrusive or involved some unnatural-looking motion (the cosmesis of using). The way the child moved while not performing any action with the prosthesis (the cosmesis of wearing) was also observed. Furthermore, the most common position of the wrist rotator and the actions for which the wrist rotator was used were observed.

In a semi-structured interview with the child and with his or her parents (mostly the mother) the following items were discussed:

- Wearing pattern of the prosthesis
- Repairs
- Wishes
- Development of the child

Finally, the pinch force of the prosthesis was measured, in most cases for an opening width of 20 mm, and the maximum opening width was measured in a number of hand positions.

5.2 CHILDREN AND PROSTHESES

The children and parents who took part in the project were recruited via the cooperating rehabilitation centers. The first group consisted of a small number of children wearing a myoelectric prosthesis or a shoulder-controlled, body-powered hook and who were known to be good users. They were particularly selected to get an idea of what good users do with their prostheses. In a later stage children were visited who tried out a new prototype of the elbow- controlled, body-powered prosthesis. In most cases they were first visited while they still wore their shoulder-controlled, body-powered prosthesis. In this way a comparison could be made between the two different prostheses as used by the same child, thus eliminating child-dependent factors.

The total group observed in the course of the project finally consisted of 26 children: 9 boys (5 with a left-sided defect, 4 with a right-sided defect) and 17 girls (11 with a left-sided defect, 6 with a right-sided defect). The age of the children was between 3 and 13 years. The following cases were observed:

- 5 myoelectric prostheses
- 6 shoulder-controlled hands
- 17 shoulder controlled hooks
- 10 elbow-controlled hands
- 1 elbow-controlled hook
- 3 cosmetic hands
- 1 child without a prosthesis.

Altogether 43 visits were paid to 26 children. The reasons for visiting children more than once were as follows:

- A change from one type of prosthesis to another (11 visits)
- A change to another type of socket (3 visits)
- To get an idea about the effects of age on the use of the prosthesis while still wearing the same prosthesis type (3 visits)

5.3 DATA HANDLING

All different or differently executed actions on a given day were noted over the day. Afterwards they were listed and stored in a specific database, one file for each visit. Actions were characterized by five fields:

- Object on which a certain action was executed
- Object specification, which could be given if necessary
- Action that was exerted on the object
- Further specification of the action if necessary
- Method of execution

Based on these files, for each visit a summary was made of the number of different or differently executed actions and their distribution over the different means of execution. A list was also made of all different actions observed over the group as a whole, including the number of visits during which a certain action was observed and the different means of execution. The number of visits was also noted. In this manner many types of information could be extracted once all actions were recorded in the databases.

Table 2   Summary results: distribution of actions

Way of execution	Number	Mean	Std. dev.	no prosth.
Without using the prosthesis	1254	36	11	
With help	135	4	4	13
Own hand only	879	25	9	23
Stump	176	5	4	
Other body part	64	2	2	1
Using the prosthesis	1910	55	15	
Direct prehension	139	4	4	7
Indirect prehension	478	14	9	14
Fixation to body	220	6	4	24
Fixation to environment	405	12	3	11
Bilateral fixation	98	3	2	7
Carrying	262	8	3	9
Pushing	205	6	4	5
Supporting	103	3	2	1
Total	3164	90	19	115

The last column gives the distribution for a child without a prosthesis. Here, the fold of the elbow was used as a prosthesis.

5.4 RESULTS

Activities observed

Based on the individual results, the average pattern for the type of use of the prosthesis and the alternative methods of doing things was determined. Because in four visits the children did not have a prosthesis with a prehension function, they were not incorporated into the calculation of the average profile. This also applies to the case of four visits during which the prehension function of the prosthesis was used only once or twice, or not at all, which sometimes could be related to the condition of the prosthesis or to the fact that the prosthesis was worn only during a short part of the day. For the resulting 35 cases on which the average result was based (and which represent a total of 3,164 actions), the use of the prehension function lies between 3 and 54 actions.

Table 2 illustrates how the 3,164 actions are distributed over the different methods of execution, with or without use of the prosthesis. The actions observed when the children did not wear the prosthesis, mostly at the beginning and end of the day, are not included in the table. The first column gives the distribution of the 3,164 actions observed over the different methods of execution. Based on the individual distribution lists of the 35 visits, the mean values and standard deviations could be calculated. They are given in the second and third columns, respectively.

For the child without a prosthesis, it is interesting to note that the functions of the forearm stump were comparable with those of a prosthesis. Even a prehension function could be observed in which the fold of the elbow acts as a terminal device. Therefore, by considering the stump in the same manner as a prosthesis, the actions of this child could be classified according to the same system. These results are given in the last column of the table.

The table shows that the total number of actions lies within the m + s range of the group. The same applies for most methods of execution. Exceptions are as follows: fixation to the body (>m + 4 s), help (> m+ 2 s) and bilateral fixation (=m + 2 s).

From the tables it follows that many actions are executed without the prosthesis, and when the prosthesis is used most activities are executed without use of the grasping function. If the grasping function is used, objects are mostly put into the prosthesis by the other hand. This is illustrated in Figure 2, which gives the values as a percentage of the total number observed. Figure 2A gives the mean values for the group, while Figure 2B gives the values for the child without a prosthesis who uses the fold of the elbow instead.

Table 3. Actions where the direct prehension function was used

Object to be grasped	Number of actions	
	Prehension	Total
Handlebars of: bicycle, tricycle, autoped, doll's carriage, wheelbarrow.	38	47
Wallbars, horizontal bar, rings.	14	50
Sleeve of a coat (withdrawing arm), socks, mittens.	10	22
Vertical bar, rope (e.g. of a swing).	7	12
Chair.	7	46
Shoelaces.	4	12
Shoes, boots.	3	8
Sum of actions in this table	83	197
Sum of actions observed less than 3 times	56	260
Total	139	457

A list of actions for which the direct prehension function was observed on at least three visits is given in Table 3. The actions are arranged according to the number of times the action was noted, together with the total number of times when the action was observed, irrespective of the method of execution, arranged according to the number of times the action was noted, together with the total number of times the action was observed, irrespective of the method of execution. In many cases the activity was also seen while it was executed in a different manner. The table therefore also shows the total number of times that the activity was observed. As the table has been arranged in such a manner that a number of similar activities are combined in one item, the numbers can be higher than 35.

The actions with the handle bars, horizontal and vertical bars, and ropes have in common that they all involve the grasping of a rotation symmetric shape that is more or less fixed to the outside world. The consequence of the rotation symmetry is that the object can be grasped from a wide range of approach angles, and thus no compensating arm movements are necessary to grasp the object (cosmesis of use). The same more or less applies to the grasping of flexible objects such as clothes.

For the bars and ropes, the activities require a symmetric action of both arms but not always at the same location in space. For the handlebars and the vertical bar or rope, direct prehension is used in more than half of the cases. The reason that the same is not

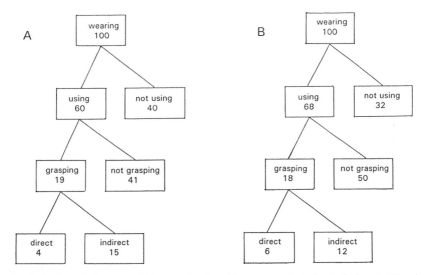

Figure 2. Indication of the use of the prosthesis while executing actions during a normal school day expressed as a percentage of all actions observed. A: Group average, based on a total of 3164 actions. B: Distribution for a child without a prosthesis who uses the fold in the elbow as a grasping device.

true for the horizontal bars is due to the fact that for a number of prostheses the opening width was too small in relation to the diameter of the bars. Also, in a number of cases a child was not able to hold the handlebars of a bicycle due to the limited opening width of the prosthesis.

A list of actions with indirect prehension function, observed four or more times, is given in Table 4. The actions are arranged according to the number of times that the use of this function was noted, together with the total number of times that this action was observed, irrespective of the method of execution. The table shows that the most frequent two-handed actions involve a number of items, such as brushing teeth, dressing, and the school activities of handling pens and pencils, cutting paper, etc. The list of 197 actions with indirect prehension in the table represents 41% of all 478 activities for which this function was used. The rest of the list consists of 18 items that were noted three times, 21 items noted twice, and 185 items noted only once. In 197 of 456 cases, i.e., 43%, indirect prehension was used. This was 35% for the 281 cases not listed in the table.

The results of Tables 5 and 6 show that in more than 50% of the cases alternatives are used for the application of the prehension function.

5.5 FACTORS AFFECTING USE OF THE PREHENSION FUNCTION

Especially in the case of experimental prototypes, it is important to determine whether the prosthesis imposes limitations that might be changed. Therefore it is important to know more about the factors that influence the use of the prehension function. In dealing with this question it should be realized that the method of use of the prosthesis only partly depends on the properties of the prosthesis and depends mostly on the wearer.

Although the primary attention of the investigators was focused on the prosthesis-dependent factors, the child-dependent factors could not be neglected. Moreover, some insight into the role of these latter factors is valuable for treatment teams because they should be taken into account in the choice of a prosthesis. Some child-dependent factors that were considered are as follows:

Table 4.   Actions with indirect prehension function

Object held	Action with own hand	Number of actions	
		Prehension	Total
Piece of paper	Cutting, tearing, sticking	30	79
Coat	Link two halves of zipper	19	34
Bag	Get object out of it	14	35
Toothpaste tube	Remove or close cap	14	49
Pen, pencil, ruler	Miscellaneous	13	15
Toothbrush	Put toothpaste on it	13	19
Socks	Pull on	12	31
Trousers	Pull on	11	38
Fork	Cut meat with knife	10	21
Pen case	Open or close zipper	10	40
Coat	Pull up zipper	8	24
Skipping rope	Swing rope	7	8
Playing cards	Take card	6	6
Toothbrush	Handle glass or tap	6	7
Toothpaste tube	Hold toothbrush	6	17
Shoe laces	Fasten	5	12
Cap	Draw pen out of it	5	9
Sharpener	Sharpen pencil	4	6
Cap	Put pen inside	4	6
Sum of actions in this table		197	456
Sum of actions observed less than 4 times		281	806
Total		478	1262

- Length and shape of the stump
- Age
- Age at receipt of a prosthesis
- Dexterity
- Dominance
- Influence of the environment; for young children this will mainly be the influence of

the parents.

Prosthesis-dependent factors are:

- Opening width
- Pinch force
- Type of control

These factors depend on the prosthesis type and the condition of the prosthesis, which in a number of cases was not found to be optimal, especially in the experimental prototypes.

In order to investigate the influence of the factors considered, the group was divided into 20 visits to a child with a prehension score lower than 15 and 19 visits to a child with a score of 15 or higher. With respect to the child-dependent factors of stump length, age, and age of fitting with a prosthesis, it was found that the group was too homogeneous to draw any conclusions. Only an encouraging attitude of the parents, as mentioned in the interviews, was found to have an influence at the 5% level in a chi-square test.

An intriguing factor is hand dominance. In the course of the investigation, a strong impression arose that some children might be dominant on the defective side. This impression was sometimes affirmed by the mother or schoolteacher. In discussions with other workers in the field of rehabilitation of children with an arm defect, they also mentioned this phenomenon. They also mentioned a number of actions with the defective side on which they based this impression, such as waving the arm, pointing at something, raising the arm, or using the arm to break a fall. A schoolteacher mentioned difficulties in children learning to write using their intact hand. Other indications might be a preference for the left or right eye when looking through the view-finder of a camera, and the use of the right or left foot when kicking a ball or starting to jump. It is known, however, that this preference need not always correspond to that of the hand. Anyway, as it seemed worthwhile in the course of the investigation an attempt was made to estimate the child's hand dominance, based on the aforementioned actions and on the opinions of the mother and the schoolteacher. The data collected did suggest some effect. However, according to a chi-square test at the 5% level, the differences were not significant.

In order to investigate the prosthesis-dependent factors, the use of the prehension function was related to the opening width, the prehension force, the type of control, and the condition of the prosthesis. Therefore, again the group of 20 visits to children with a prehension score lower than 15 was compared with that of 19 visits to children with a score of 15 or higher. For the opening width a value of 35 mm was chosen as a separation criterion. This value was found to be sufficient for holding and loosening the handlebars of a bicycle. This yielded a group of 22 with an opening width of 35 mm or more and a group of 17 with an opening width lower than 35 mm. A chi-square test showed that a small opening width indeed corresponded to a low prehension score. The differences found are significant at the 5% level.

The influence of the prehension force was investigated in the same manner. Two groups were defined. One consisted of 26 cases indicated to be "sufficient" or with a force larger than or equal to 10 N, and the other group consisted of 13 cases with the qualification "low" or with a force of less than 10 N. The results, however, did not yield any significant differences.

The technical condition of the prosthesis is given in three categories as "good," "doubtful," and "bad." However, for a separation in three quality categories the numbers are too low for a reliable chi-square test; therefore, the categories "doubtful" and "bad" were grouped together. The differences in use of the prehension function are then significant at the 1% level.

A comparison between the type of control and the use of the prehension function shows that the prostheses with shoulder control are equally divided over both categories, a pattern that is also found when the three different types with shoulder control, i.e., hand, voluntary opening hook and voluntary closing hook, are considered separately.

The differences are found in both myoelectric control and elbow control. It should be mentioned, however, that the children with the myoelectric prosthesis were chosen at the outset because they were known to be good users. Furthermore, these prostheses were in good condition. For the prostheses with elbow control, the low prehension score can partly be explained by the fact that in this category a relatively high number of prostheses were in doubtful or bad condition.

Another possible relation that was investigated was between the influence of the parents and the condition of the prosthesis. There might be a difference in attention to malfunctions of the prosthesis. Therefore the relation between these factors was investigated, but such a relation was not found.

For a comparison between the prosthesis types used in this investigation, the influence of the child-dependent factors on the outcome of the technical comparisons should be minimized. A way to cope with this problem is to observe two different types of prosthesis used by the same child. This strategy was followed wherever possible. A number of months after the new experimental prosthesis had been fitted, a second visit took place with the same program of observations. However, this comparison was not always possible. The group of potential subjects was rather limited. Moreover, the child, his or her parents, and the school had to be willing to endure having an observer following the child for 2 days. The provision of the experimental prototypes in the clinic in many cases was part of a regular checkup program wherein sometimes a new prosthesis was advised. In a number of cases the new prototype was fitted as a replacement of an already worn-out existing prosthesis, or of a prosthesis without an active prehension function. Therefore, a number of children were visited only once. This was also true for a number of children who were visited in the beginning of the project for the sole purpose of observing for what activities and in what manner a prosthesis was used.

In three of the four cases, a decrease in the use of the prehension function was found when a child changed from a shoulder-controlled prosthesis to an elbow-controlled one. However, in two of the cases the condition of the second prosthesis was worse than that of the first one. However, although the elbow-controlled prosthesis had less function than the shoulder controlled one, all nine children wearing an elbow-controlled prosthesis did not want to change back. The most important reason was the lack of a harness, although none had complained about this previously. Besides the increased wearing comfort, it was mentioned that the lack of a harness also made putting on and taking off the prosthesis much easier, because they did not need to undress to do it. Now the children could easily remove their prosthesis as desired and put it back on again if they felt it to be useful.

5.6 Discussion

As already mentioned, the feeling existed that standardized test results might not be representative of the usual actions of the child. This effect might even play a role in the methods applied in this investigation. Indeed, it was found that sometimes at the beginning of the day a child tried to do things using the prehension function, but it was done in a rather clumsy fashion. When asked afterward, the mother confirmed that the child was used to doing it in another way. However, in such cases the child very soon resumed his or her normal pattern of behavior for the rest of the day.

A valuable result of the observations made in the child's own environment is that good insight is obtained into the types of actions for which a prehension function may be really useful for a child with a below-elbow prosthesis. It was found that many actions that are often trained with application of the direct grasping function can be executed in an easier and more natural manner by using an alternative function. In practice, children determine how to do it their own way.

Picking up things from a table using the prosthesis is typically a method of task execution that is only seen in people with two arm prostheses. Yet, people with a unilateral defect or amputation are often asked to show their dexterity in this manner, as has been illustrated many times in video and slide presentations at congresses.

Consequently this type of task has often been used in laboratory studies.

It was found that a good grasping function is essential for bicycle riding. Young children who are not yet very adept in bicycle riding should especially be able to ride safely in everyday traffic. This means that it should be easy to maintain a good hold on the handlebars, but, equally important, it should also be easy to loosen one's grip. This leads to the requirement that an opening width of 35 mm should be easily attained. This is not only a requirement for the terminal device, but also for the adjustment of the cable in a shoulder-controlled prosthesis or the elbow angles in an elbow-controlled one.

An activity for which children with a sufficiently large opening width used the prehension function of their prosthesis was playing on jungle gyms. This requires an opening width of about 45 mm. Children who are not able to use the grasping function use the inside of their elbow to get hold of the bars. Although this application is not crucial, some children mentioned this as a reason for wanting a larger opening width.

For putting on trousers or socks, the pinch force in most cases was too low. However, children did not complain because they used a number of alternative techniques. For most activities for which the prehension function was used, no large forces were required. A pinch force of about 10 N seems sufficient for most applications observed.

All children had a wrist rotator, but not all children changed their wrist angle during the day. Those who did change it did not do it very frequently. In general, two positions were favored for both the hand and the hook. This means that for a lower arm defect, a passive wrist rotator is desirable. Without a wrist rotator it is possible to rotate the plane of prehension by compensatory motions of the upper arm, but this may lead to unnatural looking actions.

The cosmetic function of the prosthesis does not play a role for young children, but for some parents it is often very important. This is especially true when parents still have problems in coping with their child's defect. On the other hand, people who overcome these problems are very happy when they see that their child is very handy with, for instance, a shoulder-controlled hook. At the age of 10 to 12, the cosmesis becomes more important for the children themselves. They also become more critical of the burden of wearing. This means that some of them tend to be more selective about when, and for what activities, they wear the prosthesis.

6   Concluding remarks

The previously described activities started at a time when the thalidomide drama stressed the importance of new developments in the field of prosthesis design. However, from the technical point of view, new developments such as the myoelectric hand, looked rather promising. In the control lab at Delft it was decided not to start by doing things that were already being done elsewhere, and thus to start with design activities based on existing knowledge and experience. Because the majority of the amputees are one-sided forearm amputees, it was decided by the designers to start first with the development of a good body-powered terminal device, because existing devices were not satisfactory for a number of users. Moreover, there it makes no sense to develop an optimal elbow if the hand is not optimal.

In the beginning, the attention of the Man-Machine Systems Group was mainly focused on questions concerning the controllability of arm prostheses in terms of what were optimal input and feedback variables to choose and in what range. The laboratory studies, the contacts with the rehabilitation centers, and the amputees who participated in the treatment team discussions gradually changed this viewpoint. It was realized that the practical use of a prosthesis not only depends on the functional possibilities of the device, but also on other matters, such as the mental load required to use a grasping function. Moreover, it was found that the prosthesis was not always the most important thing that made a rehabilitation program successful or not. This gave rise to the initiative to start a field investigation to try to identify all factors that are important for the successful rehabilitation of an arm amputee. The results of this investigation gave the rehabilitation team better insight into the common problems of amputees.

It was a challenge for the designers to combine the positive properties of the different available prosthesis types while eliminating the negative ones. Design specifications now start with the idea that a prosthesis will have more of a chance of being acceptable if it looks good, either natural or agreeable, if it can be worn and used without hindrance, and if it can be easily controlled, i.e., cosmesis, comfort, and control, in that order.

The development of new ideas for a better design began at the moment parents complained that for young children only prosthetic hooks were available but no satisfactory active hands. Clinical evaluation of the first experimental prototype started more or less on an ad-hoc basis, without making use of standard lists. After the first visit to a child, it gradually gained a more structured format. Besides gaining more insight into the actual use on which to base design criteria such as opening width and grasping force, the importance of building durability into children's prostheses was also empahsized.

As already mentioned in the beginning, one of the problems in designing body-powered arm prostheses for children is the fact that forces and displacements that children can generate are much smaller than those generated by adults. Therefore a mechanism has been developed that enables a displacement with low forces until the hand makes contact with an object. Thereafter, force can be generated without much displacement. In this way the energy required is minimized (Kruit & Cool, 1989). There is a problem when a cosmetic glove is used, which has to be stretched when opening the hand. A mechanism can be designed to compensate for this force (See Chapter 8). However, the relatively large forces can introduce substantial energy losses due to friction. Therefore, a fundamental study of this problem was conducted by the design group, resulting in the development of rolling link mechanisms that look very promising (Kuntz, 1995).

The field studies provided much information on the practical application of the different prosthesis types and also the needs of the wearers and their families. They stressed the importance of offering reliable information to all the persons involved. The information obtained in field studies could make a significant contribution to a videotape produced to inform parents and children about the benefits and burdens of the available prostheses for different age groups.

References

Carlson, L.E. (1970). Below elbow control of externally powered hand. *Bulletin of Prosthetics Research* 10-14: 43-61.

Childress, D.S. (1985). Historical aspects of powered limb prostheses. *Clinical Prosthetics and Orthotics*, 9: 2-13.

Childress, D.S. and Billock, J.N. (1970). Self containment and self-suspension of externally powered prostheses for the forearm. *Bulletin of Prosthetics Research* BPR 10-14:5-21.

Cool, J.C. and Van Hooreweder, G.J.O. (1971). Hand prosthesis with adaptive internally powered fingers. *Medical and Biological Engineering*, 9: 33-36.

Dekker, J. (1981). Task evaluation of a three-grip hand prosthesis. DUT, Dept. of ME, Lab. M&C, MSc.-thesis, Report A-242 (in Dutch).

Gerbranda, T. (1974). Development of a muscle stiffness transducer. DUT, Dept. of ME, Lab. M&C, MSc-thesis, Report A-144 (in Dutch).

Holthaus, J.J.C. (1972). Analysis of hand function in relation to dominance. DUT, Dept. of ME, Lab.M&C, MS-thesis, Report A-111 (in Dutch).

Houtmam, I. (1982). Mental load in task evaluation of a three-grip hand. Free University Amsterdam, Interfaculty for Physical Education, MSc thesis.

Huisman, A.W. (1976). Analysis of arm prosthesis activities on the basis of a set of independent variables. DUT, Dept. of ME, Lab. M&C, Report N-132 (in Dutch).

Jex, H.R., McDonnell, J.D. and Phatak, A.V. (1966). A critical tracking task for manual control research. *IEEE Transactions on HFE* 7: 138-145.

Kruit, J. and Cool, J.C. (1989). Body-powered hand prosthesis with low operating power for children. *Journal of Medical Engineering and Technology* 13: 129-133.

Kuntz, J.P. (1995) Rolling link mechanisms. PhD thesis, Delft University of Technology, Dept of

Mechanical Engineering and Marine Technology, Delft.

McWilliam, R. (1964). Some characteristics of normal movement in the upper limb. *Proc. Symp. on Powered Prostheses*, Roehampton, pp. 10-16.

Meeks, D. and LeBlanc, M. (1988). Evaluation of a new design: Body-powered upper-limb prosthesis without shoulder harness. *Journal of Prosthetics and Orthotics* 1: 45-49.

Meijer, A.W.A. (1970). The behavior of the human operator in tracking experiments with tactile displays. DUT, Dept. of ME, Lab. M&C, MSc-thesis, Report A-74 (in Dutch).

Mooij, J.T. (1972). The influence of proprioception on the generation of control signals. DUT, Dept of ME, Lab. M&C, MSc-thesis, Report A-122 (in Dutch).

Moray, N. (Ed. (1979). *Mental Workload: Its Theory and Measurement.* NATO-SAD. New York, Plenum Press.

Näder, M. (1970). Erfahrungen und beobachtungen mit Myobock I und derzeitige stand der Entwicklung des Otto-Bock-Myostat-systems. *Orthopädie-Technik* 22: .337-340.

Peizer, E., Wright D.W, Mason, C.P. and Pirrello, T. (1969). Guidelines for standards for externally powered hands. *Bulletin of Prosthetics Research*, BPR 10-12: 118-155.

Pimontel, R.A. (1975). Judgement criteria for the gain in function of hand prostheses for unilateral amputees. DUT, Dept. of ME, Lab. M&C, MSc thesis, Report A-146 (in Dutch).

Pistecky, P.V. (1983). Design of arm prostheses (in Dutch). *Mikroniek* 23: 20-28.

Richter, H.J. (1987). Full cuff control. *Orthotics and Prosthetics* 40: 28-34.

Simpson, D.C. (1966). Powered hand controlled by 'muscle bulge.' *Journal of Scientific Instruments* 43: 521 -522.

Simpson D.C (1969). An externally powered prosthesis for the complete arm. *Bio-Medical Engineering*, 4: 106-119.

Simpson, D.C. and Kenworthy, G. (1973). The design of a complete arm prosthesis. *Bio-Medical Engineering*, 8: 56-59.

Simpson, D.C. and Smith, G. (1977). An externally powered controlled complete arm prosthesis. *Journal of Medical Engineering and Technology* 1: 275-277.

Soede, M. (1980). On the mental load in arm prosthesis control. PhD thesis, Delft University of Technology.

Solkesz, E.J. (1975). The influence of the control site and coordinates on the human positioning behavior. DUT, Dept. of ME, Lab. M&C, MSc. thesis, Report B-140 (in Dutch).

Tebees, W.J. (1977). Technical evaluation of hand prostheses. DUT, Dept. of ME, Lab. M&C, MSc thesis, Report A-214 (in Dutch).

Ter Maat, H.J. (1971). Analysis of elbow motions during simple repetitive tasks and during activities of daily living. DUT, Dept. of ME, Lab. M&C, MSc thesis, Report A-120 (in Dutch).

Van Ditzhuisen, G.A.J.M. (1972). Study of a possible improvement of the prehension function of the Wilmer hand prosthesis. DUT, Dept of ME, Lab. M&C, MSc thesis, Report A-115 (in Dutch).

Van Lunteren, A., van Lunteren-Gerritsen, G.H.M., Stassen, H.G. and Zuithoff, M.J. (1983). A field evaluation of arm prostheses for unilateral amputees. *Prosthetics and Orthotics International* 7: 141-151.

Van Lunteren, A. and van Lunteren-Gerritsen, G.H.M. (1989). On the use of prostheses by children with a unilateral congenital forearm defect. *Journal of Rehabilitation Sciences* 2: 10-12.

Van Lunteren A. and van Lunteren-Gerritsen, G.H.M. (1992). Field evaluation of below-elbow prostheses for children, DUT, Dept. of ME, Lab. M&C, Report N-400.

Verkaik, W. (1967). Information transmission via the skin by means of a pneumatic tactile display - part 1. DUT, Dept. of ME, Lab. M&C, Report A-52 (in Dutch).

Verkaik, W. (1968). Information transmission via the skin by means of a pneumatic tactile display - part 2. DUT, Dept. of ME, Lab. M&C, MSc thesis, Report B-52 (in Dutch).

Vlasblom, D.C. (1967). Skin elasticity. Ph.D. thesis, Utrecht State University.

Zwart, O.P.C. (1979). Control principles for an experimental hand prosthesis with two degrees of freedom. DUT, Dept. of ME, Lab. M&C, MSc. thesis, Report A-247 (in Dutch).

Chapter 8

Design of Anthropomorphic Appliances

Jan C. Cool

1 Introduction

There is need for well-designed appliances, particularly in the medical fields of rehabilitation and orthopedics. Common devices cry out for improvement, while many new instruments await development. Some essential strategies for new designs, based on mechanics, systems engineering, and human-machine interactions, are presented here.

There are three basics to design: A designer creates a concept, chooses its optimal forms, and selects appropriate materials. The designer unites the often conflicting features of these three fields. Obviously the choice of the right concept is the most important of the three.

In the conceptual phase engineers tend to design for function. In rehabilitation as well as orthopedics, this view has to be adapted to the very strong requirements set by Cosmetics, wearing Comfort, and Controllability. These three Cs describe the basic characteristics of subordinating function. Every anthropomorphic appliance at a minimum should harmonize with body contours, coordinate with normal body movements, and be controllable using minimal effort. Under these conditions prostheses and orthoses, internal as well as external, have to operate correctly, i.e., to be silent, reliable, and efficient.

2 Systems engineering

Prior to the design of a mechanical device, knowledge of systems engineering is necessary. The complete description of a physical system requires definition of both an over-variable, describing the physical quantity, acting as a potential difference over the system, as well as the through-variable, acting as a "flow" through the system. In mechanical systems, force corresponds to the flow variable and velocity (displacement) difference to the over-variable. According to Newton, force is considered to be the cause of acceleration and thus, in time, of velocity and displacement. For that reason the study of force configurations is essential for the conception of a design. The usual method of system description by displacements and rotations should be extended by a study of force attachment points, force action lines, and force configurations.

Familiarity with forces in every part of a system gives essential insight into the way the system operates. In the conceptual phase, a study of functional essential force patterns guides creativity and indicates force compensation and force control. A number of examples will be given for these three statements in the following sections.

3 Forces and creativity

Minimally necessary force patterns are powerful tools in creative engineering. The following three examples illustrate how thinking in total system equilibrium stimulates the generation of new conceptional designs.

3.1 SHOULDER ORTHOSIS

A lot of persons are handicapped by a functionless arm caused by a brachial plexus lesion or hemiplegia. In both cases often a painful shoulder joint subluxation is accompanied by the risk of edema. Especially in the case of a brachial plexus injury, an appliance is needed that suspends the arm with the elbow in the flexed position. The available appliances such as the well known mitella and the similarly functioning hemisling, prove the unsuitability of displacement oriented designs. However, a force oriented synthesis immediately shows how the goals of good function and simple design should be realized.

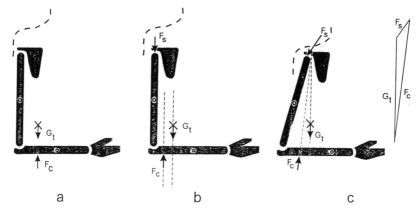

a   b   c

Figure 1.   The force configuration of a single arm suspension is unstable (a). Shifting the control force stabilizes the system (b) but leads to low wearing comfort because of shoulder loading in the cervical region. Distal positioning of the shoulder attachment increases both wearing comfort and cosmetics. The two-dimensional force configuration of the WILMER shoulder orthosis indicates the critical positioning of the textile band on the shoulder and forearm (c). G_t = total arm gravity force ; F_c = control force F_s = shoulder force.

Figure 1a shows the total system, the disabled arm in a luxated position with the forearm in the desired horizontal position. The center of mass of the flexed arm (cross) indicates the vertical action line of the weight vector. Redress of the subluxation necessitates an upward- directed control force F_c , equaling the total arm weight G_t. The applied force generates an unstable system because the attachment point $F_c = G_t$ lies under the center of mass. Shifting the attachment point proximally stabilizes the system by the generation of a gleno-humeral joint force F_s , with a simultaneous increase in the control force $F_c = (G_t + F_s)$ as shown in Figure 1b. A total system survey clarifies that the control force can be generated only if some body contact is available elsewhere. The vertical direction of F_c indicates the shoulder offers a load-bearing possibility. However, the shoulder can suspend heavy loads only distally, not in the cervical region. The inevitable distal shift of the suspension area favorably influences the stable force configuration. The shoulder and elbow joints enable the upper arm to move out of its vertical position to a more natural arm appearance. In the design a textile band transmits the control force comfortably from the distal shoulder to the proximal forearm. The two-dimensional force configuration is shown in Figure 1c with the corresponding force polygram.

The WILMER orthosis, produced in accordance with the force configuration just discussed, is shown in Figure 2. A simple force analysis led to the innovative conception of this appliance, which has a high cosmetic value because of its natural appearance and its potential to be worn under clothing. Further advantages are high wearing comfort because it does not irritate the skin and its easy donning and removal.

Figure 2.   The WILMER shoulder orthosis

3.2 ELBOW ORTHOSIS

An elbow orthosis consists of three parts: an upper arm fitting, a forearm fitting, and a joint connecting the two fittings. In normal engineering practice a dual joint is constructed in order to obtain a lightweight construction capable of resisting twisting joint torques. However, study of the force configuration reveals that twisting torques do not occur in elbow orthoses and thus a single joint is a better proposition.

Figure 3a shows a side view of the orthosis in the lateral direction. Figure 3b shows a front view of the stretched orthosis along with the forces acting on it. The external forces on the upper arm fitting and the forearm fitting act in the same plane for every elbow joint angle. The absence of a twisting moment over the elbow joint creates the innovative solution of a single-joint elbow orthosis. Wearers favor the single-joint orthosis because of its high cosmetic value, obtained by small dimensions, an inconspicuous appearance, negligible wear from clothes and low weight. For the most part, these features are realized by the single joint construction.

3.3 HAND PROSTHESIS WITH LOW OPERATING FORCE

Critical force evaluation sometimes leads to innovative design solutions. Arm amputees with a body powered prosthesis frequently wear a voluntary opening terminal device. The spring force closing of the hook or hand is the most simple solution for obtaining a pinching force at every opening width. However, this simplicity is paid for by needlessly high operation energy. If the hand/hook holds an object, the spring force productively delivers the pinching force. When opening the device, the spring force has to be counteracted by the operating force of the amputee. A method has been developed that avoids this very inefficient use of the operation force to prevent the corresponding energy loss. A two-phase operation is proposed for an innovating terminal device conception. In the gripping phase, the operation force is small because only a very weak force moves the hand/hook fingers to the closed position. In the pinching phase, a prestressed spring connected to the fingers delivers the pinching force for holding the object. Again the operation force by the user is small because it only needs to actuate the connecting

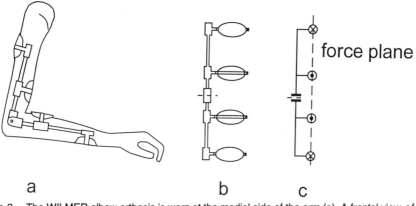

Figure 3.   The WILMER elbow orthosis is worn at the medial side of the arm (a). A frontal view of the orthosis in the stretched state (b) is converted into a mechanical connection scheme (c) with the external forces acting on it. Because all the forces act in the same plane, no twisting torque is transmitted to the single joint.

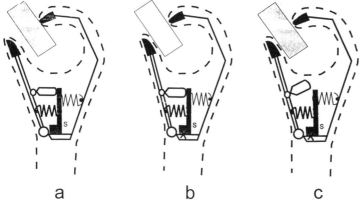

Figure 4.   The WILMER hand prothesis with WEN mechanism. The opening movement of the hand is counteracted by a weak spring between sliding part S and the frame (a). Easing of the control force decreases the hand opening until the thumb touches the object (b). By touching, the part S is clamped to the frame and the mechanism is unlocked, at the same time causing the prestressed spring force to pinch the object (c).

mechanism. The development of this idea into a reliably working design needs some other force configuration studies not presented here.

A preliminary two-phase hand/hook prototype is shown in Figure 4 in three operation stages: opened, touching the object, and pinching the object. The final design, the WILMER WEN mechanism, has an eightfold increased efficiency over simple voluntary opening devices, and thus an eightfold decreased operating force.

4   Force compensation

In the WEN-hand/hook prosthesis design of Section 3.3, the disturbing spring force has been eliminated in a new version. In other cases, in which elimination of disturbing forces is impossible, a force compensation mechanism can counteract the effect of these forces. The next examples concern appliances improved by force compensation.

4.1 FOREARM GRAVITY FORCE COMPENSATION

Movements of the forearm with an inclined upper arm are obstructed by gravity. Obviously the elimination of gravity is an illusion, so a constant force compensation mechanism has to be designed. The gravity compensation of a rotating mass can be realized very simply by a theoretical spring, as shown in Figure 5a. Mathematically exact compensation needs a spring force proportional to the length L of the spring. However, in practice springs do have an unstressed length that can only partly be reduced by prestressing. The construction of Figure 5b realizes a well-approximated solution for unstressed length avoidance at the expense of space and simplicity.

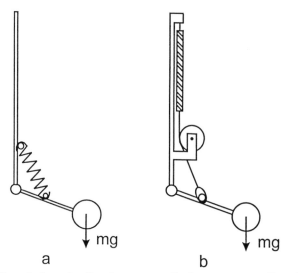

Figure 5. Two designs for the forearm gravity force compensation. The direct spring compensates gravity insufficiently because of the unstressed spring length (a). The roller and string construction induces a nearly perfect gravity compensation (b).

From a mechanical viewpoint, a rotating forearm does not resemble a rotating mass because of the unexpectedly strong influence of the weak parts around the elbow. This means that the gravity compensation has to be corrected for the influence of the weak parts. Fortunately, the characteristic of the direct spring compensation moment and the corrected gravity moment are approximately equal, meaning that the most cosmetic and simple compensation mechanism can be applied.

The WILMER elbow orthosis has been equipped with a cosmetic external force compensation. In this orthosis the compensation applied offers the possibility of harnessless body powered control from the handicapped side. See the description in Section 3.2.

4.2 GLOVE FORCE COMPENSATION IN AN ARTIFICIAL HAND

Artificial hands are almost always covered with a cosmetic glove. The various gloves available feature similar mechanical characteristics. The materials used cause an extremely nonlinear, time-dependent, force-deflection characteristic with substantial hysteresis.

Amputees do not accept gloveless hands. This means that the resisting glove force has to be compensated for. The glove hysteresis due to viscosity effects in the material cannot be compensated. Only the elastic part of the glove force can be counteracted with a compensation mechanism. The ideal direct force compensation, where disturbing force and compensation force share the same action line, requires an overlarge mechanism because of the strongly varying action line position. In the WILMER hand prostheses, less spacious moment compensation has been realized, leaving the thumb rotation joints uncompensated. See the force configurations in Figure 6a and 6b. Figure 6c shows insufficient glove compensation because of the conventional characteristics of the mechanism. The mechanism in Figure 6d, however, offers the WILMER hand prostheses a nearly glove force-free operation.

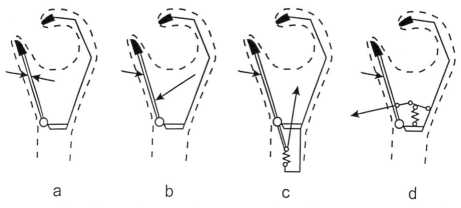

| a | b | c | d |

Figure 6. The difference between force compensation and moment compensation is illustrated by the vectors of the glove force and the compensation force (a resp. b). Two moment compensation mechanisms are shown: the direct mechanism with a compression spring (c) and the chosen four-bar mechanism with a draw spring (d).

The work of the compensation mechanisms is further clarified in energy terms. In the closed hand position the glove force is negligible, while the compensation spring is fully loaded and contains all the system's energy. By opening the hand, the compensation spring delivers the energy needed for the elastic glove deformations. At maximum hand opening, the compensation spring is released and all system energy is retained in the glove elasticity. By hand closing the system energy returns to the compensation spring. The installment of a force compensation mechanism increases the system energy; by operation of this mechanism, energy is redistributed continuously over the system, depending on its state.

5 Force control

In Section 2 study of the through-variable force was advocated. In Section 4 the force configurations led to the compensation of disturbing forces in order to improve system performance. The idea of neutralization of superfluous forces can be extended to the addition of missing forces. Examples of this technique are presented as force control.

5.1 ARTIFICIAL KNEE JOINT

A straight leg offers the possibility of standing. For sitting, the knee must flex for ergonomic and cosmetic purposes. So every above-knee prosthesis is provided with an artificial knee joint. In modern leg prostheses, the knee joint is mostly of the four-axis type: an ingenious application of a four-bar mechanism enabling stable body suspension during stance and knee bending during the swing phase of walking. This four-bar knee joint could not be used in the design of a lightweight through-knee prosthesis, mainly because of its heavy weight, bulk, and very low cosmetic value. For the new design, a small-space single-axis joint is proposed, and its axis position determined by cosmetics and available space.

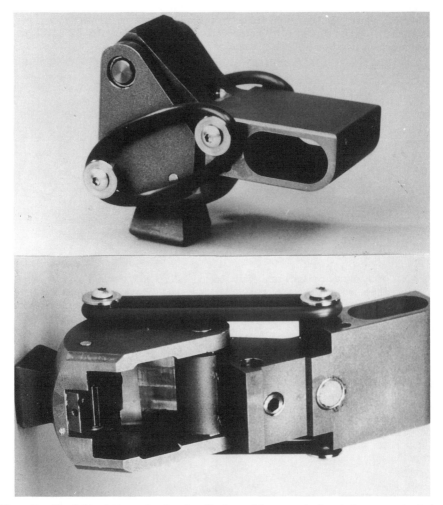

Figure 7.   The lightweight constructional realization of the new single-axis knee prosthesis in the flexed (a, above) and extended positions (b, below).

The first trials of the innovative knee joint made it clear that an anthropomorphic appliance should be designed in its entirety and not assembled from different components. The combination of the relatively proximal knee joint and the high foot mass caused an unacceptably long swing time period. The prosthesis movements were improved by the addition of simple mechanical components delivering extra acceleration forces at the start and extra deceleration forces at the end of the swing phase. Figure 7 shows the small, inexpensive, and very low mass knee joint with force-controlled swing time (patented).

5.2 CONTROL OF THE ELBOW ORTHOSIS

The WILMER elbow orthosis is designed for operation under force control. Section 4.1 describes a spring force-compensated forearm weight. The elbow orthosis operates with an undercompensated forearm weight. In a vertical upper arm position, the uncompensated part of the forearm weight forces the arm in the stretched position. The arm remains stretched unless the upper arm abducts. By abduction, the gravity influence on the forearm decreases while the compensating spring force is abduction independent. There is an abduction angle at which the influences of gravity and spring force neutralize each other, bringing the arm in indifferent equilibrium. The action of the spring force flexes the arm by further abduction. Flexion and extension of the arm are controlled by the force difference over the arm induced by abduction. A flexion controlled mechanical switch offers the wearer two fixed arm positions: stretched (particularly used for putting on and taking off the orthosis) and bent. The described force control is used to move the orthotic arm in and out of three possible modes: free stretched, fixed stretched, and fixed bent. Almost immediately after first putting on the orthosis, patients discover the orthotic arm can also be positioned very easily using inertia forces induced by small rapid upper arm movements. In practice a combination of the two effects controls the arm inconspicuously. Wearers very much appreciate the good cosmetic value, high wearing comfort, and handy controllability of the WILMER elbow orthosis. Figure 8a shows a girl wearing her orthosis, and Figure 8b shows the force control scheme of the orthotic arm.

5.3 SCOLIOSIS CORRECTION

Scoliosis is a severe local lateral deformation of the spine. Progressive scoliosis must be treated surgically. In common practice, the deformed part of the spine is forced into a straighter position. Stresses in the vertebrae limit the maximal correction obtainable. In time the correction applied decreases due to the viscoelastic properties of the spine and neighboring structures. The resulting undercorrection is initiated by the position oriented nature of the correction process. A force-controlled correction of viscoelastic materials promises better results. In the patented SKOLUT correction system, the spine is force loaded by the action of memory metal. The forces applied cause an initial correction comparable with that of position oriented systems. After implant the SKOLUT system continues correction, resulting in a less deformed and more cosmetic spine. Figure 9 illustrates the correction procedure.

5.4 ROLLING LINK MECHANISMS

Friction is a feature of mechanical systems. Often the behavior of a mechanical system is less effective than planned, because of friction. The efficiency of a glove force compensation as described in Section 4.2 is highly influenced by frictional disturbances. Because of their energy dissipating nature, frictional forces cannot be compensated for. The simplest solution is to replace sliding friction with rolling friction. Ball bearings are unsuitable for the application in hand prostheses because of their bulk, weight, and sensitivity to pollution. This consideration led to the design of the rolling link mechanism (RLM) (patent pending).

As the name indicates, an RLM is built of links that directly roll over each other in the absence of pairing elements. In essence, an RLM is only stable in a limited operating range with a high sensitivity to disturbances. Force control by the application of a thin flexible band between the rolling parts proved to be a simple and effective solution for both inconveniences. In the force-controlled mode, the RLM is an inexpensive, small space mechanism with high stiffness suitable for high load transduction. Figure 10 shows an RLM in use as the compensating system in the WILMER hand prosthesis.

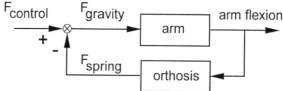

Figure 8.   The WILMER elbow orthosis is rather inconspicuous and offers a very high wearing
comfort because of the perforated flexible fittings with force-controlled suspension
(a, above). Arm abduction causes a force that controls the arm flexion/extension (b,
below).

6   Conclusions

The study of force patterns favors the synthesis of a system. From the beginning of the
conceptual phase, force patterns guide design toward a functional structure for the device
to be designed. As shown in Section 3.1, reasoning about the action line of a gravity-
counteracting suspension force points directly to the most simple design solution. A well-
functioning suspension of paralyzed arms has been created only after a basic
consideration of the acting forces. The Sections 3.2 and 3.3 show that a study of force
pattern configurations leads to design simplifications not found intuitively. In the
examples 4.1 and 4.2, the design improvements described emphasize the favorable use of
force studies. The influence of forces that disturb the functioning of mechanical systems
must be eliminated by the addition of specially designed force compensating mechanisms.
Springs and even gravity forces can be compensated for. The four examples in Sections
5.1 to 5.4 describe the improvements that force control introduces in mechanical system
performance, simplicity, and compactness.
     In mechanical systems engineering, the study of force configuration patterns is
highly recommended. Three useful generalities for design are:

Conceptualize by force patterns
Improve by force compensation
Operate by force control

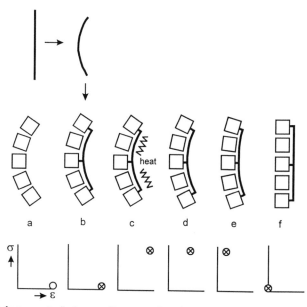

Figure 9.   The force-controlled correction procedure for scoliotic spines. The top row indicates
the correction rod shapes; in the middle row the relative positions of five scoliotic
vertebrae in six (a to f) recovery states are given; the bottom row indicates the
corresponding stress-strain relation. The original deformed spine (a) is provided with
a precurved correction rod (b). Heating the memory metal correction rod (c)
introduces a nearly constant correction force, which continuously redresses the
spine (d, e). Correction ends if the memory metal rod no longer generates force in
returning to its original straight shape.

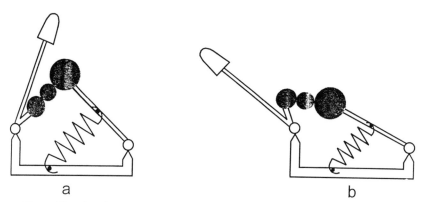

Figure 10.   The glove force RLM (rolling link mechanism) compensation system in closed
(a) and opened (b) hand position.

7   Suggested Reading

Cool, J.C. 1989. Biomechanics of orthoses for the subluxed shoulder, *Prosthetics and Orthotics International,* Vol. 13; 90-96.

Kruit, J., and Cool, J.C. 1989. Body powered hand prosthesis with low operating power for children, *Journal of Medical Engineering & Technology,,* 13; pp. 129-133.

Kuntz, J.P. (1995) rolling link mechanisms. PhD thesis, Delft University of Technology, Delft.

Leerdam, N.G.A. van (1993). The swinging UTX orthosis. PhD thesis, University of Twente, Enschede.

Lemmers, L.G. (1994). The UFITT prosthesis. PhD thesis, University of Twente, Enschede.

Sanders, M.M. (1993). A memory metal based scoliosis correction system. PhD. Thesis, University of Twente, Enschede.

Chapter 9

Rehabilitation of Persons with Upper Extremity Defects in a Multidisciplinary Treatment Team

Werneck van Haselen

In 1967 a collaboration was established between the De Hoogstraat rehabilitation center and a number of technical institutes, among which was the Delft University of Technology, resulting in the foundation of the Working Group on Orthoses and Prostheses (WOP). The activities of the WOP not only concerned the individual patient but also resulted in the design of improved arm prostheses and orthoses, which changed treatment strategies. The treatment to be discussed in this chapter is for patients with a paralyzed arm caused by damage of the plexus brachialis, the nerve bundle between the spinal cord and arm. Provision of an elbow orthosis and, if necessary, a shoulder arthrodesis, a fusion between the shoulder blade and upper arm bone, in most cases leads to successful rehabilitation.

1 A short history

The birth in the early 1960s of a number of children with severe limb deficiencies in several European countries, mainly England and Germany, gave rise to increased activity in the fields of rehabilitation and prosthesis development. In the Netherlands only a small number of children were involved. Nevertheless this resulted in closer cooperation between the medical and technical sciences in the development and application of prostheses and orthoses. It was soon determined that insufficient expertise was available in the Netherlands for this small population of patients with upper limb defects. It was also recognized that much more knowledge and experience were necessary for adequate treatment.

In 1967 a collaboration was established among De Hoogstraat, one of the oldest rehabilitation centers in the Netherlands, formerly at Leersum and now at Utrecht, the Delft University of Technology, and the Medical Physical Institute TNO at Utrecht. A working group, the Working Group on Orthoses and Prostheses, was founded for the benefit of patients with defective upper limbs, such as congenital deficiencies, amputations, and paralyses. Among the first collaborators, Ms. Mien Luitse, occupational therapist; Dr A. Verkuyl, at that time the medical director of De Hoogstraat, and Henk Stassen should be mentioned.

The aims of the group were the treatment of patients and the advancement of research and development. An important aspect was the collection, recording, and interpretation of the data with respect to the treatment methods and aids provided. The group consisted of collaborators in the rehabilitation center, such as a physiotherapist, an occupational therapist, a social worker, a psychologist, a prosthetics technician, and a rehabilitation medical doctor, and external members from the DUT and MFI-TNO.

The group met twice a month. In the morning individual contacts were arranged between group members and patients, and in the afternoon the information, treatment goals, and practical advice were discussed with the patient and, in the case of children, the parents. This method of working was a breakthrough in medical research in the field.

The group has been in existence for more then 25 years, although the memberships have changed over time. However, this method of working also had its influence on other treatment groups of the rehabilitation center and in other centers in the Netherlands. Moreover, a number of research projects described in the other chapters can trace their origins to this group. Over the years, a certain specialized knowledge in this field has grown. As a result many patients come from other parts of the country for treatment. To give an indication of the type of treatment resulting from this method of working, one of the categories of patients will be discussed: patients with a brachial plexus injury.

2   Rehabilitation of patients with a lesion of the plexus brachialis

An example of the type of patients who were seen in the WOP is the group with a damaged arm nerve bundle. This group consists mainly of young people who have suffered an automobile or moped accident. In general the clinic has about 10 patients. These are mostly those individuals with more severe injuries, who are sent to a specialized center. This means that the patients often have other injuries that require primary attention, such as a contusio cerebri, i.e., a type of brain injury, limb fractures, and/or spinal cord and pelvis injuries, damaged blood vessels, and damaged internal organs. In about 75% of cases these other injuries dominate the first treatment phase, so that in many cases the plexus lesion is recognized rather late.

2.1 DIAGNOSIS

In recent years improved imaging techniques have allowed for earlier diagnosis. It is necessary with all possible methods that within 1 week after injury a clear distinction is made between an avulsion, i.e., when a nerve is pulled out of the spinal cord, and/or damage to the nerves in the plexus, i.e., at the outside of the spinal cord. If a roentgenogram diagnosis shows fractures of the collarbone, then there may also be an avulsion. A number of tests exist that together enable a reliable diagnosis.

Especially in the first years after the start of the WOP, patients often were referred to the center at a rather late stage. In the early 1980s the mean time between the injury and the start of the rehabilitation treatment was 20.4 months. Such a late start clearly has a negative influence on the final results.

2.2 NEUROSURGICAL TREATMENT

From the beginning, neurosurgical intervention was favored by the treatment team, although in the ideal case it should be undertaken within a half-year of the accident. The peripheral part of a nerve always degenerates. However, the part that is still connected to the cell body can grow again. In the latter case the nerve ending should grow out into the remaining tubelike structure of the peripheral part. Such a connection might be restored by surgery. If there is a lot of damage, a nerve graft from another part of the body may be used to create an interconnection. However, if the operation is performed more than 300 days after the accident, recovery seldom occurs.

2.3 REHABILITATION

After the medical condition has become more or less stable, a rehabilitation program can be started. The first step is making a diagnosis with respect to disorders and disabilities. The diagnosis not only refers to dysfunctions of the motoric system but also to neuropsychological and social problems. Rehabilitation involves a laborious learning process for these patients, who often do not originally possess an optimal attitude for learning. Besides the specific training programs, multidisciplinary conduct oriented coaching is necessary. Learning to cope with disorders and disabilities is often an essential part. In many cases this involves learning to cope with remaining handicaps, such as pain, motion disorders, one-handedness, etc. The main goal is to gain more independence in the activities of daily life, hobbies, work, and mobility.

2.4 TECHNICAL PROVISIONS

An important part of the attention of the group is focused on adaptations and aids. Contrary to the case with amputations of the upper extremity where the most attention is focused on hand function, the starting point in plexus lesions is the function of the shoulder. In the course of evolution, the shoulder has undergone a drastic change. Motions in only the sagittal plane have evolved to a mobility in all directions. This has had significant consequences for the multifunctional role of the arm and hand. Therefore, the shoulder is one of the most complex motion systems of the human body. Its large mobility due to its small articular surface and wide capsule of the glenohumeral joint do not contribute to the stability of this joint. The stabilization is mainly performed by muscle activity. For a more detailed description of the shoulder mechanism, the reader is referred to Chapter 4.

In the case in which the muscles that bear the weight of the arm have lost their function, a subluxation occurs, i.e. the upper arm hangs on the passive structures, which then overstretch, resulting in reduced bloodflow to the arm, pain, and loss of control. In this case a special lightweight orthosis that can be worn under clothing, the Wilmer wearing sling, is used. Using the principle of a lever, wherein the weight of the forearm generates an upward force, the upper arm is lifted, so that the subluxation is ended (see also Chapter 8). When, over the course of time, the function of the muscles mentioned at least partially returns, this device is no longer necessary.

In the case in which there is a stable shoulder joint but loss of the flexion function of the elbow, with or without remaining hand function, a lightweight elbow orthosis is used (Cool, 1976). This orthosis can also be worn under clothing. Two braces are fitted to the upper arm and forearm, respectively, and pivot at the axis of rotation of the elbow joint. A spring is fastened with one end to the upper arm brace while the other end stresses a string that is guided over a roller and fastened to the forearm brace.

There are two ways to move the forearm from a stretched position to a bent position. The first is by moving the arm sideways. In this case the spring force is larger than the component of the gravity force that stretches the arm and thus the elbow bends. The second method is to throw the arm upward using the shoulder muscles. When the motion in the shoulder joint is stopped, the momentum of the moving mass of the forearm then causes a flexion in the elbow. The pivot is provided with an automatic locking mechanism in the 90 degree position. It can be loosened when the forearm is flexed upward by one of the two mentioned methods.

The elbow orthosis has a number of advantages:

- The patient has a more stable feeling, because he or she regains control over the forearm.
- It functions as a training device for cases in which some muscle activity returns.
- If some hand function is left, it can be used again.

The importance of the role of the orthotist should be mentioned. Although the orthosis consists of a number of standard components, the value for the wearer is determined by a careful fitting. Therefore, this work should only be done by well-trained specialists. If, for instance, the fitting is uncomfortable or the mechanism is not well adjusted, the orthosis will be rejected.

The elbow joint is complicated, because it also allows for axial rotation of the lower arm. This type of motion is no longer possible with the available elbow orthoses. In some cases hand function can be improved by surgical methods such as displacement of a tendon so that a grasping function can be created.

2.5 FIXATION OF THE SHOULDER

When after some time, usually 1 or 2 years after the accident, it becomes clear that subluxation of the shoulder is permanent while the muscles that control the motions of the shoulder blade still function, fixation of the upper arm bone (humerus) to the shoulder blade (scapula), shoulder arthrodesis, can be considered. As an orthopedic surgeon once noted: The problem of a person not having enough muscle function can be reformulated as "having too many joints." By performing this operation, the patient again has some control over the upper arm, although range of motion is limited. It should be mentioned that an arthrodesis has a number of disadvantages:

- The operation is irreversible.
- After the operation, some form of temporary fixation has to be applied while the bony structures grow together, e.g., plaster of Paris or some type of internal or external fixation. This brings a lot of discomfort and vulnerability.
- The choice of the optimal angle between the scapula and humerus is not easy. Moreover, the accuracy with which the chosen angle can be achieved is not very high. As a consequence, the patient may no longer be able to reach his or her pants pocket, or the arm may be in an uncomfortable position when lying in bed.
- Especially when a lot of time has elapsed between the accident and operation, the quality of the bone has degenerated, which makes good fixation very difficult.

The last point offers an argument for early surgery. However, the irreversibility of the operation is a psychological threshold that is difficult for many patients to cross. Moreover, many patients still hope for a return of function of the involved muscles. In the past, particularly, when diagnostic methods were of limited nature, rehabilitation teams also were not certain on this point.

Many methods for fixation of the shoulder have been described in the literature (Crenshaw, 1971; Post, 1978; Depalma, 1983; Rowe & Leifert, 1988). This in itself is an indication of the fact that this operation is not without difficulties. The bony elements are difficult to fit, due both to their shape and their structure. Furthermore, there often is serious decalcification. Therefore it is difficult to use plates and screws. A frequently applied method is to use of plaster Paris after internal fixation with a bone chip, to secure position of the arm with respect to the thorax. This results in a lot of discomfort to the patient due to perspiration. For this reason this operation is only executed in winter in the Netherlands.

In a cooperation between the Academic Hospital at Utrecht and the Delft University of Technology, a new method has been developed that makes use of a special external fixator (Nieuwenhuis & Pronk, 1989; Pronk, 1991). The operation is executed in two stages. In the first stage three pins are inserted into each of the bony elements, the scapula and humerus. Afterwards two T-shaped frames are fixed to the three pins in each of the bones. The frames are mutually connected by means of the fixator, the angle of which can be adjusted in the rehabilitation center by having the patient try out a number of activities. After wearing the fixator for a number of days, the patient decides whether or not he wants to continue with the procedure. If the patient chooses not to, the pins are again removed. If the patient decides to continue with the arthrodesis, the second stage is performed. The fixator is removed from the pins while the adjusted angle is preserved. Then the actual operation is executed. After removing the cartilage from the joint surfaces, the bony structures are pressed together using a force that is adjusted by means of a screw and the wound is closed. A further advantage of this method is that the shoulder muscles can be trained by the physiotherapist, so that a much shorter rehabilitation period is necessary with the classical method.

In the case in which there is not only subluxation, but also loss of function of the lower arm, the combination of a shoulder arthrodesis with provision of an elbow orthosis can restore a lot of arm or even hand function. In a number of cases a fracture of the fused joint has occurred, which, however, did not result in subluxation, but which introduced a few degrees of mobility. Patients were not unhappy with this situation because it offered them some more comfort. In the future it might be desirable to create such a situation intentionally.

2.6 AMPUTATION

Some authors mention plexus lesions as a reason for amputation and provision with an arm prosthesis (Yeoman & Seddon, 1961; Wynn Parry, 1974; Lauman & Schilgen, 1977; Malone, 1982). In our opinion this should only be done in the case of a serious crush injury with fractures and soft tissue pathology. The loss of a limb is far reaching, both in a physical and a psychological sense. Especially in the case of an upper arm amputation, no functional improvement is obtained by provision with a prosthesis. Moreover, a stump is cosmetically less acceptable than an atrophied arm or hand.

It was found that the combination of a shoulder arthrodesis and an elbow orthosis restores some function with a complete plexus lesion. Even in the case of an arm that has lost its sensibility, positioning of the lower arm and hand are again possible under visual guidance. Objects can be fixated with respect to the environment. With a 90 degree flexed elbow, the lower arm can be used to carry things or to open a door. Therefore a lame arm supported by an unobtrusive orthosis with good wearing comfort that allows some easy active positioning can be considered a better prosthesis.

2.7 ROLE OF OTHER DISCIPLINES

Physiotherapy

As already mentioned, an accident often results in a number of injuries to the musculo-skeletal system. For instance, most patients also suffer from shoulder and back pain. Often a scoliosis, i.e., a sideways curvature of the back, is present. Thus, besides exercises to maintain the condition of the affected muscles and to preserve mobility in the joints of the injured arm, these additional complaints are also treated.

Occupational therapy

As also mentioned above, the main goal is to regain independence. Even in a case in which there is partial restoration of some of the muscle function of the arm, this may take quite some time. Thus, in all cases the patient is trained to do daily activities one-handed. This often means doing things differently, for instance, tying shoestrings with a special knot or making use of special aids to handle pots and pans, etc. It is important to find solutions for hobbies that the patients are enjoying. These activities often involve doing all kinds of repair jobs, as well as sports. Furthermore, attention is paid to bicycle riding and car driving. In cases in which the car controls have to be adapted, patients can apply for a restricted drivers license on which the required adaptations of the car are indicated.

Social work

It is important that patients have future prospects for work. On the one hand, patients have to be encouraged to use their remaining abilities and to learn to cope with their handicaps. On the other, it is important to determine whether the patients can return to their work; if not in the same job as before, then perhaps in another one. Therefore at an early stage their contacts should be made with employers and, if necessary, with other organizations that can assist in this matter. Finally, it should be mentioned that a rehabilitation team should not only be multidisciplinary but should function interdisciplinarily, i.e., there should be many mutual influences on thinking of the team.

References

Cool J.C. (1976). An elbow orthosis. *Bio-Medical Engineering*, 11 (10):344-347.

Cool J.C. (1989). Biomechanics of orthoses for the subluxed shoulder. *Prosthetics and Orthotics International*, 13 (2):90-96.

Werneck van Haselen

Crenshaw A.H. (1971). Arthrodesis. In Crenshaw A.H, (Ed.),*Campell's Operative Orthopaedics*. Vol 2, 5th ed., pp. 1185-1191, C.V. Mosby Company, St Louis.

Nieuwenhuis F.J.M., and Pronk G.M. (1989). An adjustable external fixator to perform a glenohumeral arthrodesis. In Paul J.P., Barbenel J.C., Courtney J.M., and Kenedi R.M.(Eds.), *Progress in Bioengineering*, pp. 170-173. Adam Hilger, Bristol-New York.

Pronk G.M. (1991). The shoulder girdle; analysed and modelled kinematically. PhD-Thesis, Delft University of Technology, ISBN 90-370-0053-3, 244 p.

Post M. (1978). *The Shoulder; Surgical and Nonsurgical Management*. Lea & Febiger, Philadelphia.

Depalma A.F. (1983). *Surgery of the shoulder*. 2nd ed., JB Lippincott Co, Philadelphia.

Laumann U., and Schilgen L. (1977). Varisierende subcapitale Osteotomie in Verbindung mit Schulterarthrodese und Oberarmamputation bei Plexusparese. *Orthopädie Technik*, 28 (5): 66-67.

Malone J.M., Leal J.M., Underwood J., and Childers S.J.. Brachial plexus injury management through upper extremity amputation with immediate postoperative prostheses. *Archives of Physical Medicine and Rehabilitation*, 63 (2): 89-91.

Rowe C.R. and Leffert R.D. (1988). Advances in arthrodesis of the shoulder. In: *The Shoulder*, Ed. Rowe C.R., pp. 507-520, Churchill Livingstone, London.

Wynn Parry C.B. (1974). Management of injuries to brachial plexus. *Proceedings of the Royal Society of Medicine* 67: 417-420.

Yeoman P.M. and Seddon H.J. (1961). Brachial plexus injuries: Treatment of flail arm. *Journal of Bone and Joint Surgery,* 43-B: 493-500.

Chapter 10

Development of a Lightspot-Operated Communication Aid

Jan Goezinne

1   Introduction

Research into communication aids at the Delft University of Technology started in 1969 when Wilhelmina Luitse, occupational therapist at the De Hoogstraat rehabilitation centre, asked the Man-Machine Systems Group to investigate the communication problems of severely handicapped people. At the time, the development of conservative treatment for people with a traumatic lesion of the cervical spinal cord, which reduced the mortality rate from more than 90% to less than 10%, resulted in a steadily growing number of severely handicapped people. De Hoogstraat could offer the patients comprehensive treatment but few suggestions with regard to sensible and rewarding activities after rehabilitation. One of the possibilities was the ability to operate a typewriter. The best available option was the use of an electric typewriter with a headstick or a mouthstick. At less than 20 characters per minute, typing speed proved to be disappointingly low and the required physical effort limited use to an hour a day for most patients. The Man-Machine Systems Group, under the inspiring leadership of Henk Stassen, took up the challenge to find a better way, thus initiating a research program that continues to this day.

Several operating principles were considered. The most promising made use of small movements of the head to point a lightspot at a matrix of characters, each provided with a photodetector. A system of this kind, the PILOT, was available at the time but it proved to be very unreliable. In 1970, after some basic research concerning the optimal size and layout of the character matrix, a first prototype of the Lightspot Operated Typewriter or LOT was built in Delft and evaluated in De Hoogstraat. The evaluation was very successful (Soede et al., 1973). Users reported an average speed of about 60 cpm after a few days training, often rising to more than 100 cpm after a few weeks. Moreover, lightspot pointing proved to be much less tiring than typing with a headstick. People in a poor physical condition were able to use the system for several hours a day without excessive fatigue. Following these positive results, five improved prototypes were built and extensively evaluated in several Dutch rehabilitation centers (Stassen et al., 1982). For many patients, lightspot operation proved to be superior to any available alternative.

In 1975, a commercial equipment manufacturer was found to take care of development and production. The LOT Mk3, as the final product was called, became available in 1980 (Fig. 1). An independent foundation, OSKAR, was established to take care of distribution. Over the next 5 years, only 10 machines were sold. The few people lucky enough to receive a LOT Mk3 were enthusiastic about the speed and ease of operation, and used their machines intensively, but the small number of units sold made it impossible to continue the existing scheme of development, production, and distribution. The reasons for the lack of success of the LOT Mk3 were high cost, large size, and limited flexibility.

Motivated by the positive reactions from the users, Delft University decided to redesign the system from scratch. One of the problems that had to be solved was finding a short, recognizable, and internationally acceptable name for the new system. LUCY, derived from the Latin word *lux* (light), made its debut in 1990 (Fig. 2). Cost and size were drastically reduced while flexibility was much enhanced. In addition to lightspot control, other control methods could be used, making LUCY suitable for a

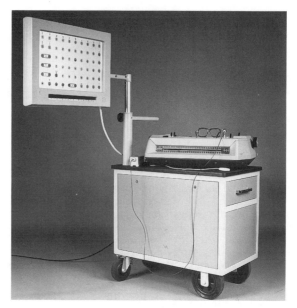

Figure 1. The LOT Mk3 was the first commercially produced lightspot-operated typewriter. Introduced in 1980, it had limited success because of high cost, large size, and limited flexibility.

Figure 2. LUCY appeared in 1990 and was an instant success. It is often used as a lightspot-operated keyboard for a personal computer, but it supports many other input methods and output devices.

large group of handicapped people (Stassen et al., 1993). Among its various output modes was a built-in keyboard emulator for an MS-DOS personal computer, which proved to be an important feature. By the end of 1994, about 300 units had been produced and sold, making it the most successful communication aid of its kind in the Netherlands. Several design changes have taken place since 1990, the most important being a much improved lightspot detection system based on a small laser pointer. Development of LUCY is an ongoing process that will no doubt continue for many years to come.

2 Communication aids and their users

Communication is a basic human need. Not being able to speak or write is a condition that most nonhandicapped people find hard to imagine. Yet, not being able to operate a personal computer because the keyboard is inaccessible seems equally frustrating. Modern communication aids can do much to alleviate these problems. In the Netherlands, the number of severely motor disabled people that could benefit from a communication aid is estimated to be between 150 and 250 per million population. Most of them will probably never use a communication aid because of age, unfamiliarity, or psychological barriers. Based on experiences with LUCY, the most important user groups are those with cerebral palsy (spasticity), motor neurone disease (ALS), cervical spinal cord lesions, traumatic brain lesions, muscular dystrophy, and multiple sclerosis. The resulting motor impairments are various. For a disease like ALS in the final stages, available motor control may be as little as the blinking of an eyelid. In other cases, some hand and arm movements may be possible, but not enough to operate a standard keyboard. Patients with a cervical lesion of the spinal cord usually have good control over their head motions but very limited control over the upper and lower extremities.

The simplest possible input device for a communication aid is a single switch. In some cases, for instance, ALS in its final stages or severe spasticity, this is the only available option. The simplest communication aids consist of a panel with a string of letters or symbols. By activating the switch, a scanning operation is started and a pointer, often a lamp, indicates the letters in sequence. Activating the switch once more selects the symbol at the pointer. These simple devices are often used with young children or mentally handicapped people. More efficient than the linear scan is the two-dimensional scan. In that case, the characters are arranged in a matrix. Selection is performed in two stages: first the rows and then the columns. Matrix scanning devices were the first commercially available communication aids, and they continue to be very popular. A modern approach is to implement the system as a software program for a PC and to use part of the screen for the character matrix. Most manufacturers offer a variety of input switches, ranging from miniature touch switches for small eyebrow movements to rugged buttons for spastics. The advantage of single switch scanners is that they require minimum motor ability. The disadvantage is low speed. The combination of having to wait passively until the pointer has reached the symbol and then having to respond quickly is ergonomically unsound and makes operation slow and rather tedious. Typical input speeds are between 10 and 20 cpm, and few people are prepared to use matrix scanning for more than 1 hour a day.

Several methods have been tried to improve scanning systems. Three-dimensional scanning, wherein a part of the matrix is selected first, which is then scanned in row-column fashion, can offer some improvement in speed but makes the device more complicated to operate. Another approach is to offer word prediction. In that case, the system contains lists of probable words that are presented to the user once the first one or two letters have been selected. The user can then finish the word with just one selection. Although these systems can offer a reduction in keystrokes of up to 50%, the corresponding gain in operating speed is much lower. There seem to be two reasons for this. First, the operation of a communication aid by an experienced user is an almost automatic, rhythmic action that can be performed without much mental effort. Any interruption of the rhythm, such as looking at the list of predicted words, deciding if the desired word is in the list, and then selecting it, takes time and mental effort. Second, the most frequent words in a language are also the shortest, which means that most predictions offer low gain.

A better way to improve the speed and ease of operation is to find an input method that makes optimal use of the available motor functions. Dual switch operation, one for the rows and one for the columns, makes it possible to get rid of scanning by pressing a switch as many times as required to move the pointer. Although more motor actions are required, the dual switch approach seems particularly suitable for spastics who find it difficult to perform precisely timed actions. With available communication aids, the next step is a five-switch arrangement. Four switches are used to move the pointer up, down, left, and right, while the fifth switch selects the character. Five-switch operated matrix systems have become quite popular, offering speeds of up to 40 cpm with reasonable operating ease. Switches can be in the form of a keypad or a joystick, making these systems suitable for a large group of people with different degrees of motor impairment. In some cases, a standard PC input device such as a mouse or a trackball can be put to good use.

For many severely handicapped people, movements of the head are the best way to control a communication aid. The problem then is to translate head movements into an electrical signal. Three systems are currently commercially available. The first uses an optical sensor that is coupled to a matrix panel with light emitting diodes (LEDs) as position indicators. The sensor, which can be attached to a head band or a pair of glasses, picks up signals from four infrared sources in the corners of the panel. From the relative intensities of these signals, the pointing direction of the sensor can be calculated. The matrix LED at the pointing position is then switched on and the corresponding character is selected after the LED has been on for a small, predetermined period of time. The problem with this system is the low resolution discrete visual feedback to the user. Small, involuntary movements of the head have no effect on the indicated pointing position. During selection, the head has to be kept as steady as possible because there is no indication of any position drift until it is too late and the wrong LED lights up. This makes operation of the system rather slow and tedious for anyone with less than perfect head control. Compared with other possibilities, this system is clearly inferior.

Much better are systems that use a computer screen for visual feedback. A popular, commercially available system uses a "head mouse" to translate movements into electrical signals. This system has three ultrasound microphones, mounted in a triangle on a frame that can be worn on the head. An ultrasound generator is placed near the screen, and the system can calculate the pointing position from the relative intensities of the signals from the three microphones. Head movements are then translated into pointer movements over a software-implemented character matrix on the screen. The visual feedback can have good spatial resolution, but temporal resolution is limited by the refresh rate of the computer screen. Fast movements, common for experienced users with good head motion control, do not result in unbroken lines of movement on the screen, but rather in a series of dots. This effect decreases operating ease and results in eye fatigue when the system is used for long periods of time. Other negative factors are time delay, noise, and nonlinearity in the sensor system. All of these effects are clearly noticeable in current systems and tend to make operation of the system tiring when used for prolonged periods of time. However, for people who do not want to use their systems for many hours a day, this might be a good solution. Also, these systems can be produced relatively inexpensively because, apart from the head movement sensor, they consist of just software and a standard PC.

Finally, there is the optical pointer and photosensor matrix. In this case, visual feedback is optimal: unlimited spatial and temporal resolution, no time delay, no noise, and high linearity. This results in a system that can be operated at high speed for long periods of time. Experienced users can attain top speeds of over 120 cpm and sustained speeds of 90 cpm for many hours. Even people with limited head control, for instance, spastics, can routinely attain up to 60 cpm for almost as long as they like. The operating speed and the time during which the system can be used without excessive fatigue is 3 to 6 times better than when using a headstick. However, compared with the PC based systems mentioned earlier, it is inherently more expensive to produce. Lightspot-operated systems are the best choice for many handicapped people who want a high-performance communication aid that can be used for long periods of time.

3   Development of the optical pointer

The first step in the development of a lightspot-operated communication aid is the optical pointer. The basic design objectives are as follows:

- It should produce a clearly visible, well-defined lightspot on the character screen.
- It should be small, lightweight and very rugged.
- Its shape and color should not attract attention.
- It should not generate heat or stray light.
- It should be possible to attach it firmly to a headband or glasses.
- Its pointing direction should be easily adjustable and remain fixed after adjustment.
- It should be economical to produce.

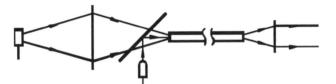

Figure 3. Optical system of early pointers. The lamp unit contains an ordinary lamp for visible light, a modulated infrared LED lamp for reliable detection, and an optical filter to add the two components. A flexible fiber optic cable, about 2 meters long, carries the light to the pointer. Pointer length is 60 mm, diameter 20 mm, and weight 30 grams.

Back in 1970, the only light source with sufficient brightness was the halogen-filled tungsten filament lamp. The optical system of the first LOT consisted of a stationary lamp unit, about 2 meters of flexible optical fiber cable, and the actual pointer. The light from the lamp was focused on one end of the optical fiber and the other end was attached to a pointer with a lens to project an image of the fiber end onto the screen. This resulted in a round, white lightspot of about 20 mm diameter. Evaluation showed that reliable detection of this lightspot could not be achieved. To make reliable detection possible, the lightspot must have some characteristic that makes it different from all other sources in the environment. This was achieved by adding a high-frequency modulated component to the light from the pointer.

The basic configuration is shown in figure 3. The modulated light is generated by an infrared LED. An optical filter, transparent to visible light but reflecting infrared light, was used to add the two components. The main problem with this system was the cable, because the glass fibers made it quite heavy and inflexible. Also, the fiber cable was easily damaged by being bent over a sharp edge or being stepped on, events that happen in real life. In 1972, much time was devoted to fitting the entire optical system into a lightweight pointer, thus avoiding the use of optical fiber. To keep the hot bulb from scorching the user's hair was quite an engineering feat in itself. A useful pointer was developed, but the idea had to be abandoned because of the high cost of production. Although costly, complex, and not very reliable, the optical fiber pointer remained the only available option until about 1984, when very high efficiency LEDs became available.

Figure 4. Optical system of the LED pointer which is much better than earlier pointers because of its simple construction, high reliability, and low cost. Pointer length is 110 mm, diameter 25 mm, and weight 20 grams.

Before 1984, LEDs could not be used as a source of visible light for a pointer because they were much too weak. The new LEDs were 20 times brighter. Finally, it was possible to design a pointer that was much better than the fiber system. Because the actual surface brightness of this LED is much lower than that of a tungsten filament, care had to be taken to use all available light from the LED. The basic construction is shown in Figure 4. The top of the LED is projected by the lens as a round, red lightspot of 15 mm diameter on the character screen. An ordinary mirror is used to halve the length of the tube. Compared with the fiber system, the LED pointer has many advantages. The construction is very simple and much less expensive to manufacture. Due to the low weight of the components, it can be made very rugged and will easily survive rough treatment or a fall on the floor. Power consumption is less than 1% of the fiber system, and the light from the LED can be fully modulated, resulting in a stronger signal from the photodetectors and consequently more reliable detection. Finally, the electrical cable is lighter, more flexible, and more reliable than a fiber optic cable. The only disadvantage is its size, which results from the requirement of high optical efficiency. Other optical arrangements, resulting in smaller size, are possible but they would be much more expensive to manufacture. Still, the LED pointer was a big improvement and has given good service for many years.

Figure 5. Optical system of the laser pointer. Light is emitted from a very small dot, one-tenth the diameter of a human hair and a thousand times brighter than the sun. Pointer length is 30 mm, diameter 12 mm, and weight 13 grams.

Visible light laser diodes appeared around 1987, but they were much too expensive to be useful for the optical pointer. By 1992, prices had dropped to an acceptable level and a laser pointer was developed. The optical system is extremely simple, as shown in Figure 5. The laser diode is made from a very pure crystal of gallium, aluminum and arsenic about 1 mm long. The light is generated in a very small channel inside the crystal, about 5 microns wide and 3 microns thick. At the end of the channel, the light is diffracted and fans out as a cone with a top angle of about 30 degrees. What makes the laser different from any other light source is the extremely high intensity at the emitting surface, typically about 100,000 times higher than a LED, 10,000 higher than a tungsten filament, and 1,000 higher than the surface of the sun. Even a tiny lens will focus the light into a small, bright spot, easily visible in sunlight. Power consumption is low and the light can be fully modulated. This makes the laser an almost ideal light source for an optical pointer.

The laser pointer became available in 1993 and is now rapidly replacing the older LED pointers. It comes close to fulfilling all the basic design objectives mentioned earlier. A disadvantage is the possibility of eye damage. If the beam from the pointer enters the pupil and remains focused on the same spot of the retina for a few seconds, a small dot may be permanently damaged. However, accidental eye damage is highly unlikely. First, the human eye has a natural reflex: When it is suddenly exposed to intense light, the eyelid closes within a few tenths of a second. Second, looking straight into the beam is an unpleasant experience and people quickly look away or close their eyes. The resulting short, fleeting exposure does no harm to the retina. The optical power from the pointer is about 0.2 mW, well below the maximum level of 1 mW required by international safety standards for lasers.

4   Lightspot detection system

The basic design objectives for a lightspot detection system are as follows:

- Each cell should respond quickly and reliably each time it is hit by the lightspot.
- It should be insensitive to variations in the ambient lighting.
- It should remain sensitive to the lightspot in conditions of very high ambient lighting.
- It should be insensitive to stray electric or magnetic fields from the environment.
- It should be economical to produce.

One of the first choices to be made is the number of cells in the matrix. Clearly, the more cells the matrix has, the more expensive it is to manufacture. To emulate a keyboard from a typewriter or a computer, all the often used keys should be directly accessible. Less often used keys can be made accessible through a two-step procedure, activating a special cell first, which changes the functions of all other cells in the matrix. The LOT, which was designed to operate like a typewriter, has 60 cells in a 6 by 10 matrix. With LUCY, designed as a PC keyboard, this was increased to 88 keys in an 8 by 11 matrix.

Table 1:   Characteristics of low-cost photodetectors

	Relative sensitivity	response time microseconds	relative cost
Photodiode	1	0.1	1
Phototransistor	100	1	2
LDR (see below)	10,000	10,000	0.3

The values are for comparison only. Actual devices may vary considerably.

The first step in designing a detection system is the selection of the photosensitive element. Basically, there are just three alternatives that meet the requirement for economical production. Their characteristics are summarized in Table 1. At first sight, the light dependent Resistor (LDR) seems to be a good choice. The cost is minimal, which is important because the system uses so many of them. The sensitivity is extremely high, simplifying the design of the signal processing circuits. The response time of about 1/100 of a second is well within acceptable limits. For these reasons it was used in the first LOT prototype in 1970. Each matrix cell had an LDR, an amplifier, and a small pulling magnet mounted under the keyboard of an electric typewriter. Because of the high sensitivity of the LDR, the amplifiers were simple and inexpensive. However, evaluation showed that a more reliable detection system was needed. Sometimes just opening a door changed the lighting to such an extent that the typewriter produced a string of characters all by itself. Reliable detection could be achieved by adding a high-frequency modulated component to the light from the pointer and band-pass filtering the signals from the photodetectors. Because common light sources, in particular fluorescent tubes, can produce fairly strong modulated components up to several kilohertz, the modulation frequency was chosen to be much higher, at 64 kHz, making reliable detection relatively easy.

The response time of a LDR is much too large to detect a 64 kHz modulated component. Consequently, the choice of photodetectors was narrowed down to photodiodes or phototransistors. Because the modulated component in the lightspot was very small, the phototransistor was selected because of its relatively high sensitivity. Even then, the available signals were very small. To bring the signals up to the level required for further processing, the bandpass amplifier must have high amplification and tightly controlled characteristics. Such an amplifier is necessarily complex and expensive. Using a separate amplifier for every cell in the matrix, like the prototype LOT, would be much too expensive. The solution was a matrix of switches, one for each photocell, as shown in Figure 6. The actual photocell circuit is kept as simple as possible. Each phototransistor can be connected, in turn, to the common bandpass amplifier. By rapidly cycling through all cells in the matrix, the processing circuit can determine which cell is illuminated by the lightspot. This system performed well and was used in several types of LOT from 1972 until 1990, when the first LUCY appeared.

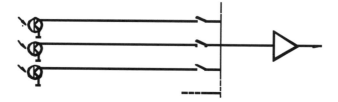

Figure 6.   Detection system with switched phototransistor matrix. This system requires a minimum number of components but may be vulnerable to external electromagnetic fields because of the relatively long distances between the phototransistors and the amplifier.

Although the detection system of the LOT did not produce errors when lighting conditions changed, it did become insensitive to the lightspot at lighting levels above 500 lux, which is about the recommended value for an office. For the LOT, this was not much of a problem, as it was used in a fixed location inside a controllable environment. LUCY, however, had to be suitable for use outdoors on a wheelchair, where lighting levels can easily reach 5000 lux or more. The problem of loss of sensitivity could be traced back to the phototransistors. Due to their inherent sensitivity, even moderate levels of light will produce large photocurrents through the transistors. At the lighting level expected outdoors, this current becomes unmanageable, saturating the transistor and making the detection system insensitive to the lightspot. Consequently, from the three photodetectors mentioned earlier, just the photodiode remained. Due to its lower sensitivity, the photocurrent remains manageable and saturation can be prevented, even in very bright light.

In the early stages of the development of LUCY, it was decided to add a LED indicator to every cell of the matrix. This would make the system much more versatile, as the indicator matrix could provide the necessary visual feedback for input methods other than lightspot control. After adding a LED to every cell, the next step was trying to use this LED as a photodiode. Although a LED is designed to emit light, not to detect it, the internal construction of a LED is basically the same as that of a photodiode. Most types of LEDs are indeed photosensitive, although their sensitivity is at best an order of magnitude lower than a typical photodiode. Experiments showed this idea to be feasible, but turning the idea into a reliable and reproducible system took more than 2 human-years of intensive effort. The very low signal levels generated by the LED detectors were the main cause of the problems. The design of the common bandpass amplifier and the conductor layout on the printed circuit board proved to be extremely critical and had to go through many iterations. However, the final version did meet the requirements, and more than 50 first-generation LUCYs were produced and sold during 1990 and 1991.

The LED detection system performed well and required only a small number of inexpensive components, but there were problems. The variations in the electrical characteristics of the LEDs as supplied were quite large, which meant that they had to be carefully tested and selected before they could be used. This, and the fact that the sensitive bandpass amplifier required an elaborate and critical adjustment procedure, made production quite labor-intensive. Another problem was the directional sensitivity of the LED detectors, limiting the angle under which LUCY could be used to about 20 degrees from normal. This meant that users had little freedom of movement when working with LUCY. There was also a problem with working outdoors. Although the detection system did operate perfectly, the lightspot from the LED pointer became almost invisible in bright daylight. Finally, magnetic fields from some computer monitors could upset the detection system. Clearly, this left something to be desired.

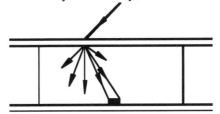

Figure 7.   Principle of laser spot detector. The screen can be made to fill the whole area of a cell, minimizing the required pointing accuracy.

The development of the laser pointer prompted a completely new approach to the detection problem. In earlier systems, the lightspot was fairly large and the photodetector very small. The small lightspot from a laser cannot be used with a small detector, as this would make hitting the target very difficult. Instead, the detector should have as large a photosensitive surface as possible. The solution found is shown in Figure 7. A semitransparent, diffusing screen is placed at some distance from a photodiode. Part of the laser beam is transmitted through the screen and fans out diffusely. Some light hits the diode wherever the spot hits the screen. The amount of light falling on the diode is almost independent of the angle of incidence of the beam, which solves the problem of directional sensitivity of earlier systems. Because just a small part of the light from the pointer reaches the diode, the available signal levels are quite low.

One of the lessons learned from earlier designs was that it is very difficult to transport these small signals over the circuit board to the common amplifier without them being affected by the inevitable stray electric and magnetic fields. For this reason, a simple preamplifier was put in the immediate vicinity of the photodiode, amplifying the signal about 100 times. The resulting high-level signals are no longer vulnerable to disturbances from outside. An independent LED indicator was added to every cell, increasing the total number of components per cell from 3 to 7. As this number is multiplied by 88, the total number of components in the new detection system is quite high, but because the components are inexpensive and do not have to be specially selected, the total cost of the detection system is only slightly higher. The preamplifier approach made the design of the printed circuit board much easier, and the common bandpass amplifier did not need any adjustment at all. This detection system has been in use since 1992. It has proved to be very reliable and easy to use because it requires minimal pointing accuracy. It can be used in bright daylight, from any direction, and is insensitive to electric and magnetic fields. Moreover, it is easy and economical to manufacture. Compared with earlier systems, it leaves little to be desired.

5   Optimal size and layout of the control panel

In the early stages of the project, Matthijs Soede investigated the motions of the head when subjects moved a lightspot quickly to a pointlike target (Soede et al, 1973). The diameter of the lightspot and the matrix pitch were variable. Usually, there was an overshoot when reaching the target. If the lightspot leaves the target briefly during the overshoot, a detection error may result. This means that the larger the lightspot is, the smaller is the number of errors. In actual fact, the diameter of the lightspot is limited by the amount of light available from the pointer and the required brightness of the spot. In early versions of the LOT, the overshoot effect was clearly noticeable. Sometimes the machine produced two strikes instead of one. After the lightspot detection system was made to ignore any interruption of less than 0.1 unwanted double strikes were effectively eliminated.

The average movement times for various combinations of pitch p and lightspot diameter D are plotted in Figure 8. The horizontal scale was chosen to illustrate Fitts' law. This empirical law states that the average movement time for rapid, cyclical movements between two targets of diameter D at distance p depends linearly on the logarithm of the ratio 2p/D. The figure shows that the values obtained by Soede fit this law quite well. To minimize the movement time and the effects of overshoot, the lightspot should be made as large as possible, but not larger than the matrix pitch to avoid overlapping two photocells.

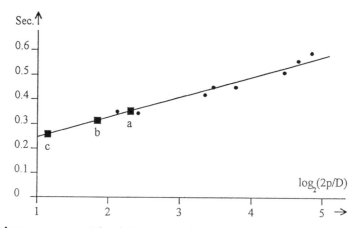

Figure 8. Average movement time between two characters for various combinations of matrix pitch p and lightspot diameter D. The dots indicate values measured by Soede in 1973. The squares indicate the operating points of the three generations of detection systems. a = LOT, b = LUCY 1, c = LUCY 2.

The LOT had a matrix pitch of 50 mm, a lightspot diameter of 20 mm, and a recommended operating distance of 1 metre. The corresponding 2p/D ratio is shown in Figure 8. In the first generation LUCY, the matrix pitch and operating distance were halved, keeping head motions the same. With this matrix pitch and a lightspot diameter of 14 mm, the 2p/D ratio is lower, resulting in shorter movement times and improved ergonomic qualities. Finally, in the second generation LUCY, with the laser pointer and large-area photosensors, the logarithm of the ratio almost reaches the theoretical minimum of 1, where targets overlap. Enthusiastic reactions from people who have used both systems intensively leave no doubt about the superior ergonomic qualities of the current detection system.

To maximize typing speed, the matrix characters must be arranged in such a way that the distance traveled by the lightspot is minimal. To calculate the average distance between characters when typing normal text, a table of digram frequencies is necessary and sufficient. These tables are available for most languages. When contemplating optimization procedures, it soon becomes obvious that simply trying all permutations is impossible. For 26 letters plus space, the total number is 27 factorial, or about 10 to the power 28, far beyond the capabilities of available computers. A suboptimal procedure was used. First, the position of the space, the most frequently occurring symbol, was fixed. Then, the four next most frequent letters were arranged around the space in random order. All pairs of letters were then exchanged, each time calculating the average distance, and the exchange pair with the shortest distance was retained. This exchange procedure was repeated until no pair with a shorter distance could be found. Then, some of the next most frequent letters were added and the whole procedure was repeated until all the letters had been used. By varying the number and initial placement of the letters, many different matrices could be found with average distances differing no more than a few tenths of a percent. From these, a matrix was selected that did not obviously show any undesirable words. The results for four different languages are shown in Table 2.

Table 2. Character layout for different languages. The number shown is the average distance between characters when typing normal text if the matrix pitch is 1.

Dutch (1.75)	German (1.70)	French (1.70)	English (1.73)
XPORDHC	QVOLBKX	KBMPJWX	JBFRUVX
YBT-EGU	JWTIRCY	FALERVY	KCOEHWY
ZJANILQ	FUS-EHZ	GIS-DUZ	PLA-TSZ
WFVSMK	PMANDG	HCTNOQ	QMDNIG

Other symbols on the layout were not optimized but were added in a logical way. It has been suggested that a standard typewriter keyboard or an alphabetical layout would be better because users would not have to learn a new layout. For English, a standard QWERTY typewriter layout resulted in an average distance of 2.9 and an alphabetical layout, in a 7 by 4 matrix, resulted in 2.7. When compared with 1.73 for the English matrix, the advantage of optimization seems clear. Interestingly, the space symbol and the eight letters surrounding it account for about 65% of all text.

6   System design

Of the many considerations that influenced system design, ergonomics, reliability, flexibility, portability, and cost will be discussed briefly.

6.1   ERGONOMICS

At the start of the project, the decision to develop lightspot operation instead of other possible control methods was based on ergonomic considerations. The decision proved to be right. Even today, almost 25 years later, lightspot operation still is the best choice for many handicapped people. After evaluation of the first LOT, a character display was added to the control panel, enabling users to read and correct a line of text before it was printed. Also, to obtain smooth operation, a beep generator was added to give instant audible feedback once a character had been accepted by the machine. This configuration has not changed over the years. Handicapped people want to be as independent as possible and communication aids should be designed accordingly. To adjust internal settings and parameters, many available aids have special knobs or switches that are inaccessible to the user. With LUCY, the handicapped user can adjust all internal settings and parameters over a wide range and with high resolution.

6.2   RELIABILITY

A communication aid has to be reliable. The average user will soon reject an unreliable electronic aid and revert to more basic methods, like using a headstick. Reliability has always been a very important design objective. In the days of the LOT, high quality components and expensive construction techniques were used to make the equipment shockproof and to protect it against the occasional coffee spill. Cost was not a first consideration then. It was designed like professional medical equipment, where reliability and performance are more important than cost. When LUCY was being developed, much effort went into finding inexpensive construction techniques that avoided mechanical strain on the printed circuit boards and provided ample shielding against mechanical and electrical disturbances from outside. Striving for high reliability makes good economic sense too. Because communication aids are not distributed in large numbers, a repair service has to be centralized. Even in a small country like the Netherlands, fixing a problem in a remote location may take a technician the better part of a day.

6.3   FLEXIBILITY

One aspect of flexibility in a communication aid is the range of input devices and control methods it supports. While the LOT offered just lightspot operation, LUCY was designed to be more flexible. Adding a LED indicator to every cell of the matrix provided visual feedback for other control methods. The current version of LUCY supports one-, two-, or five-switch control and a joystick, mouse, or trackball input. Flexibility has many advantages. In the first place, the number of potential users is much larger. It is estimated that for every lightspot user, there are at least two others who have to use other input methods. Eventually, this will lead to larger production runs and consequently lower prices. In the second place, the possibility of using different input methods is particularly important for patients suffering from progressive diseases such as ALS. When these patients start using a communication aid, they often have sufficient motor control to benefit from the speed and ease of lightspot operation. In later stages, a trackball or joystick may be more appropriate. In the last stages, a switch-controlled scanning aid might be the only way to communicate. This gradual downward option is beneficial to both the patients, who are spared the psychological and practical problems of having to adapt to different aids, and health care authorities, who are spared the costs of providing new aids.

Another aspect of flexibility is the range of output devices. The LOT was designed for printer output only. Although the LOT Mk3 did have provisions to add other output functions, this option was never used because it proved to be too expensive. Part of the design philosophy of LUCY was not to add extensions later but to build flexibility in from the start. A careful study was made of the various input and output devices that had to be connected now and in the future. The current version of

LUCY supports a printer, a speech synthesizer, an external character display, a keyboard emulator and a mouse emulator, for an MS-DOS PC. The keyboard interface can emulate every key or key combination of a standard keyboard, which means that any PC software can be used unmodified. Also, LUCY was designed to be compatible with existing speech synthesizers. The combination of a wheelchair mounted LUCY and a speech synthesizer has been particularly valuable to many users.

The last aspect of flexibility is the range of internal functions of a communication aid. LUCY was designed to accept any combination of control method and output device. This means that there is a large number of internal settings and adjustable parameters. All of these can be changed by the user giving yes/no answers to questions that appear in the built-in display. An important feature of LUCY is the possibility of storing arbitrary sequences of characters or control commands in memory and recalling them by selecting just one or two symbols. This so-called macro capability proved to be very popular with users. LUCY can remember more than 900 different macros. The complete set of macros can be stored in a PC and reloaded into LUCY, which is important in circumstances where different people use the same machine. Finally, several exchangeable character layout screens are available, each optimized for a particular control method.

6.4 PORTABILITY

The first LOT prototypes were designed for use in a rehabilitation center where many patients had to share the same machine. Because some of the patients were immobilized, the equipment had to be transportable. With the bulk and weight of the then available electronics, this was not an easy design task. The final configuration consisted of a trolley, containing most of the electronics, and a fairly large control panel, fully adjustable in height and orientation. In later versions of the LOT, this configuration was retained. All this required an elaborate mechanical construction, adding much to the overall size and cost of the equipment.

Reduction of size and weight were primary design objectives for LUCY. The latest version measures 30 x 25 x 5 cm. Due to its relatively low weight of 2.5 kg, it can be mounted on a wheelchair or placed near a computer monitor using only simple, inexpensive mechanical attachments. Another advantage of small size is that the equipment looks more pleasing and draws less attention. The importance of this fact should not be underestimated, as there have been several cases of people not accepting a LOT Mk3 just because of its rather imposing appearance. Finally, there is the factor of power consumption. The LOT needed a mains power connection to operate. During the development of LUCY, great care was taken to keep the power consumption as low as possible. When used with a PC, no external supply is needed because it can be powered from the keyboard connector. On a wheelchair, it can be connected directly to the wheelchair battery. Finally, it can be operated from a small battery pack for several hours, making it truly portable.

6.5 COST

In the days of the LOT, cost was not a primary design objective. Compared with the costs of rehabilitation, housing, and transportation of severely handicapped people, the LOT did not seem particularly expensive. However, the lack of success of the LOT Mk3 made it clear that its cost was a major obstacle to widespread use. A completely new, low-cost system was needed. When designing a microprocessor based system, there are basically two approaches. One is using modular circuits, available from many manufacturers, and building a complete system by placing various modules in a rack. The other is starting from basic components and designing the system and the printed circuit boards oneself. The modular approach takes much less development time but the resulting system is usually more expensive and more voluminous than a custom designed system.

The LOT Mk3 was a modular system, consisting of many modules, racks, connectors and cables. This, and the requirement of transportability, which meant lots of specially manufactured mechanical components, made it an expensive machine. LUCY was designed as a custom system, using basic, inexpensive, mass-produced electronic components. It consists of two printed circuit boards, one for the lightspot detection system and one for the microprocessor and input/output circuits. The mechanical construction is very simple. A polycarbonate front panel, an aluminum back panel, an injection moulded ABS case, and four screws are all that is needed to assemble a LUCY. To reduce the cost of production still further, Jan Goezinne and Bernard Visse from Delft University decided to start a part-time commercial enterprise, Shannon Electronics, which has taken care of the production of LUCY since 1990. All this resulted in a reduction from US $19,000 for the LOT Mk3 to US $1,400 for LUCY, not including distribution costs.

7   Distribution

However useful a communication aid is, its success depends critically on the quality of the distribution organization. Distribution has many aspects. The first step is promotion, which includes writing articles, mailing information leaflets, and showing the equipment at exhibitions, rehabilitation centers and schools. Useful as these activities may be, experience has shown that giving potential users ample time to try out the equipment by themselves is the most effective way to go. Once a potential user has been found, individual needs and motor impairments have to be carefully assessed . Different control methods and parameter adjustments have to be tried to match the aid to the individual. The eventual success of a communication aid depends very much on the skill of the assessor.

When an aid has been selected, financial support must be found because the cost of the equipment usually puts it out of reach of individual users. The distributor must be well aware of the different possibilities for obtaining funds and guide the user through the often complicated and time-consuming application procedures. Once funding has been obtained, the aid must be installed. As individual needs and circumstances vary widely, installation of the equipment often calls for considerable technical expertise and inventiveness of the distributor. Next comes training, to make sure the handicapped people will be able to make optimal use of the equipment. After sales, the distributor must provide a telephonic help desk and a repair service that can fix technical problems in a reasonable time. Finally, the distributor must provide feedback to the development team. This includes registration of equipment failures and evaluation of complaints and questions from users.

Successful distribution of a communication aid requires many different skills, which makes a good distributor hard to find. Distribution costs will substantially influence the price of an aid. With LUCY, distribution accounts for about 60% of the end-user price. Before LUCY, distribution was organized by the OSKAR foundation. A comprehensive training and support program had been developed, involving up to 20 hours of instruction. Training was offered separately from the hardware. This scheme met with a surprising amount of opposition from the financiers, who considered it unnecessary and too expensive, even though the cost of the program was just 7% of the hardware cost. For this reason, the price of LUCY now includes about 4 hours of training, which has been found sufficient for the average user. Extra training by an experienced therapist is available at extra cost. After the disappointing sales of the LOT Mk3, the OSKAR foundation had to end its distribution activities. Fortunately, Revalidatie Techniek het Dorp, a distributor of many kinds of aids for the handicapped in the Netherlands, was willing to take up the distribution of LUCY. In this larger organization, it proved possible to provide all the services mentioned earlier in an economical way. Over the years, they have done a splendid job and have contributed much to the success of LUCY in the Netherlands.

The rising trend in sales of LUCY is an illustration of the rising popularity of communication aids in general. In many European countries, the number of manufacturers and distributors has more than doubled over the last 5 years. The quality of communication aids is improving steadily. This has been stimulated by the PC revolution, as the combination of a communication aid and a PC is obviously very powerful. About 60% of all LUCYs are used as keyboards for MS-DOS PCs. Another popular output device is the speech synthesizer, often mounted on a wheelchair, serving the needs of about 30% of all LUCY users. Lightspot operation is the most often used input method, estimated at 60%. Other popular input devices are the five-switch pad, the joystick, and the trackball. Single- and double-switch operation is used in about 10% of all cases.

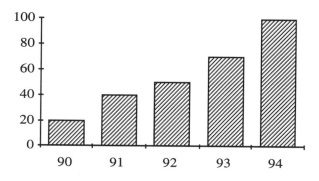

Figure 9.   Number of LUCYs sold since 1990. Export accounted for 10% in 1993 and 30% in 1994.

The development of LUCY is an ongoing process. Current activities include the development of picture overlays for young children and a wireless link to control a PC and other devices from a wheelchair. Recently, much time has been devoted to making LUCY suitable for other countries and languages. Full sets of control software, character overlays, and user manuals for Dutch, English, French, and German are now available. Distributors have been found in Germany, France, and the United Kingdom. Other countries will follow in the near future, among them the United States and Canada.

References

Soede, M., Stassen, H.G., van Lunteren, A., and Luitse, W.L. (1973). A Lightspot Operated Typewriter for severely physically handicapped patients. *Ergonomics*, 16; 829-844.

Stassen, H.G., Soede, M., and Bakker, H. (1982). The Lightspot Operated Typewriter: A communication aid for severely bodily handicapped patients. *Scandinavian Journal of Rehabilitation Medicine*, 14; 159-161.

Stassen, H.G., Goezinne, J., and Visse, B. (1993). A new communication aid for children and adults. *Journal of Rehabilitation Sciences*, 6; 20-24.

Chapter 11

STAP, a System for Training of Aphasia Patients

Lenie de Vries
Ben Wenneker

1   Aphasia and its treatment

Aphasia is a language disorder caused by brain damage, in the majority of cases a rupture or an occlusion of a blood vessel, known as a cardio-vascular accident (CVA) (Darley, 1982). As a result, one half of the body can be totally or partially paralyzed, but language impairments may also occur. Several types of impairment may result, dependent on the localization and extent of the brain damage. People can lose the ability to speak, write, or read in a normal way, or to recognize other people's words. This means that it is difficult to have contact with other people, a dramatic change in the life of the persons involved and their families.

In many cases a certain degree of spontaneous recovery takes place. However, in a number of cases therapy is necessary. Therapy aims to decrease the limitations on communication (Stachoviak, 1991). A main aspect of the therapy is repetition. This, however, requires much patience of therapists, family, and patient. Especially for the patients, who often know exactly what they want to say but cannot find the right words, the many unavoidable failures in expression are very frustrating. Moreover, the number of therapists in relation to the number of patients is such that the amount of therapy is limited to only a few hours per week. To decrease the language disorders of aphasia, automated training programs such as System for Training of Aphasia Patients (STAP) can be used. Such programs can be used independently after a short period of instruction.

In the aphasia therapy of the De Hoogstraat rehabilitation center, the training is based on the use of all modalities available for communication. This means that besides written and spoken language, use is made of pictures, objects, gestures, and special training materials. In addition to individual training sessions directed at the special disorder of the individual patient, group sessions are held in which patients are encouraged to use their abilities of mutual communication. The family of the patient is also involved in the training program. As a special communication aid, the "Taalzakboek" (Language Pocket Book) was developed, which, in modified form, is presently being published by a nationwide support organization for aphasia patients. It consists of a number of pictures with their matching nouns or verbs, all arranged in context groups such as clothes, food, time, and travel. Thus, therapists and family can use it together with the patient. It also has a central role in therapy sessions.

2   Starting points of the STAP project

The introduction of the first microprocessors and contact with the Man-Machine Systems (MMS) Group headed by Henk Stassen gave rise to the question of whether it might be possible to develop a system with a number of programs based on the existing treatment philosophy with the following requirements:

- The patients must be able to work independently with the system.
- The patients must get help whenever asked for .
- The patients must get feedback on their actions.

- The system must offer a new element to the existing treatment methods.

For the MMS group, these requirements formed an interesting human-machine problem: The machine must be controlled independently by someone who has problems with reading and comprehension. Thus in 1981 the STAP project was born as a collaboration between the rehabilitation clinic and the university.

It was decided to start with a flexible system in the clinic with the ultimate goal being the development of a training aid that could be connected to a home television set and could be used without the help of a speech therapist. The following specifications for the experimental system in the clinic were formulated:

- Communication between patient and system, and between speech therapist and system, should be user friendly.
- The software should enable the generation of new exercises in a relatively short time.
- The therapist must be able to interrupt running programs.
- Connection of a variety of input-output devices should be possible.
- Maintenance of the system should be simple.
- Extension of the system in the course of the project should be possible.

The first exercises were designed for the most severe cases. They were based on existing material consisting of a set of boxes, each containing a picture and a set of letters or words, which the patient has to combine. For example, the first program consisted of a number of pictures that could be presented in random order. The patient then had to write down the name of the presented item on a piece of paper, because the use of a keyboard was too complex. Therefore the first system had a single input device, i.e., a large button they could press. The programs made use of pictures, letters, words, and sentences, which the patient could call onto the display by pressing this button.

For example, a picture of a cat was presented on the screen (Fig. 1). If the patient pressed the button, the first letter was presented. In this way the patient could verify whether his first letter was correct, or if he did not know the first letter, he thus obtained some help that might enable him to generate the next letters himself. Each time the button was pressed, a letter was presented until the whole word was on the screen. Pressing the button then produced the next picture and so on until the whole series had been processed. In the first program a set of simple words were used, such as cat, in Dutch kat, which is an example of a CVC word, where CVC stands for Consonant Vowel Consonant. Series with longer words were also made.

A trial of the first program with a group of 7 patients showed that for 3 of them the program was too easy and for one of them it was too difficult, which illustrated the need for a more extended program package, containing more difficulty levels. Another program presented word anagrams, i.e., the letters belonging to the picture were mixed up, and the same was done with syllables in a word in a sentence. Finally, sentences had to be judged for correctness, either grammatically or with respect to their contents.

Figure 1.   Example of an exercise

These programs were all run in a menu structure on an Apple II with 48+16 kByte RAM, a 5 1/4-inch floppy disk drive, a video display, and a graphics tablet, which was used to generate the pictures. These pictures were all adapted from the Taalzakboek. The software was written in Basic. Because the standard characters were too small, a special character set was generated.

3   First evaluation

In order to get an indication of the therapeutic effects of STAP, in 1983 a small-scale evaluation was carried out with patients from three different aphasia treatment centers. In addition to the rehabilitation center that was involved with the development of the system, there was an outpatient center and a nursing home. At the beginning and end of a given period a group of patients was tested with respect to their proficiency in naming pictures and words, both verbally and in writing. The time spent with the STAP system was recorded in the computer for every user. Thus the relation between the time spent using STAP and the changes in the test score could be determined. A positive correlation was found. Moreover, both the patients and the speech therapist involved were very positive about the programs. Furthermore, the speech therapists involved recommended extending the software with programs of a higher level for patients who could use a keyboard.

4   From Apple II to IBM-PC

In order to fulfill the wishes of the therapists, a more powerful computer was necessary. In the meantime IBM had introduced its first PC, which offered more possibilities in terms of memory capacity and standard software. Therefore the existing STAP programs were converted to the IBM format, and the software was also extended with exercises for patients who could use a keyboard. In the first version, a deliberate choice had been made to use a one-button input system. Although this made it much easier for the user, it put severe limitations on the possibilities for the program package. Initially it seemed that these patients were not capable of typing. However, when using STAP it was found that patients, upon seeing letters appear one by one, wanted to try to type words by themselves. The possibility of utilizing typing enormously extended the training prospects. The insights that resulted from the evaluation of the first prototype led to the development of a second version of STAP. Additional features were as follows:

* Use of color (limited)
* Recording training times for each of the exercises
* Use of function keys
* Typing of words and sentences

5   A pilot study in the home environment

In 1986, PCs were installed for a half-year period in the homes of 10 CVA patients. These patients had left the rehabilitation center and received only a limited amount of therapy in an outpatient setting. Before and after this period, a number of tests were used to measure their language disorders. Moreover, interviews were held with patients, their partners, and their therapists. The tests showed a general improvement in their ability to name words, although for only 4 of 10 patients was this improvement significant at the 5% level. Patients were very happy that they could train independently. They could choose their own pacing and got feedback and help whenever they wanted. Another positive aspect was the fact that they could work with computers, which were not yet very common in those days, giving them more self-confidence. The patients experienced the exercises as useful because this had a therapeutic value for them, and also because work or other pastimes were no longer possible.

6 Toward a flexible treatment system

As already mentioned, aphasia is a general name for a collection of different language disorders. Therefore, a speech therapist starts with a number of diagnostic tests to ascertain the nature of the language disorder and to determine the capabilities of the patient. Based on this information, a therapy plan is made. Depending on the amount of progress made, the plan has to be adjusted over time.

In order to adapt STAP to this mode of treatment, it was decided to develop a programming system in which the therapist generates a number of exercises on a floppy disk for the patient as homework. The patient works with these exercises on a PC-based individual training system at home. After each session feedback about results is obtained. The time spent on each program and the scores obtained are recorded on disk. Based on these data the therapist can adjust the therapy and homework for the next period.

The treatment system consists of a number of clusters that can be extended. Each cluster consists of a database and a number of programs. All exercises can be executed on several difficulty levels and a number of items can be selected from the database with the help of a menu.

The main clusters are as follows:

- P (Picture) Cluster. The programs in this cluster aim at recognition of words and the generation of parts of words based on the presentation of pictures.

- W (Word) Cluster. The programs enable the recognition, remembering, and production of word forms. The database contains about 6500 basic words. For the nouns, the plural form is also available and for verbs the root and the past participle, so that altogether the database contains 14000 items.

- CW (Category Word) Cluster. This cluster is meant to train in e.g., word finding and generation of words that have a certain characteristic in common, for instance a set of animals or a set of words that start with the letter "a."

- S (Sentence) Cluster. This is used to generate words in the context of a sentence, to learn to distinguish between syntactically correct and incorrect sentences, and to formulate grammatically correct sentences.

Following is an example of a program structure:

- A stimulus word, e.g., a verb, appears on the screen.
- The user has to type a reaction, e.g., the corresponding past participle.
- With every letter typed, additional information is given, e.g. a mouth image.
- Wrong letters appear in a different color, after which they are slowly removed. Then the user gets an adjustable number of retrials.
- In general, a program also contains some adjustable help functions, such as a repeated presentation of the stimulus word or the presentation of three letters, one of which has to be chosen.

Results of the exercises can be used on three levels:

- Feedback to the patient
- Information for the therapist about the exercises executed in a certain period and the scores attained

- Information for evaluation studies

An important property of STAP is that it does not depend on a specific treatment method. All speech therapists can select exercises based on their own vision that are tailored to the needs of a specific patient.

7   Availability of STAP

The system has been available commercially for the Dutch language market since 1989, and 65 copies have been sold. The firm that until recently had the distribution rights initially intended this product as its first step into a new market. However, over the course of time it changed its policy, and another company, which was already involved in communication systems for handicapped persons, has now taken over the distribution rights. Based on the information obtained in a survey of present users, the company intends to produce a new version of STAP. In addition to making some improvements in the software based on present technology, the number of exercises will be extended and other changes will be made, such as inclusion of synthetic speech.

References

Darley, F.L. (1982). *Aphasia.* Philadelphia, WB Saunders Company.
Stachoviak, F.J. (1991). Developments in the Assessment and Rehabilitation of Brain-Damaged Patients. Tübingen, Gunter Narr Verlag.

Chapter 12

Some Unexpected Results of a Study on Information Processing for Use in the Rehabilitation of Patients with Spinal Cord Injuries

Han Bakker
Ton van Lunteren

1 Spinal cord injuries and their rehabilitation

A spinal cord lesion is a breaking of the continuity of the spinal cord by an accident or disease. The result of this broken continuity is the neuronal disconnection of the lower part of the body: There is no direct connection between the brain and the upper part of the spinal cord, and between the lower part of the spinal cord and the nerves in the lower part of the body. No voluntary movement is possible in the lower part of the body. Sensation has been lost. The voluntary part of micturition, bowel movements, and sexual function have also been lost. Damage of the spinal cord in the neck region will result, apart from the difficulties just mentioned above, in loss of function of the arms. If the breaking of the cord is complete, no healing can be expected at all.

The intention in rehabilitation is *not* healing, but rather teaching the patient to use the remaining functions, if necessary, with the help of an orthosis. A team of experienced workers is necessary to reach this goal. Such a team should consist of the following (Bakker,1977):

1. Rehabilitation doctor for the necessary medical checkup
2. Physiotherapist for strengthening the still available muscle power
3. Social worker for job adaptation and solving financial problems such as insurance, etc.
4. Occupational therapist for training in the activities of daily living (ADL) and for advice on house adaptation
5. Sport teachers
6. Psychologist to guide the patient in the assimilation process
7. Technicians of different standards to construct adaptations
8. Last but not least, the nurse

A point of importance in the treatment of these patients is the prevention of complications. The most serious complications are as follows:

1. Pressure sores, especially in parts of the body without normal sensation
2. Urinary tract infections, especially when the urinary bladder is not emptied completely
3. Spasticity as a result of loss of movement and incorrect positioning in a bed or chair
4. Airway infections due to insufficient breathing
5. Serious obstipation due to insufficient emptying of the intestines

At first, the preventive measures are to be taken by the therapist, but later on, and especially at home, if possible the patient has to be responsible for his or her own treatment. You are always safer in your own hands than in the hands of others. Even if the proper measures cannot be taken by the patient, he or she should learn to ask a friend or relative for assistance. However, the patient should be in charge.

2   Information processing project

2.1   MOTIVE FOR THE PROJECT

One of the results of the contacts between the De Hoogstraat Rehabilitation Center and the Delft University of Technology was a project to investigate the possibility of reducing the duration of the treatment period of patients with a spinal cord lesion without loss of the quality of treatment. The average duration in this period was on the order of 1 year. For the patients it is a long time to be away from home. For society a shortening of their stay in the rehabilitation center would mean a reduction in costs. Home adaptations are often necessary, for instance, for wheelchair accessibility. The procedure to obtain financial support takes some time, so it has to be started at an early stage of the rehabilitation process. However, in the early 1970s, the present diagnostic methods were not yet available, so it was not always sure whether the patient had a complete lesion or a partial one. In a number of cases, it was not clear whether some restoration of function would occur. Therefore, the staff waited to start procedures for home adaptations until they were more sure about the final state of the patient. Furthermore, the patient also hoped for a certain degree of recovery and would not accept any actions that were based on the expectation of a negative outcome.

   As a consequence, home adaptations were not always completed when the patient was ready to go home. A first analysis of patient files showed that in most cases information on the diagnosis, the course of therapeutic activities, the state of the patient, and the reasons behind decisions was very scarce. Therefore, Henk Stassen proposed to start systematic data collection in order to develop a model that could predict the future state of the patient. This project was started in 1973 and continued until 1979. The following description is based on earlier publications (Stassen et al., 1980; Hoogendoorn et al., 1983).

   From a control engineering viewpoint, the treatment of patients with a spinal cord injury can be described like any control system (Fig. 1). The treatment activities of the team can be considered as the input to the rehabilitation process. In addition to activities of the treatment team, often unpredictable external forces, which cannot

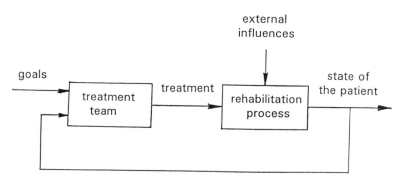

Figure 1.   The rehabilitation process with the variables of interest

be influenced by the team, also act on the rehabilitation process. The state of the patient can be considered to be the output. This state is compared with the desired state, i.e., the goals of the treatment that are thought to be attainable by the treatment team. In addition to executing treatment activities, the treatment team also acts as a controller. Based on the differences between the desired state and the actual state, the team decides what activities are necessary in the coming period as input to the rehabilitation process.

   A description of the rehabilitation process can be obtained by recording the inputs and outputs of the rehabilitation process for a group of patients over the duration of their stay in the rehabilitation center. To complete the description of the controlled system as a whole, it is also necessary to record the goals of the team. Here it should be mentioned that the final goals cannot be stated beforehand because many factors are still unknown, so these goals have to be seen as short term goals that have to be reformulated during the treatment period. Thus besides the inputs and outputs of the rehabilitation process, the goals also have to be recorded. A study of the rehabilitation process for this category of patients was started based on this concept.

The study could be divided into four main phases, i.e.,

- Data collection
- Data reduction
- Development of a prognosis model
- Analysis of the total rehabilitation process

2.2 DATA COLLECTION

In the period of 1973-1975, data on the treatment and rehabilitation states of 58 patients with spinal cord lesions were recorded on a weekly basis. Data collection was carried out by team members. Altogether 74 treatment variables (inputs) and 153 rehabilitation state variables (output) were recorded. The latter can be classified as follows:

- 39 medical variables, e.g., urinary tract infections, pressure sores
- 19 ADL variables, e.g., dressing, eating, wheelchair use
- 73 psychosocial variables, e.g., emotional state, relations with treatment team members, partner, or relatives
- 22 support variables, e.g., state of adaptations of car and house, and provision of orthoses

The average duration of the stay in the center was 43 weeks, so altogether about 500,000 pieces of data were collected and stored on the central computer of the university. All data were scored on a scale between 0 (maximally disabled) to 5 (functionly normal).

2.3 DATA REDUCTION

Data reduction started with the elimination of variables that hardly varied or even remained constant. Thereafter, those variables that had a high correlation were combined into one new variable. In this way the 74 treatment variables were reduced to 4 input variables: nursing, occupational therapy, physiotherapy, and bladder training. The 153 state variables were reduced to 12 new variables: pressure sores, bladder control, para-articular osteopathy, pain, spasticity, self-care, wheelchair riding, ability to stand, time out of bed, attitude of family, mood, and relation with therapists.

2.4 DEVELOPMENT OF A PROGNOSTIC MODEL

Two prognostic models were developed. The first one was the Semi-Markov model. This is a probabilistic model that describes the probabilities for a given state variable that had a certain value over a given number of weeks to change to another given value in the next

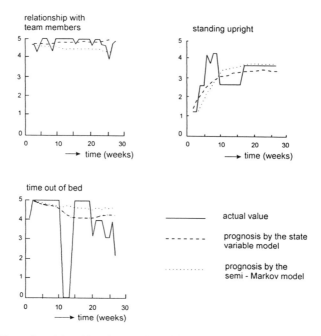

Figure 2.   Examples of time histories together with predictions made by the two models

week. This model is rather complex because a lot of probabilities have to be estimated. It is also rather limited because it does not take into account the possible interaction between variables.

The second model describes the input-output relation between the 4 therapeutic activities as input variables and the 12 state variables as outputs. This model includes the feedback relation in which the state of the patient influences the therapeutic activities and also a possible interaction between the state variables. Nevertheless, it has a more transparent character and the parameters can be estimated more easily. Figure 2 gives an illustration of predictions obtained with the two models versus actual developments. The predictions are generated based on the state in the second week and, in the case of the input-output model, a planned treatment, based on an average strategy. The figure shows that both models have a smoothing effect, i.e., they predict the general trend. The predictions are somewhat better for the input-output model than for the Semi-Markov model.

2.5 ANALYSIS OF THE TOTAL REHABILITATION PROCESS

The parameters of the input-output model consist of two matrices. The 12 by 12 matrix A describes the mutual relations between the present values of the state variables and those of the previous week. The 12 by 4 matrix B describes the relations between the therapy and the state of the patient. In matrix A, the diagonal elements have a value between 0.79 for "time out of bed" and 0.99 for "attitude of the family," while the off-diagonal elements have values between -0.19 and 0.11, which means that the present state is mainly determined by the value it had the week before. The state of other variables in the previous week has only a small influence, which shows that the reduced state variables are more or less independent.

If the input variables are also normalized so that they vary between 0 and 5, then it is found that the elements of the input matrix have values between -0.025 for the relation between "physiotherapy" and "pressure sores," and 0.067 for the relation between "physiotherapy" and "time out of bed." As already mentioned, no conclusions can be drawn from these figures about causal relations between therapy and state of the patient. The matrix describes the net result of both the effect of the amount of therapy on the state of the patient, and the effect of the state on the amount of therapy. In the first case it does not seem plausible that an increase in the amount of physiotherapy caused a worsening of the pressure sores. In this case the most likely explanation would be that an improvement of the variable "pressure sores," meaning that these sores decreased, enabled an increase in physiotherapy. For the positive relation between "physiotherapy" and "time out of bed," however, cause and effect can work both ways and probably do. In principle, system identification methods exist that enable separation between causal effects. However, these methods require more data and less disturbances than in the present case and thus could not be applied.

Although the results obtained did not give a direct answer to the original question, they did provide an indirect one. The data gathering activities, and the feedback provided by the investigators, made team members aware of the importance of keeping a systematic record of the changes in the state of the patient. Moreover, they got more insight in the overall rehabilitation process, so that in an earlier stage they could decide on necessary adaptations and appliances.

In this respect, the goal list, which was filled out at regular intervals, played an important role. This led to helpful discussions, especially when team members had different expectations. Here it should be mentioned that the patient also filled out this list and was involved in the discussions. For instance, when the patient's goal was to walk independently, while team members had already excluded this possibility, the first step was to learn to accept that the future would be different from what had been hoped.

Thus the study as such had already contributed to the goal for which it had been executed, independent of its direct results. It had triggered a process that is described in more detail in the following section.

3   Effects on the working method of the treatment team

At the start of the project the team members were the heads of the departments. They had to disperse the information obtained during the team meeting to their coworkers. When data gathering started, it was felt that more direct contact was needed between these coworkers because their activities could benefit from the information provided by the others. Direct information from worker to worker is faster and less apt to produce mistakes or misunderstanding. So we changed the practice of holding team meetings with heads of departments and had our meetings with the therapists who were directly involved in the treatment of a particular patient. In this way, a group of more or less independent therapists started to function as a team.

For the patient this method is not without disadvantages. All persons involved in the patient's treatment form a group. If the patient has a question, one of the team members will be approached, and usually this team member first discusses matters in the team meeting before answering the patient. The patient has to deal with a group of experts. This does not result in an easy exchange of opinions between the patient and therapists. So, the team membership was enlarged to make the patient a member. The team meeting is thus a round table meeting of experts to discuss the problems of the patient, who is an expert by virtue of experience.

In the beginning, some team members were afraid to offer negative information. Now we are convinced that the discussion of negative points is beneficial for the outcome of the rehabilitation process. This can be illustrated by the following example:

A patient did not turn up at the physiotherapy department to do his exercises. The physiotherapist called the patient "lazy." In discussing the matter during the team meeting, the patient told us that he did not understand why he had to do these exercises. They were rather useless in his opinion, and he decided to read a book instead. The lesson to us was as follows: Always explain, time and again, why you expect seemingly useless actions.

Sometimes the therapist himself is the cause of the lack in progress in the rehabilitation process. Teaming with the patient gives the opportunity to improve the attitude of the team members. One of the difficulties we came across when things were discussed with the patient was one of language. Every discipline has its own language, which is often difficult to understand even to other disciplines. The use of such a discipline-bound language increases when the speaker is uncertain. Difficult words cover up for lack of knowledge. Therefore much attention was paid to providing a clear formulation.

Furthermore, it was thought important to give patients more background information about their disabilities and the corresponding treatments. It already was our custom at the rehabilitation center to give a lecture about a particular disease at regular intervals. These lectures were attended by patients, their relatives, and staff members. All information was also provided in writing, offering the opportunity to read it over again and to show it to relatives. These lectures were made possible because patients with the same diagnosis were admitted into the same department. An incidental advantage of this group formation is that patients who are more advanced in the rehabilitation process teach the new ones from their experience.

4   Activities of Daily Living

The most important data collected to describe a rehabilitation process are data regarding Activities of Daily Living (ADL), because they determine the amount of independence the patient has reached in spite of physical limitations. But are these data always representative? The following case shows that sometimes other factors are more important for a patient's future.

A Moroccan man was admitted to the rehabilitation centre with a traumatic tetraplegia. The man was married and had two children. After 8 months he was independent in self-care in a wheelchair. Because his house was not adapted for wheelchair use, we arranged a new house for him that was accessible. The man was dismissed and booked out, being ADL-independent. We were proud of the satisfying results. After 3 months he was seen for a checkup. It turned out that all that time he had been dependent on his wife and children. It was impossible to change the attitude of the members of his family.

The reason of this disappointing outcome was a cultural misunderstanding. In the Moroccan culture it would have been very impolite, even a sign of a lack of love, for the family to leave the man doing things with difficulty. The family members were able to help make things easier.

We would not have made this mistake if we had been aware of the fact that ADL involves not merely one facet, but three (Vreede, 1993), that is,

- ODL (Operations for Daily Living), the mental and physical functional activities applied in ADL in so far as they can be performed and experienced consciously
- ADL (Activities of Daily Living), the actual intentional activities, usually referring to an individual or a group of individuals
- IDL (Ideas in Daily Living), the pursuits that presuppose a common value or social purpose in so far as they are described concretely and can therefore be analyzed in terms of ADL

Our mistake was to train our patient in ODL. But, although he succeeded in doing these operations, they did not become his own way of living. His ideas of daily living differed completely from ours. He preferred to receive love from his family instead of difficulties and hardship for himself. To summarize, we can perform an ADL only when the necessary ODL is available, but it will be performed only if it is part of an existing IDL. In IDL there are important differences, especially between people from different cultural backgrounds. If these differences are neglected, the rehabilitation process will fail.

The multidimensional structure of ADL is also determined by the context in which an activity takes place. The components describing the context are as follows:

- *Performance*, the ability to dress, to transfer from a bed to a wheelchair, to bathe, etc.
- *Product*, e.g., is the ability to move, properly clothed, among other people
- *Place*, either the rehabilitation center or the adapted house
- *Persons* involved, nurses at the center and the family at home
- *Period*, the duration of the process of clothing or the hour of the day, in this case, the morning

With reference to the above-mentioned case, it becomes quite clear in studying these components that the change should have warned us. We should have paid special attention to this fact. Without these *persons* being available, there was no ADL. In the case of illness or absence of the wife, the patient would have to stay in bed. This shows the importance of the *person* component.

The last aspect of ADL to consider is the difference between usual daily performances, "usual" or "own" to a certain person, and unusual performances, "alien" or "strange" to a person. Examples are as follows:

Squatting: For Asian and African people this is easy, but it is impossible for most European people.
Eating habits: Use of chopsticks, fingers, or fork and knife.

It is much easier to learn an action that is already common to you but temporarily lost, due to a disease or accident, instead of a completely new action.

It is always important to establish how far the patient has got in the processing of accepting the loss of abilities. Is the patient still denying the seriousness of the situation, or is the patient depressive? In both cases our therapeutic activities will have an outcome that is different from what we expected. *We* expect a certain result, but the patient, still in the denial phase, expects a complete recovery.

It is a well-known fact that most therapists are content with the result of their actions, while the patient remains discontented. This is understandable, as the patient compares the present situation with a healthy state before the accident or disease, whereas the therapist remembers a severely ill patient, rushed into the hospital.

5 Concluding remarks

The direct benefits for the rehabilitation center, which were obtained during the execution of the study, can be summarized as follows:

- By recording data in an orderly manner, information about the experience became available to therapists.
- This information provided more insight into the rehabilitation process and increased

the expertise of the therapists.

- The formation of a group of specialists into a cooperating team increased the effectiveness of treatment.
- The formation of the team triggered a change in the relationship between the patient and therapists. The patient had a more active part in decisions concerning the treatment program. The discussions of the goals of treatment especially made the patient more aware of the future situation.
- The growing experience of the treatment team also brought a growing awareness of other factors that had not been considered previously, such as cultural patterns.
- The term activities of daily living means far more than a list of activities. It is our relation to the surrounding world, changing in time, depending on place, shaped by ourselves and our abilities.

References

Bakker, H. (1977). The rehabilitation team (in Dutch). In: Verkuyl A. (ed.): *Gehavend en wel*. Samson, Alphen aan de Rijn. ISBN 90-14-02509-2, 188 p.

Hoogendoorn, R., Bakker, H., Van der Kolk, G.J., Balk, P., Morsink, G., Stassen, H.G. and Van Lunteren, A. (1983). Information processing for the use in the treatment of severely physically handicapped persons. In: Van Lunteren, A. (Ed.) *Fourth Progress Report of the Man-Machine Systems Group 1976-1982*. Delft University of Technology. Dept of Mechanical Engineering, Report WTHD-161, pp. 125-151.

Stassen, H.G., Van Lunteren, A., Hoogendoorn, R., Van der Kolk, G.J., Balk, P., Morsink, G. and Schuurman, J.C. (1980). A computer model as an aid in the treatment of patients with injuries of the spinal cord. In: *Proceedings of the ICCS*, Cambridge (Mass.), IEEE, pp.385-390.

Vreede, C.F. (1993). *A Guide to ADL*. Eburon,, Delft. ISBN 90-5166-316-1. 149 p.

Section B

HUMAN CONTROL OF VEHICLES
AND MANIPULATION

Chapter 13

Trends in Human Control of Aircraft, Railway Trains, and Highway Vehicles

Thomas B. Sheridan

1   Introduction

The way humans control their transportation vehicles is changing. These changes are characterized by higher speeds, tighter headways relative to other vehicles, increased attention to traffic regulation and safety in general, increased optimization to conserve resources, and more use of computers, artificial sensors, and communication with other vehicles and with system infrastructure. With respect to the human driver, these changes signal a move toward more automation and human supervisory control, in which the human instructs a computer and the computer performs control automatically through electromechanical sensors and actuators. The chapter describes these changes in each of three major modes of transportation: aircraft, railway trains, and highway vehicles. While the implementation of new technology and the form of interaction with the human operator is necessarily somewhat different in each transport mode, the similarities are greater than the differences.

2   Aviation

New technologies for commercial aviation (Wiener, 1988; Billings, 1991; Sarter and Woods, 1994) include TCAS (Traffic Alert and Collision Avoidance System), GPS (Global Positioning System) to take fixes of latitude and longitude from orbiting satellites, and various aids to detect windshear (in which neighboring air masses have significantly different velocities), etc. The "glass cockpit" was first introduced several years ago with Boeing's 757 and 767, in which information that heretofore was presented on separate displays was integrated by computers and displayed on several cathode ray tubes (CRTs). These have replaced the multiple independent mechanical flight instruments and have permitted simplification of the instrument panel. Autopilots have been provided with multiple control modes, e.g., for going to and holding a new altitude, for flying to a set of latitude-longitude coordinates, or for making an automatic landing when an airport has the supporting equipment. In the Airbus A320 the primary flight mode is fly-by-wire through smaller side-sticks, in dramatic contrast to the old control yokes. In the cockpit, computer-based expert systems give the pilot advice on engine conditions, how to save fuel, and other topics. Performance management systems are now available to optimize fuel and time.

A recent addition to the flight deck is the Flight Management System (FMS). The FMS is the aircraft embodiment of the human-interactive computer intrinsic to the concept of supervisory control — a computer system designed to talk to the pilots in their own language and in turn to interact with multiple special-purpose control computers scattered throughout the aircraft. The FMS permits pilot selection among many levels of automation, provides the pilot advice on navigation and other subjects, and detects and diagnoses abnormalities. The typical FMS Control and Display Unit (CDU) has a CRT display and both generic and dedicated keysets. More than 1,000 modules provide maps for terrain, navigational aids, and procedures. A closely associated Engine Indicating and Crew Alerting System (EICAS) presents synoptic diagrams of various electrical and hydraulic subsystems.

A proposed electronic map that would show flight plan route, weather, and other navigational aids is illustrated in Figure 1. When the pilot enters a certain flight plan, the FMS can automatically visualize the trajectory and call attention to any way-points that appear to be erroneous on the basis of a set of reasonable assumptions. (This might have prevented the programmed trajectory that allegedly took KAL 007 into Soviet airspace.)

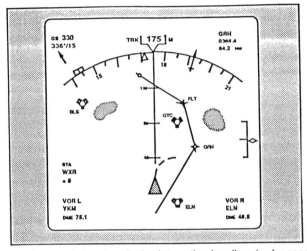

Figure 1.   Proposed plan view display integrating heading, track speed, way points, radio frequencies, course prediction, weather, and other information. (From Billings, 1991, with permission.)

The FMS extends a trend in aircraft control that began with purely manual control through a joystick (and a later a yoke) to control ailerons, and elevators and pedals to control the rudder (and brakes). This control became power aided (like power steering in a car), and still later became fly-by-wire (with no mechanical connection to transmit forces and displacements).   Purely manual control was replaced, first by simple autopilots to hold an altitude and/or heading, later by autopilot modes to achieve an altitude or heading in a transient maneuver, and finally to automatically take off or land the aircraft. At first there were essentially no buttons to push, and gradually piloting has become to a far greater extent a matter of button pushing.   This trend started as a matter of pushing buttons on the panel that are uniquely dedicated to specific functions, many of which are still there, now collected into the Mode Control Panel (MCP). However, more and more the pilot uses the general alphanumeric keyset of the FMS CDU. Typically one CDU is at the most convenient location for the captains's right hand, the other is similarly located for the first officer's left hand, in such a way that either can back up the other.

The move from sticks and control yokes to button pushing is more than what meets the eye (or more literally, what meets the hand!). More important is the move away from direct body image control (which the psychologist might call "psychomotor skill") to supervisory control (which the psychologist might call "cognitive skill"). It is the FMS that more than any other technology forces this move, since the FMS forces the language of communication (and probably more critical, the language of the pilot's thinking) to include alphanumeric streams (words, mnemonic codes). The pilot's mental models still include the physical space of the aircraft, and the lift, drag, thrust, and wind components of force that propel the aircraft in its Newtonian trajectory, but must now also include some of the "if -- then -- " logic computer technicians use.

Control of the aircraft, of course, is not only determined by the pilot but is also determined by air traffic control (ATC). The same kinds of technological changes that have been taking place in the aircraft itself have been happening in ATC. Traditionally ATC is accomplished through a network of stations around the world that are connected to the aircraft by two-way radio and "see" the aircraft by means of radar. In most countries not only commercial carriers but also general aviation (GA) aircraft are required to carry transponders that identify them to ATC with a simple code. If they plan to fly into the airspace of major airports, these transponders must have the capability to transmit aircraft altitude as well. This provides an identification tag next to the blip seen on the ATC operator's radar display.

Airports with control towers (the smaller GA airports are an exception) typically have one ATC operator for takeoffs and landings, and one for ground (taxiway) operations, both of whom operate using visual observation. Somewhat larger airports have an additional operator for approach and departure operations, who works primarily from a radar screen. The most critical operations are usually found in the terminal area radar control rooms (TRACONS) associated with major airports. These are the ATC operators who must ensure that aircraft arriving from various directions are ordered with proper spacing into the landing pattern and out of the way of other aircraft that are leaving that airspace or that are just passing through. A final category of ATC operators is found in the "enroute" facilities (one for each major sector in a large country such as the United States) which monitor commercial traffic on major airways ("highways in the sky") laid out between radio navigation beacons. Some recent technology permits every aircraft in the United States to be displayed on a single large screen. During major weather problems, when, for example, airports must be closed, it is important to monitor and control the cascading effects of backups and delays from one sector to another.

New and better weather radar, weather prediction technology, and communication links between ATCs and supporting weather and other facilities have evolved. As international air travel has increased, the problems with communication have required rulings that pilots and ATC operators speak a common language, which usually is English (but, in practice, these rules are not always adhered to). Controllers and pilots who share a non-English native language often defer to their own language over the radio.

One major innovation about to occur in some form is "datalink," a two-way communication with aircraft by digitally encoded messages. Data link allows much more data communication with the ground than possible with traditional voice channels. This enables new capabilities but also poses new problems. Such capabilities include sending a variety of information about aircraft conditions to the ground for recording and analysis in real time, but there are commercial rights to privacy that inhibit this possibility. Such capabilities also include making available to the pilot any or all of the radar information about other aircraft in the pilot's area, which might seem to be an advantage. The problem is that if the pilot sees all of what the ATC operator sees, and the pilot is ultimately in charge of the aircraft, does ATC thereby lose control?

Perhaps even more serious is the question of how communication between ATC and the pilot is accomplished. With datalink ATC instructions, landing clearances, etc., after being sent to the aircraft digitally, can be put on visual displays. So too can pilot queries and responses, thereby (perhaps) obviating the need for voice communication. But pilots like the voice communications and have confidence in it (it is immediate, flexible, can be as informal as necessary, offers immediate confirmation that the other party is paying attention, and, by listening to the "party line," they have a sense of what is going on in the surrounding air traffic). Communicating by datalink may be more reliable in a narrow technical sense (voice messages get cut off and essential information can be lost, require repeating, etc.), but datalink communication could turn out to be less reliable in a broader sense. What about having both? That may increase workload. So the form in which datalink will be implemented remains under active discussion.

Another innovation now being seriously considered in the United States is "free flight," under which aircraft would be free to fly enroute by direct great circle routes and to deviate from those as they wish, either vertically or horizontally, depending on winds aloft, weather conditions, and other air traffic, all of which they now can observe for themselves. (They still would have to abide by ATC instructions as they converged on their destination airport). Current efforts in the author's laboratory at MIT are evaluating pilot decision aids for vertical free flight and suggest that significant fuel savings are possible in this manner (Patrick, 1995).

3   Rail transport systems

While the aircraft is surely the most "high-tech" mode of transportation (and also the newest, the one with tightest selection and training requirements for human operators, and the most regulated in other respects), it is interesting to consider how similar changes are occurring in other transport modes. Consider railroad trains, the oldest of the three modes. Though slow to follow, railway systems are copying much aircraft technology, such as use of the GPS for determination of latitude and longitude, and the use of new sensing technology to measure critical variables (in this case, e.g., the temperature of wheel bearings). Locomotives are even using "fly-by-wire" controls, which in the case of trains combine the brake and throttle adjustments into a single control. (Existing systems have separate controls for the brake and throttle, evolved from air pipe valves and locomotive steam valves that the driver used to operate directly.)

A train driver's primary task is to control speed, and as part of this to anticipate the uphills and downhills that tend to decelerate or accelerate the train by gravity; to know the fixed speed limits at different locations along the track where there are curves, or switches, grade crossings, heavily populated areas, and to stay current with information about temporary speed limits due to weather or problems with the track or road bed, or maintenance crews currently working at certain locations. The driver must know the train momentum and friction effects of the wind and track as a function of speed for different consists (number of cars and type of load), and most know the limits of train thrusting and braking (both normal and emergency). The driver is responsible for reading and comprehending the wayside signals at the entry to each block or section of track (which becomes more and more difficult as speeds increase), for communicating with the central dispatcher as necessary, for knowing or being able to access the scheduled time of arrival at each station, and for braking the train to a stop at precisely the correct point at each station. The driver must also monitor brake pressure and other variables of train and track condition, change the pantograph connection to the overhead electrical power cable from DC to AC and reverse, and operate diesel motors, as required. The driver must integrate all of the above so as to maintain safety, arrive at the next station on time, and minimize the use of fossil fuel or electrical energy.

In some sense the train diver's job looks easier than that of the airplane pilot, since one might think of a train as a one-dimensional airplane. A special problem with the train is that it has very large momentum (it takes up to 3 km to stop a modern high-speed train traveling at 300 km per hour, even under emergency braking). Yet, if an obstacle lies ahead — a truck stalled on a grade crossing, a bridge or section of track that is not in good condition, or a large rock or rock slide from a mountain — the train does not have "maneuvering room" as the aircraft does; it must stop. Some trains are equipped with automatic speed controllers that can be set much the same as cruise control systems in automobiles. But, much like the automobile's cruise control, this does not take care of necessary braking. Some newer trains do have automatic train protection (ATP) systems to signal the driver to put on the brakes if a train is detected immediately ahead (trains have been easier to detect than other obstacles by means of electromagnetic sensors) and apply the brakes automatically if the operator does not respond in time. However, the existing infrastructure of slow and stop signals are on fixed posts along the side of the trackbed, and not only can these not be seen from a distance, they are also very difficult to read as the trains reach higher and higher speeds.

Flexible, computer-driven displays for monitoring that are located on the operator's console (the "glass cockpit") are gradually finding their way into railway systems. What might such displays be used for, and how might they assist the driver in speed control? One obvious use is for displaying in the cab the same information that now must be gleaned from the outside signals, and providing this in an easy-to-read manner right in front of the driver. This of course depends on digital communication between the wayside and the locomotive cab, which means an expensive overhaul of the present signaling system.

An advanced version of such an in-cab display has recently been proposed by our group (Askey, 1995). Its purpose is to help the driver anticipate by (1) previewing speed constraints, (2) predicting the effects of alternative throttle and brake settings that might be set momentarily, and (3) providing an optimal throttle setting.

Currently, even though speed signal information from the wayside is being transferred into the cab itself, this does not help if the operator cannot see trouble far enough ahead. The driver certainly cannot see through the windscreen for more than 1 km ahead, and much less that at night (headlights on trains are a signal to outsiders rather than providing much illumination for the driver). Experienced drivers use a mental model of speed constraints set by curves in the track, grade crossings, and population densities, which are fixed and can be learned, but, as noted earlier because of track maintenance, rock slides, snow, etc. there may be other speed constraints that are not as easily anticipated. For these reasons our proposed display previews the track for a number of kilometer, showing curves, speed limits, and other features, both fixed and changing (Figure 2).

It also shows prediction curves of the speed, determined from a computer-based dynamic model. These curves show how the speed will change as a function of the track distance ahead if the current throttle setting is maintained. There are also predicted speed curves for maximum service braking and maximum emergency braking (a different braking system). Finally, there is a continuous indication of throttle settings that will get the train to the next station on schedule (assuming the train is at or near schedule currently and the winds are known), meeting the known speed limits, and under these constraints minimizing fuel. This indication can be updated iteratively by a dynamic programming algorithm. The latter display is akin to the flight director in an aircraft. This system was tested in a dynamic human-in-the-loop train simulator with a number of trained driver subjects and was shown to improve performance significantly over driving with conventional displays.

Rail systems have their equivalent of air traffic control in the dispatch or traffic control centers, usually located in major rail junctions. Here a crew of several people set schedules, set switches, communicate with trains by telephone as necessary, and monitor rail traffic for large areas. Up to now their principal tools have been the telephone, large sheets of paper with schedule tables and/or plots of location versus time for every train for the main line sections of track, and wall-mounted diagrams or maps showing (usually with a small light bulb for each block of track) which blocks are occupied by trains. They have few if any computer-based tools to predict the cascading effects of late trains or accidents.

4   Highway vehicles

Highway vehicles — including trucks, buses, taxicabs, special vehicles such as fire-trucks, police cars, and ambulances, as well as private automobiles — have also followed the trend toward use of new sensor, communication, computer, and control technology. The European Community countries were probably the first to organize a major development project. One such project, called PROMETHEUS, began roughly a decade ago and has recently been completed. Its aims were to introduce new computer-

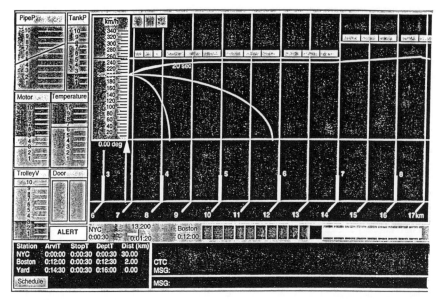

Figure 2. Askey's experimental preview/ prediction/ advisor display. At the left are conventional instruments. At the right, referenced to the speed indicator (vertical) and distance from the present position (horizontal), are (1) bars at the top ("preview"), indicating speed limits for upcoming kilometer sections; (2) upper curve ("advisor"), indicating ideal speed trajectory to satisfy speed limits, schedule constraints, and energy minimization; (3) curve almost coinciding with first part of the upper curve ("predictor," displayed in contrasting color), which indicates prediction of train speed if current throttle setting were held constant; (4) middle curve ("predictor"), indicating maximum service braking; and (5) lower curve ("predictor"), indicating maximum emergency braking. The curves are continuously recalculated based on present state and computer models or train dynamics and track configuration.

based navigation and control systems, to look ahead to common signing, speed control, and other infrastrucure developments for European countries, and to enhance the position of European automotive manufacturers in world markets. A second program, called DRIVE, with similar aims but a different organizational structure, is still ongoing. In Japan similar national projects instituted cooperation between vehicle manufacturers, suppliers, and government agencies. In about 1990 the United States Department of Transportation (USDOT), in concert with state agencies, vehicle manufacturers and suppliers, insurance companies, and universities, founded a unique advisory and educational organization initially called Intelligent Vehicle-Highway Society of America. Its name was later changed to Intelligent Transportation Society (ITS) of America after metropolitan rail interests complained, but it is still largely highway vehicle oriented. From the outset human factors considerations were seen as important to the success of ITS.

At roughly the same time, General Motors, Avis, American Automobile Association, the State of Florida and the City of Orlando undertook a demonstration project in passenger vehicle navigation. One hundred rentable Oldsmobiles were outfitted with GPS transducers, simple inertial transducers, 386 PCs, CD ROMs to hold street map information for Orlando, speech generation devices, and CRT graphical display units. The rental agency customer could indicate his or her destination to the computer, and the system would then guide the driver there by a combination of voice and graphical displays. This included, for example, advice to move to one side of the road or another in preparation for a left or right turn at a forthcoming named intersection. The GPS was accurate enough in most cases to detect whether the driver stayed on the correct road, and if not would tell the driver how to get back on the correct course. While the author has himself ridden in one of these vehicles, evaluation results of this multiyear project have not been made fully public as of this writing.

Controversy over putting such navigation technology into passenger cars mostly centers around safety issues. Does use of such a system detract significantly from the driver's attention to the road, or does it result in greater confidence and greater safety? Can older drivers accommodate their visual focus between the road outside and the internal graphics or other displays sufficiently rapidly? Is this yet another addition to the driver's workload, which in the future may include not only cellular telephone but also fax machines, e-mail, etc.? Are map displays too distracting (currently some states in the United States claim yes and legally prohibit such front-seat map displays, classifying them the same as television sets). If maps are used, should the map direction always correspond to the vehicle direction (the vehicle then becoming a fixed symbol) or should the map remain fixed with north (or south) up, and the vehicle becoming a moving symbol? The answers to these questions may be settled to some extent by research, but the final arbiter will be the marketplace.

The major safety objective is to avoid collisions. New technology provides a different approach to collision avoidance. Instead of accepting the "first crash" (of the vehicle with environmental obstacles) and focusing only on technological means to ameliorate the "second crash" (of the driver and passengers with the interior of the vehicle), there is now the opportunity for technology to prevent the first crash. In this regard, perhaps the second major new development for "intelligent vehicles" is "intelligent cruise control." Current cruise control systems do not know when one vehicle is about to collide with another. Intelligent cruise control systems make use of microwave and optical sensors to detect the presence of a vehicle in front, and the resulting signal can be used either to warn the driver or to brake automatically. Interestingly, current demonstration projects have mostly stayed away from actually applying the brakes, choosing instead to decelerate and downshift. The reason for this seemingly irrational choice is a concern that once claims are made that an automatic system will brake, the driver may become less inclined to apply the brakes himself, and the manufacturer will be wide open to tort litigation.

Human-in-the-loop simulation is a promising method to test driver and vehicle response under circumstances deemed too unsafe to test on the highway or on test tracks with actual vehicles. Human-in-the-loop simulation technology has been used for many years in aviation, nuclear power, and the military for research and engineering development, for verification and quality assurance for new systems, and for training and marketing. Crude automobile and truck driving simulators have also been used, but the costs of such devices have precluded very extensive use, except for a very few such systems (e.g., a moving-base simulator at Daimler Benz in Berlin).

In our laboratory at MIT we have used a driver simulator to test braking reactions and following distances of drivers under various conditions, and have run comparable car-following measurements on major highways at rush hour. What we have found (Chen,1996) is that following distance correlates surprisingly poorly with the speed of the traffic stream, to the point that most drivers could not stop in time to avoid collision were the vehicle ahead to suddenly apply its brakes. In association with this work, we have evolved a quantitative model with which, assuming the initial speed of the traffic stream, the stopping distance of the lead vehicle, the braking characteristics of the following vehicle, and the pavement conditions, random (or biased), draws are made from density functions (obtained experimentally as described earlier) for the following-car driver's reaction time and following distance). This is then used to calculate the probability of collision (Fig. 3).

Currently the most ambitious U.S. project is the Automated Highway System (AHS). This project seeks to demonstrate how vehicles may be driven by computers on special lanes designated only for properly equipped vehicles. A variety of automatic steering and longitudinal control techniques have been demonstrated, including platooning of vehicles at high speeds with only 1 meter headway, which is controlled

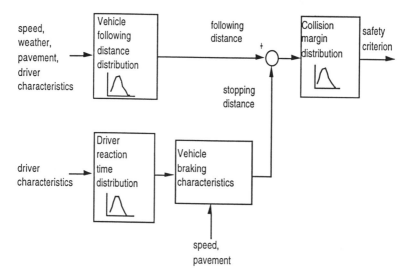

Figure 3.   Monte-Carlo procedure for inputting test parameters and sampling from experimental distributions of vehicle following, driver reaction time, and vehicle braking characteristics to obtain the probability of collision.

electronically. Allegedly, according to the University of California (Berkeley) PATH Program researchers (Hedrick, personal communication), it can be shown that it is actually safer to have high speed vehicles in close proximity to one another than spaced at longer distances, since in this manner large velocity differences can never build up and hence large impact collisions cannot occur.

The intelligent highway system also has its counterpart in air traffic control, called ab advanced traffic management system. This is not a unique concept, because in fact many new highway and tunnel projects have built accompanying traffic management centers. The newest such system is associated with the largest current construction project in the United States, Boston's eight billion dollar Central Artery /Tunnel. In this project's control room, several operators observe traffic by means of roughly 400 video cameras, and a comparable number of magnetic and other sensors (not only of traffic but also of heat, smoke, carbon monoxide, and toxic chemicals of various kinds). They use a sophisticated computer system that infers where and how serious any abnormality is and make a recommendation for action (e.g., send in a tow truck from a particular location, or certain firefighting equipment, police, etc.). They can change variable message signs, block traffic lanes, initiate visual and auditory alarms, and activate fire extinguishers.

Our laboratory is currently participating in helping to design this "incident response system." In this regard one supervisory control problem that has not received the research attention it is due (or even as much as teaching and monitoring) is intervention. In this system the supervisory operator is expected to decide within seconds whether to accept the computer's advice (in which case the communication to the proper responding agencies is automatically commanded), or to reject the advice and generate his or her own commands (in effect to intervene in an otherwise automatic chain of events). An operator simulation of this system is currently being run to evaluate what one can expect of a trained operator under different conditions.

5   Who is in charge?   The problem of authority in new transport systems

The problem of authority between human and computer is one of the most difficult in new transportation systems (Boehm-Davis et al., 1983). Popular mythology is that the human operator is (or should be) in charge at all times. In the past when a human turned control over to an automatic system, it was mostly with the expectation that circumstances were not critical, and that he or she could do something else for a while (as in the case of setting one's alarm clock and going to sleep). Even when continuing to monitor the automation (as would occur in supervisory control) people are seldom inclined to "pull the plug" unless they receive signals clearly indicating that such action must be taken, and unless circumstances make it convenient for them to do so. Examples of some current questions being debated in the aviation field follow:

- Should there be certain states, or a certain envelope of conditions, for which the automation will simply seize control from the operator? In the Airbus, for example, it is impossible to exceed critical boundaries of speed and attitude that will bring the aircraft into stall or other unsafe flight regimes. In the MD-11, the pilot can approach the boundaries of the safe flight envelope only by exerting much more than normal force on the control stick.

- Should the computer automatically force deviation from a programmed control strategy if critical unanticipated circumstances arise? The MD-11 will deviate from its programmed plan if it detects windshear.

- If the operator programs certain maneuvers ahead of time, should the aircraft automatically execute these at the designated time or location, or should the operator be called upon to provide further concurrence or approval? The A320 will not initiate a programmed descent unless it is reconfirmed by the pilot at the required time.

- In the case of a subsystem abnormality, should the affected subsystem automatically be reconfigured, with after-the-fact display of what has failed and what has been done about it? Or should the automation wait to reconfigure until after the pilot has learned about the abnormality, perhaps been given some advice on the options, and had a chance to take initiative? The MD-11 goes a long way in automatic fault remediation.

One can ask further, if all the information required for full automation is available, why not automate fully? In the railway sector this question has been asked, and in some special cases the decision was to remove the driver entirely (e.g., trains on rails or rubber-tired guided vehicles in some new airports, and trains on dedicated tracks such as that between Orly Airport and Paris). Many engineers assert that automatic control is essential for modern high-speed trains, and there is simply nothing to debate. However, history has shown with regard to automation that we are not always as smart as we may think. Although automation is now widely accepted in aviation by pilots, airlines, and regulators, there remain accidents that have been blamed on the automation itself. In the case of the Airbus there is now an active discussion about whether automation has gone too far. It is salient to note that Charles Stark Draper, the "father" of inertial guidance and the Apollo navigation system that took the astronauts to the moon, was observed by the writer to proclaim at the outset of the Apollo Program that the astronauts were to be passive passengers and that all the essential control activities were to be performed by automation. It turned out that he was wrong. Many routine sensing, pattern recognition, and control functions had to be performed by the astronauts, and in several instances they "saved" the mission.

Assuming sufficiently accurate models of vehicles and their interaction with their respective media, and sufficiently accurate state measurements, optimal automatic control is quite feasible. Alternatively, as noted earlier, an optimization calculation can be made continuously and used, not for automatic control, but to display to human operators a best profile for human control action. Proponents of this view maintain that if the humans keep precisely to such a profile, they can in any case do a better job at control than trying, in effect, to perform complex calculations in their heads.   In addition the humans are there if some totally unpredictable events were to occur. They can allay public fear of riding in an unmanned vehicle, and would inhibit litigation on the basis that if an accident occurred both automation and human were present and doing their best.

Automation is sure to improve and to relieve the human of more tasks. However, we must guard against unsubstantiated claims, as are now being made for how "intelligent automobiles" will increase safety and reduce congestion. Empirical system demonstrations speak louder than words.

6   Conclusions

Demands on modern transportation systems include higher speeds, tighter headways relative to other vehicles, increased traffic regulation, and increased emphasis on safety. In an effort to meet these demands, there is a trend toward more use of computers, artificial sensors, and communication devices, and with respect to the human driver, a move toward automation and human supervisory control.   Current research and development in the three areas of aviation, railroad trains, and highway vehicles illustrate common themes regarding human control.   The proper extent of automation is undergoing active debate.   The answer must lie with empirical demonstrations of performance and reliability.

References

Askey, S. (1995). Design and evaluation of decision aids for control of high speed trains: Experiments and model. PhD thesis, Massachusetts Institute of Technology , Cambridge, MA.

Billings, C.S. (1991). *Human-centered aircraft automation: A concept and guidelines.* NASA Technical Memorandum 103885. Moffet Field, CA: NASA Ames Research Center.

Boehm-Davis, D., Rurry, R., Wiener, E., and Harrison, R. (1983). Human factors of flight deck automation: Report on a NASA-industry workshop. *Ergonomics* 26: 953-961.

Chen, S. (1996). *Estimation of Car-Following Safety: Application to the Design of Intelligent Cruise Control*, PhD thesis, Massachusetts Institute of Technology, Cambridge, MA.

Hedrick. K., personal communication.

Patrick, N.J.M. (1996). PhD thesis, Massachusetts Institute of Technology, Cambridge, MA.

Sarter, N. and Woods, D.D. (1994). Decomposing automation: Autonomy, authority, observability and perceived animacy. In Mouloua, M. and Parasuraman, R. (Eds.) *Human Performance in Automated Systems: Recent Research and Trends.* Hillsdale, NJ: Erlbaum, pp 22-27.

Wiener, E.L. (1988). Cockpit automation. In Wiener, E.L. and Nagel, D. (Eds.) *Human Factors in Aviation.* San Diego, CA: Academic Press, pp. 433-461.

Chapter 14

Human Factors in the Gossamer Human-Powered Aircraft Flights

Henry R. Jex

1   Introduction

The First Kremer Prize for human-powered flight went unchallenged for 20 years until Paul MacCready's Gossamer Condor flew the 1-mile-plus figure-8 course in 1977. Then the second Kremer Prize, for crossing the English Channel via a human-powered aircraft, was awarded a mere 2 years later to the Gossamer Albatross. These were team efforts involving, in a microcosm of volunteers, many of the human factors problems faced in any large project, plus some unique ones. This presentation gives us an insider's view of the human factors challenges met in the design, development, test, and operation of these pioneering human-powered flights.

Henk Stassen is a master of the art of making simple, handed-lettered, multicolored presentations. He often developed a student's first presentation transparencies, introducing his art by example, to serve the student for a lifetime. This presentation was originally inspired following such a Stassen lecture. Although it cannot match Henk's deceptively simple style, this lecture is offered as a holographic ode to his didactic expertise. (Henk's copy was in four colors, and this black copy suffers from lack of color).

2   Background

Numerous attempts at building an entirely human powered aircraft (HPA) during the 1930s through 1950s in Europe were only partially successful, in that they could sometimes travel about 1 km in straight flight but could not be turned around to return to their launch site. In 1960 Henry Kremer, a British industrialist, proposed a £50,000 ($86,000) prize for the first HPA to fly a figure-of-eight course around two turn pylons

1/2 mile (804.6 m) apart, and pass over a 10 ft (3 m) height hurdle at the start and end of the course, between the pylons markers. Various attempts, mostly in the United Kingdom, failed until Dr. Paul B. MacCready, an experienced indoor-model-aircraft builder and hang glider pilot, realized in 1976 that the correct approach was to use hang-glider technology to an extreme: a long wing of high aspect ratio for low span loading, an ultralight structure with a wire braced single spar, and a lifting canard (elevator) to help carry weight and to tilt for yawing moment controls. This ultralight concept was successful, and the resulting Gossamer Condor won the First Kremer Prize at California in August 1977, almost 17 years after the announcement! (For a superbly documented background on human-powered flight and the Gossamer HPA, see Grosser, 1981.)

Henry Kremer then raised the challenge drastically: to win a £100,000 (then $210,000) a HPA would have to cross the English Channel from England to France, a distance of more than 23 miles (38 km). With the technology for a successor already under way, a skilled and enlarged Gossamer team took on the task of traveling 20 times the figure-of- eight course, across the dangerous English Channel.

This chapter describes the ergonomics and human factors engineering involved in accomplishing that task, crossing the Channel in just under 2 years, on June 12, 1979.

SOME COMPARISONS

CONDOR	vs	ALBATROSS
Aug-77	Date	June-79
Figure-8	Course	Cross channel
1.3	Distance-mi	23 (35 air mi)
7.5	Duration- min	167
71 + 141 = 212 A/c Pilot Total	Weights - lb	64 + 141 + 10 = 215 A/c Pilot Eqpt. Total
96	Span - ft	94
.33	Power - hp	.25
10+	Gestation -mo	18
<100,000	Cost -$	>300,000
86,000	Prize -$	210,000

2.1 Some comparisons between the Condor and Albatross flights

While the distance and duration of the cross-channel flight in 1979 was represented a 20-fold increase over the 1977 flights, the comparisons to follow show a remarkably similar size and weight. However, significant technical improvements were made, such: double surfaced airfoils, carbon-fiber/epoxy tube structure, minimum induced-loss propeller, tapered wing planforms, and body streamlining. Cumulatively these led to 25% reduction in required power, which was the key to better than 2-plus hour endurance.

Further, the need to fly over rough water in the presence of moderate surface wind turbulence required continuous control corrections while pedaling, so numerous control-system refinements were demanded, as discussed later. These developments, plus more human factors related issues, led to a much longer "gestation" period and a quadrupling of costs.

HUMAN FACTORS WERE INVOLVED IN:

• **Project Operations**: Management
                             Design, Fabrication
                             Logistics
                             Flights

• **Pilot - Powerplant**: Selection
                             Training
                             Conditioning

• **Aircraft Design**: Configuration
                             Power system
                             Stability, Control
                             Handling

3   Human factors

The Gossamer Albatross mission to cross the English Channel posed many unique issues with challenging human factors implications. Long-distance project operations, in England and over open water, were new to the team. Pilot (power-plant and control) selection, initial training, and conditioning were aggravated by the different environment of food and weather. The new aircraft design had to have displays and controls to allow continuous corrections while pedaling at maximum effort over a sea surface with confusing visual cues from its semi-transparent surface, erratic wave swells, and wave-surface induced turbulence. The latter turned out to be much worse than expected from flights over runways and often put maximum power and control workload on the pilot, simultaneously.

4   Albatross development personnel interactions

Due to the innovative aspects of the Albatross mission, the configuration, and human factors, a very tight team effort was essential. It was not unlike the famous Lockheed "Skunk-Works" which developed the Mach-3 Blackbird, but on a much smaller scale. The next figure shows the key interactions of interest and the number of personnel involved. Near the flight operations, the AeroVironment Gossamer Team expanded to nearly 30 persons, and for the cross-channel flight there were over 200 news personnel on boats crowding the flimsy aircraft's air space in an often dangerous manner (Groser, 1981).

    Note the strong ties shown between the pilot-training and aircraft-design and flight-test activities. More than for any modern aircraft, the various Gossamer pilots had a major impact on every detail of the propulsion system and controls. Most of the pilots were champion-level bicyclists, and many were hang glider pilots as well.

ALBATROSS OPERATIONS

5   Available pedal power versus duration

The generic trend for available power from human pedaling against a pure torque load at a near optimum pedaling rate is shown in the figure below. Noted there are the important distinctions between the "normal willing" efforts used in lab tests, "highly motivated" efforts seen in competitions, and the "physiological limits" seldom used, except in emergencies. Due to unexpected head-wind conditions in the channel, this emergency margin had to be used, to reach shore, with only 1 minute remaining of Bryan Allen's 170 minute physiological limit!

The actual test data below show that only .25 hp could be achieved for the estimated 2.0 hr Albatross flight, versus the .33 hp achieved for the 7 minute Condor flight. Also shown are the tested 2 hour maximum efforts point of Brian Allen (cross) along with that needed for the planned and actual flights.

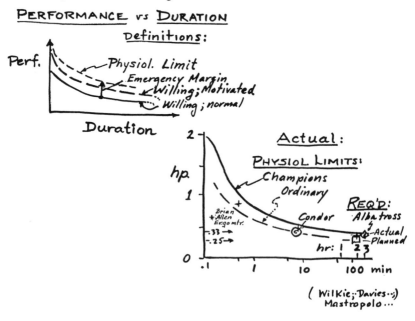

6   Optimizing the power system

Further improvements resulted from optimizing the gearing and required propeller speed. In the figure below the measured short term (15 s.) maximum power available from pedaling at various rpm is shown. The dots show the power, while the triangles show the pedaler's efficiency in terms of shaft power delivered per caloric power available (Davies, 1979). It is apparent that for a maximum endurance channel crossing one must pedal much slower than for maximum power generations. The lower aircraft drag and higher efficiency of the Albatross slender "toothpick" propeller enabled the pilot to operate at an efficient 70 rpm instead of the 110 rpm required for the earlier Condor. Additional endurance and better control resulted from a more bicycle-like upright posture.

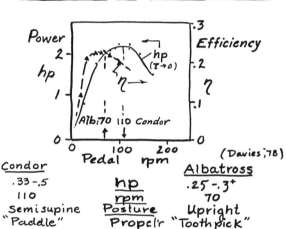

7   Training factors

Training the pilot was greatly complicated by the fact that he had two demanding tasks to perform simultaneously: to exert near maximum propulsive physical workload for a constant airspeed, and to continuously monitor and control a slightly unstable 96-foot-span aircraft about its pitch, yaw, and roll axes, all in the presence of surface wind gusts. Consider, first, the physical training required. The seminal research of Calvert et al. (1976) on the dynamics of training for swimming and running competitions provided great insight, as diagrammed in the next figure. The roles of genetic makeup (e.g., Brian Allen consumed just the amount of water he needed), negative feedback from fatigue (due to excess practice), and positive feedback from achieved performance in tests all serve to modify the strength and cardiovascular training required to convert preflight "fitness" to actual "condition" during the event. Note that the different "time constants" estimated by Calvert and others interact in a complex way, such that there was an optimum wait of a few days for fatigue recovery before the final flight.

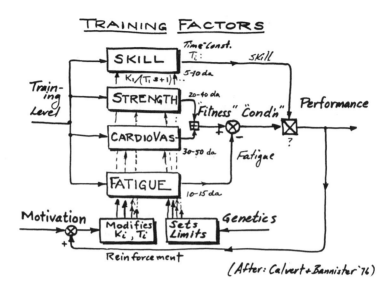

TRAINING FACTORS

(After: Calvert + Bannister '76)

8   Aerodynamic design implications

Here, as in the first successful engine powered aircraft, the 1903 Wright Flyer of 66 years earlier, the primary objective was to minimize the power required. This was obtained by using the three methods described later, with the following implications: *light weight* via a hang glider and indoor model type of structure having a monospar wing and external wire bracing; *low drag* via an efficient airfoil, a long span, a high aspect ratio wing, and minimal vertical surface; *low airspeed* via a high lift coefficient airfoil, lifting canard.

The optimum configuration developed by MacCready was uncannily like the 1903 Wright Flyer. Both were canard aircraft with a pusher propeller and wing warping for turn control. Like the '03 Flyer, the wing load wires also provided good wing torsional

DESIGN   FACTORS

<u>Primary objective</u>:   <s>Minimize power</s> req'd.
   <u>via</u>: Light weight   (wire braced ; composites)
         Low drag   (≈ all wing ; long span; fly low)
         Low speed   (high c_L airfoil ; large wing)

<u>Optimum Configuration</u>:   Canard w/ pusher prop
         Front wing supports 8% wt. ; provides trim + control
         80% of surface is wing (more efficient)
         Wire bracing carries main loads + stiffness

<u>Some consequences</u>:
         Brought within normal athlete capabilities
         Marginal (negative) "weathervane" stability
         Very precise control of speed and altitude req'd.
         Air "apparent-mass" >> inertial mass (heave, roll, pitch)
         Cannot fly in winds > 5-7 kts.

stiffness, a feature lacking in many other HPAs. Some of the consequences (to be discussed) included: There was marginal weather-vane directional stability, precise control of airspeed and pitch trim was needed, and the apparent mass reactions of air displaced by aircraft/air relative motions in heave, roll, and pitch were not negligible, as in normal aircraft. This meant that the Gossamer aircraft could not be flown in winds over about 7 knots.

AERODYNAMIC CONTROLS

- Lifting canard for pitch trim + control
- Pedal-power for altitude control
- Tilting canard "rudder" for yaw + roll control
- Wing warp (against turn) for trim   ($V_o/V_i = 1.2$)
  - Higher drag of washed-in "wing creates turning moment

FOR REGULATION AGAINST DISTURBANCES:
Use canard controls for holding attitude + yaw rate + roll any
Use pedaling to control altitude

FOR MANEUVERING THE FLIGHT PATH:
Step in wing warp, coordinate via canard tilt.
Coordinate pedaling inputs with canard retrimming

PILOT ALWAYS IN THE CONTROL LOOP:
No inherent stability
High mental-workload

9   Unusual aerodynamic controls

The controls of the Gossamer aircraft are deceptively simple, and they are quite unlike the similar-appearing 1903 Wright Flyer.

- The lifting canard provided pitch control moments and trim (the '03 Flyer actually had a downloaded canard!).

- Altitude gain was strictly from pedal power, because the Gossamers were flying near their minimum power speed.

- Tilting the canard's lift in roll acted as an effective forward rudder to yaw the aircraft.

- The three-position wing warp control creates a wing twist opposite to the direction of turn, whereby the extra induced drag of the "washed-in" wing causes a turning moment, while the twist balances the rolling moment due to the outer wing going 20% faster than the inner wing in the standard turn.

The rules for maneuvering the Gossamers are summarized later. Even though these were gradually refined to simplify the steering and pitching guidance, the pilot had to be in the control loops continuously. The high mental workload of this situation added to the fatigue of long flights.

10   Cockpit design factors

Seldom has the human factors engineer had to combine the pilot and the powerplant inside the cockpit. The demand for efficient pedaling posture resulted in a modified hi-tech bicycle seat and pedals, and custom-designed hand grip canard pitch and tilt controls. The other hand worked the three-position wing warp device and controlled the microphone. Pilot vision was impaired by the main structural downpost in front of his eyes.

As in any heat engine, much of the pilots effort went into waste heat, here cooled by a sophisticated transpiration system — sweating! The sealed cockpit often fogged up and a special set of cooling ducts was devised to cool the pilot and clear the fog. For the Channel flight a reflective foil sunshield was essential to shade the pilot at the expense of forward-left visibility.

Only 2.0 liters of water were carried, enough for the estimated 2.0 hours of flight. Unfortunately the headwinds slowed the crossing to 2.75 hours, and Brian started to collapse as muscle cramps and heat prostration symptoms developed. With his radio and sonic altimeter dead, he started to abort the flight at 1.5 hours. Miraculous, as he flew higher to receive a towline from the Zodiac boat below, the air became calmer so he pushed on towards his physical limit. This unexpected meteorological effect really saved the flight.

Safety was a key priority with the American sponsor, the Dupont Corporation. Due to the low and slow flight over water, crash injuries would be unlikely, but drowning while trapped in the cockpit was a real concern (Brian could not swim). So, using a crash helmet, inflatable life preserver, and careful rehearsals of ditching procedures, a safe rescue was assured. Brian's concern over this situation merely helped his motivation to not give up.

COCKPIT FACTORS

Minimal Structure:
- Downpost + ribs + mylar
- Bike seat + pedals

Pilot Temperature Controls:
- Hot + steamy ; sealed cockpit
- Air vents ; breath exhaust
- Sun shield
- Water supply: 2 liters (2 hrs)

Safety:
- Fly low + slow
- Energy absorbing downpost
- Crash helmet ; life preserver

11   Displays, controls and handling qualities

The pilot displays were rudimentary but essential. Just below the horizon to the pilot's left side were presented: true airspeed from a propeller-anemometer on the boom, propeller rpm from a shaft tachometer, and altitude above the surface from a sonic altimeter developed especially for the Gossamers by Polaroid Corporation. The pilot used the near horizon for pitch and roll attitude reference, while the aircraft angles of attack and sideslip were made visible by plastic streamers from the canard trailing edge. During the flight, the altimeter (crucial for over-water surface detection) was lost as the battery faded at 2 hours, so voice transmissions from the Zodiac crews were used: "clearance 5 feet"... "3 feet"... "1 foot"... "climb"... "CLIMB"! The simple control law was as follows: Pitch the canard to keep it on the horizon, and tilt it to keep the vehicle level in roll (Jex and Mitchell 1982).

The handling qualities of this ponderous but gossamer-like aircraft were well damped despite very slight longitudinal and directional instability. Gentle "standard" turns were made by toggling the warp control knob to the left or right, and using the canard tilt for turn coordination, as explained earlier. Because of its docile characteristics, the Albatross was easy to fly in calm air, but coping with the wavy turbulence over the channel became a fatiguing mental workload.

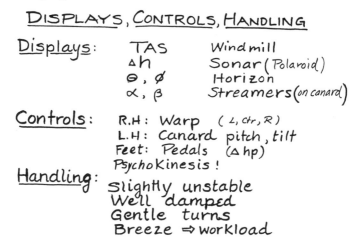

DISPLAYS, CONTROLS, HANDLING

Displays:
TAS	Windmill
Δh	Sonar (Polaroid)
Θ, ∅	Horizon
α, β	Streamers (on canard)

Controls:
- R.H: Warp (L, ctr, R)
- L.H: Canard pitch, tilt
- Feet: Pedals (Δ hp)
- PsychoKinesis !

Handling:
- Slightly unstable
- Well damped
- Gentle turns
- Breeze ⇒ workload

12 Conclusions

The final version of the Gossamer Albatross was a much more sophisticated design than the Gossamer Condor, with a carbon-filament and Kevlar structure, a 1 mil Mylar skin, 400,000 psi steel wire bracing, and state-of-art high L/D airfoils and propeller. With the carefully trained and Olympic-quality pilot/powerplant's improved ergonomics, duration was increased from 7 minutes to 170 minutes. The design was optimized for minimum weight and a high lift coefficient for low-speed flight, so it resembled the similarly designed Wright Flyer of 1903 in many respects. Even the employment of a somewhat unstable vehicle that was relatively easy to control matched the Wright's bicycle-based paradigm. Further improvements in the handling qualities of this design concept would allow its use for pylon speed events, longer flights, and operations in more typical calm weather.

CONCLUSIONS:

- Minimum power paramount; human factors secondary.

- Opt. design for Kremer courses: canard "C.C.V." (like Wright's!)

- Trained bicyclist for T > 1 minute
  > 160 minutes

- Improvements req'd for:
  sport flying
  typical weather
  longer flights

13   Epilogue

In the decade following the first Gossamer flights, a number of new speed and distance records were set. The table below summarizes some of those known to the author. The short-term speed of smaller vehicles has been increased to over 25 miles per hour (40 km/h) by careful drag reduction, under the impetus of further Royal Aero Society prizes. The Gossamer aircraft technology and team also designed the solar-powered aircraft Solar Challenger. In 1981 it took off unaided near Paris, France and flew Louis Bleriot's famous route across the channel to a landing field south of London. So far, this is the only case we know of this feat's accomplishment. Therein lies a new challenge! Later, an M.I.T. team under the direction of John Langford, Mark Drela, and Guppy Youngren, built a Gossamer-like craft with an an aft-tail, the Daedalus, which flew 74 miles (119 km) from Crete to Santorini in April 1988.

EPILOGUE:
CURRENT STATUS OF KREMER EVENTS

Team:	Date:	Plane:	Pilot:	Score:
"FIGURE-8" - USA:		(Foci .5 mi apart)		
MacCready	August '77	"Gossamer Condor"	B. Allen	7.5 min £48K
- EUROPE				
G. Rochelt (Ger)	June '84	"Muscular I"	H. Rochelt	4.1? min £20K

CROSS ENGLISH CHANNEL: (Human power only)

MacCready	June '79	"Gossamer Albatross"	B. Allen	$2^h 47^m$ £100K
(Solar plum) MacCready	'81	Solar Challenger	S. Ptacnik	163 mi Paris-London

TRIANGULAR SPEED COURSE: (1.5 Km, 10 min pwr store, +opp. dir); +5%

M.I.T.	Apr '84	"Monarch"	F. Scarabino	19.1 mph £20K
MacCready	July '84	"Bionic Bat"	P. MacCready	20.5 mph 5K
Rochelt	Aug '84	"Muscular-II"	H. Rochelt	22.2 mph 5K
MacCready	Dec '84	"Gossamer Swift"	B. Allen	23.5 mph 5K
MacCready	Apr '85	"Bionic Bat"	P. MacCready	25+ mph(?) 5K

References

Calvert, T.W., E.W. Banister, M.V. Savage, T.Bach (1976). A systems model of the effects of training on physical performance, *IEEE Tranactions on Systems, Man, & Cybernetics*, SMC-6: 94-102.

Davies, C.T.M. (1979). The selection, fuel supply and tuning of the engine for man powered flight. In *Third Man-Powered Symposium Proceedings*, The Royal Aeronautic Society, London.

Jex, H. R. and D.G. Mitchell, (1982). *Stability and Control of the Gossamer Human-Powered Aircraft by Analysis and Flight Test*, NASA CR-3627.

Grosser, M. (1981). *Gossamer Odyssey, The Triumph of Human-Powered Flight,* Boston, Houghton Mifflin Company.

Chapter 15

Navigation Simulation

Robert Papenhuijzen

Computer programs that simulate the behavior of the complete system of a ship under human control can make powerful tools for investigations in the field of navigation and fairway design. Hence, two different models have been developed to simulate the behavior of the helmsman. Further, two alternatives were developed for a model of the behavior of the navigator, the human in charge of the navigation process. On the basis of a large number of experiments, an in-depth evaluation of the various models has been performed. It appears that under most conditions a stand-alone configuration of a navigator model, i.e., not complemented with a separate helmsman model, generally results in realistic simulation of the operator output and the overall performance of the system under human control. Only if realistic simulation of the helmsman's behavior under less critical conditions is a prerequisite is the combination of a navigator model with a separate helmsman model considered necessary.

1   Introduction

Within the framework of research into human control and supervision of slowly responding systems, a comprehensive study has been devoted to the simulation of the behavior of a vessel under human control. The reason for this is that fairway dimensioning and safety issues require increasingly sophisticated research tools. Larger vessels that are more difficult to handle are being built, vessel traffic is intensifying, and there is an ever growing number of ships that transport dangerous cargo. Furthermore, there is an increasing need for safe, cost-effective, and environmentally acceptable alternatives to road transport.

Since the early 1970s, fairway dimensioning and safety research has generally been performed using a ship bridge simulator, for accomplishing realistic sailed tracks, given a combination of a professionally steered vessel, and any generally confined, hypothetical or existing waterway. But, in fact, many highly relevant studies are carried out only partially, or are not carried out at all, due to the high cost of operating a ship bridge simulator. Therefore, it was decided to attempt to develop simulation programs that cover all elements of the navigation process, including the human operator components.

In general, it is possible to discriminate between two major human operator functions: those of the *navigator* and *helmsman*. Being the supervisor of the complete navigation process, the navigator sets the engine speed. It is the helmsman who actuates the rudder. Depending on the situation the ship is in, the helmsman is supplied by the navigator with either heading or rudder commands for this purpose. If a major course change is imminent, or if the waterway that is navigated is relatively confined, the navigator gives rudder commands and the helmsman acts merely as a command control element. Otherwise, the helmsman is given heading commands.

On a seagoing ship, the role of the navigator is played by a mate or, particularly in coastal waters, by the captain or a pilot. Traditionally, rudder control was performed by a helmsman. Nowadays, however, an automatic course-keeping system (autopilot) is more generally employed. In inland waterway navigation and sometimes on board a small seagoing ship, the functions of navigator and helmsman are combined into one position.

During the early 1970s, a research project was conducted that was intended to devise a model of the behavior of the *helmsman* (Veldhuyzen, 1976). As a result, two models were developed:

- A linear model including a gain, lag, and lead term, describing the helmsman's behavior in controlling a ship sailing in calm water
- A nonlinear model, describing the helmsman's behavior in controlling a ship sailing in either disturbed or undisturbed water, with the model being based on the internal model concept

Then between 1985 and 1993, a *navigator* model was developed. As a result, a model of the state estimation behavior of the navigator was proposed, based on Kalman filtering. A simple prototype was built to demonstrate its feasibility (Papenhuijzen, 1988). Next, due attention was paid to modeling planning and control behavior. For this two basically different approaches were chosen, one rooted in linear optimal control theory (Papenhuijzen and Stassen, 1987, 1989) and the other one based on fuzzy set theory (Papenhuijzen and Stassen, 1992). Both approaches have yielded an operational navigator model that was coupled with suitable models of an environment and a ship.

In this chapter, an overview is given of the background and developing path of the project. Further, a brief description is given of the various helmsman and navigator models. Relatively more emphasis is put on the presentation of experimental results, on a discussion of these results, and on a more general evaluation of the project outcomes.

2 Helmsman models

It was decided to focus on the course-keeping task of the helmsman, because modeling rudder command control would be insignificant because of the large time constants of the system to be controlled compared with the helmsman's response. Before actually modeling the helmsman's behavior, a number of ship bridge simulator experiments were executed in order to establish the major characteristics of the course-keeping task. Some notable outcomes of the experiments are summarized in Section 2.1. Next, two principally different helmsman models were developed: an uncomplicated linear model, as described in Section 2.2, and a more sophisticated nonlinear alternative, which is described in Section 2.3.

2.1   ORIGINAL MODEL

Preliminary ship bridge simulator experiments were performed on the simulator of the TNO Institute of Mechanical Engineering (TNO-IWECO) by trainees of the Amsterdam School of Navigation. For the experiments, models of large seagoing vessels with either stable or unstable stationary course-keeping characteristics were used.

Two major conclusions were derived from the experimental results:

- Changes in the steering wheel position were generally applied as discrete steps.

- The implementation of a heading command involves four basically different phases.

During the first phase the helmsman generates an output in order to start the ship rotating. During the second phase, the rudder is kept amidship. During the third phase, the helmsman stops the rotating motion of the ship. When the desired heading is achieved with only a small rate of turn remaining, the fourth phase starts (rudder angle approximately zero). If the rate of turn is not small enough, there is an overshoot, and the four phases are repeated again until the desired state is finally reached. This procedure can be shown clearly by means of the phase plane: the rate of turn of the ship $\dot{\psi}(t)$ plotted against the heading error $\psi_e(t) = \psi(t) - \psi_d(t)$. An example of such a phase-plane plot is shown in Figure 1.

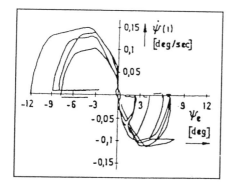

Figure 1. Phase plane plot

2.2 LINEAR MODEL

Even though it is not possible for a linear model to feature a discrete rudder setting character, the development of such a model was attempted. It was considered attractively simple, since it was expected that a linear model might provide an option sufficient for some categories of basic research. Various models were considered, and the three-parameter model of Stuurman (1969), which includes a gain, a lag, and a lead term, was adopted.

2.3 NONLINEAR MODEL

A first version of the nonlinear model was built, basically consisting of the following two components:

- An *internal model* of the ship dynamics, which was used to make predictions of future headings and rates of turn of the ship
- A *decision making element*, to interpret the predictions and to base the subsequent control actions upon the predictions

For the internal model, a suitable second-order differential equation was chosen. The decision making element was directly based on the phase-plane concept (Fig. 1). In a phase plane, four phases can be indicated, with the boundaries between the first phase regions and the second phase regions representing the objective being pursued during the first phase. Approximation of the region boundaries using straight lines results in the phase plane indicated in Figure 2. The rudder angle is kept at zero during the second and fourth phases. During the other two phases, a rudder angle is selected to achieve the corresponding objectives after a given time interval, while also considering that the rudder angle to achieve an objective should not be too large.

When a ship is sailing in disturbed water, the heading signal shows a considerable high-frequency component, in addition to the intended low-frequency components, due to waves acting upon the hull. The helmsman is trained not to try to compensate for this high-frequency phenomenon, as otherwise fuel consumption would be excessive and instabilities might occur. As a consequence, models for the helmsman to cope with disturbed water as well as calm water, should include an estimation algorithm to determine the low-pass filtered heading and heading rate (yaw). Because, in practice, the bands of the high-frequency phenomenon and the intended heading signal component almost overlap, such an extension to the *linear* model was considered impossible. A heading and yaw estimation algorithm was designed for the *nonlinear* model based on top

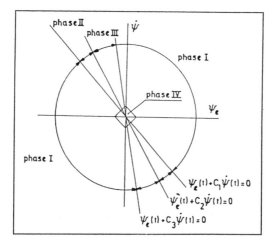

Figure 2.   Phase plane plot resulting from straiht-line approximation

detection and averaging. In this model, estimated values of $\hat{\psi}_s(t_t)$ and $\hat{\dot{\psi}}_s(t_t)$ for the low-pass filtered heading and yaw are determined whenever a top has been detected. Up-to-date information on the heading $\psi_s(t)$ and the yaw $\dot{\psi}_s(t)$ during the interval between two tops is available by resetting a copy of the internal model to the calculated values, after which that model is integrated for every subsequent time step until a new top has been passed.

Finally, a block diagram of the complete nonlinear model is shown in Figure 3. It may be clear that the variable $\delta_d'(t)$ in this figure represents any value of the rudder setting, which allows a prediction of the heading and yaw to be made based on the time interval $(t, t+\tau)$. The prediction process may be meant to evaluate a control action already in progress, but it may also serve to probe a given value of the rudder setting in order to adequately achieve a new desired state of the vessel.

3   Validation of the helmsman models

After completing an extensive parameter estimation program, the performance of both helmsman models, given calm water conditions, was compared with the outcomes of preliminary experiments on the ship bridge simulator. This is described in Section 3.1. Next, the extended model was tested under wave disturbance conditions, discussed in Section 3.2. In addition, the possible contribution of different types of navigation instruments was investigated, as well as the suitability of the extended nonlinear model to describe such effects.

3.1 SHIP MANEUVERING IN CALM WATER

The linear model and the non-extended nonlinear model were used to re-run the experimental program described in the Section 2.1. An example of the results for such an experiment is given in Figures 4 and 5. It may be clear that both models manage to yield a fairly realistic representation of the actual heading signal. Only the nonlinear model, however, is capable of producing a realistic description of the rudder setting signal as well.

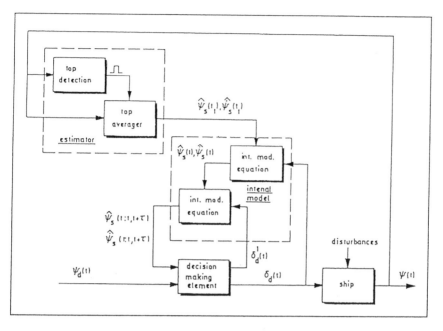

Figure 3. Linear model

3.2 SHIP MANEUVERING IN WAVES

The opportunity was seized to investigate the effect on the helmsman's steering performance on supplying either a rate of turn indicator or a predictive display showing a prediction of the heading of the vessel compared with the desired heading. A concise experimental program was carried out on the ship bridge simulator and was duplicated with the extended nonlinear model, augmented with models of the additional navigation instruments. Considerable wave influences were exerted for all experiments.

The major conclusions of the experiments were as follows:

- The extended nonlinear model is capable of reproducing the outcomes of the corresponding ship bridge simulator experiments.

- A rate of turn indicator improves the performance of the helmsman.

- A predictive display leads to better performance with respect to the heading error and rudder amplitude. However, the number of rudder calls may increase, without increasing the mental workload.

4 Navigator models

As was the case for the helmsman project, two basically different models have emerged from the navigator project, differing in the way in which the navigator's planning and

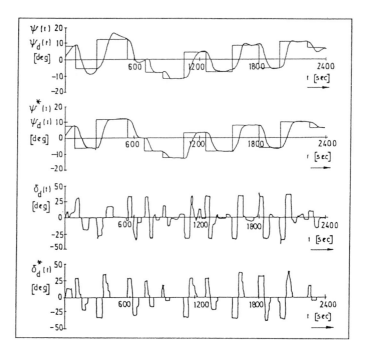

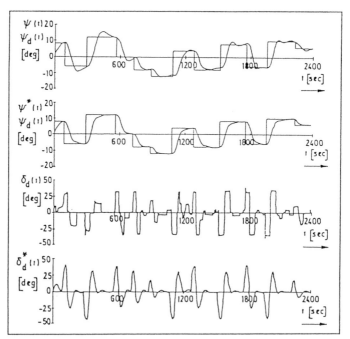

Figures 4 and 5. Comparisons of results from two models with actual measurements.

control behaviors are modeled. For the first model, planning and control simulation is based on linear optimal control theory, whereas the other model employs a total of 10 fuzzy logic controllers (Lee, 1990) to simulate the various decision-making processes. Both models may either include a Kalman filter to simulate state estimation or draw on perfect state information as a basis for planning and control actions. The working principles of the state estimation model and the planning and control models are summarized in the remainder of this chapter. An in-depth treatment of the various concepts has been given by Papenhuijzen (1994).

4.1 STATE ESTIMATION

The major elements of the navigator's state estimation behavior that should be reflected in the state estimation model in some way are as follows:

- The interpretation of on-line information (observations) by referring the navigator's knowledge of the geometry of the surroundings.

- Use of earlier observations as an additional source of information. In particular when the amount of on-line position-related information is limited, this component of state estimation, called dead reckoning, is very important.

- Manner which dead reckoning is interwoven with the taking and processing of new observations.

In fact, in a discrete-time Kalman filter, all these elements are represented in a very straightforward way. Further, the Kalman filter concept provides an elegant framework for modeling learning effects that relate to understanding the ship's dynamics and adaptation of the navigator's internal representation of the ship as the ship's dynamics change due to bank suction or shallow water effects.

Consequently, a prototype state estimator was built, supplied with suitable empirical relationships for relating the observation error variance terms to the values of the corresponding observed variables. For testing purposes, the prototype was coupled to a sophisticated dynamic model of a container carrier, which was used in the ship bridge simulator of the TNO-IWECO. The results of two experiments are shown below. In both experiments, the same turning circle was sailed. At regular intervals, typically 4 seconds, a fixed combination of observations was taken, applying a normally distributed white-noise error signal to each separate observation. During the first trial, the compass reading and relative bearings with respect to three buoys were taken into account. The second trial was used to investigate the influence of leading line information by adding to the combination of observations the subtended angle between the two leading-line beacons. As the output of the experiments the actual position and heading of the ship have been indicated. In addition, both the actual and the estimated track of the ship are plotted for each minute, and the 90% confidence region has been drawn around the corresponding estimated position. A confidence region is defined here as the smallest area that surrounds the estimated position and in which the actual position is expected to be with some given probability. Since the Kalman filter concept implies the assumption that the position component estimation errors are characterized by a bivariate normal distribution, the confidence regions take the form of an ellipse.

The result of the first experiment is shown in Figure 6a. The estimated position is indicated by the discontinuous line. Over periods of 4 seconds the state estimator determines the position of the ship only on the basis of a priori information. In between, readjustments of the estimated position are performed on the basis of observations, which account for the sudden jumps in the estimated position path. The influence of advance knowledge on the accuracy with which the observations are taken is clearly demonstrated. Figure 6b shows that, in close accordance with reality, the fixes became much more accurate in a direction perpendicular to the leading line, and the confidence regions

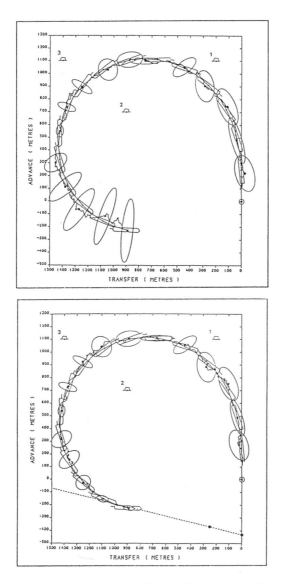

Figure 6. Experimental results (a, above). Influence of advance knowledge (b, below).

changed accordingly. Furthermore, in this experiment applying leading-line information yielded very little improvement in position estimation accuracy in a direction parallel to the leading line, which is only natural.

4.2 CONTROL THEORETIC APPROACH

Conceptually, the complete planning process may be considered to comprise both long-term and short-term planning. Long-term planning is aimed at determining a desired route and preliminary time planning, whereas short-term planning determines the exact trajectory to follow. Long-term planning is, in fact, the voyage preparation process, yielding just a few target states, corresponding time instants, and an indication of how important it is to realize the various components of the target states. Given the field of application of the program, it was not considered necessary to simulate long-term planning as such. As an alternative, a very small number of target states, generally one, is to be specified by the user in order to define the navigation task.

The problems of track planning and track following are basically interrelated. In defining the short-term track to be followed, not only the attractiveness of a track in terms of acceptable risk but also the dynamic properties of the ship have to be considered. For that reason it has been found logical to develop a submodel in which the description of the navigator's short-term planning behavior and track following behavior are integrated.

The ultimate goal to be achieved by the track-planning and track-following submodels includes the following aspects. First, a target state, an outcome of the long-term planning process, has to be reached within a given span of time. Furthermore, the risk that is involved in sailing to the corresponding position and arriving there with a given velocity is to be kept sufficiently small, as are the necessary control inputs. Linear optimal control theory provides an elegant framework for generating realistic control inputs on the basis of the combined subgoals mentioned. The three distinctive aspects of short-term planning reappear in a control theoretic context as terminal error weighting, state weighting, and control effort weighting, respectively.

Unfortunately, deriving an optimal control strategy in terms of a linear optimal control law is not straightforward. In this particular case, the optimal solution of the regulator problem, which includes the short-term track to be sailed, is not known. On the contrary, the optimal combination of a trajectory and a control signal should be a *result* of the algorithm, rather than an *input* to it. Consequently, the control problem to be solved is the generalized discrete-time linear optimal regulator problem (Papenhuijzen and Stassen, 1987).

Solution of this control problem is achieved by iterative application of the linear optimal control law. To start with it applies anything but the optimal solution, but improved solutions result eventually. Two additional loops are nested within the resulting loop. One of these is supplied in order to achieve absolute convergence of the algorithm by damping the convergence speed as much as necessary. The inner loop is meant to avoid instabilities, which may be rooted in the control law prescribing larger control inputs than those that are possible from a physical point of view. An overview of the complete track planning algorithm is shown in Figure 7.

As soon as the track-planning algorithm has duly converged, the latest control law is saved to be used to simulate track following, featuring multivariable system control, i.e., vessel control by applying both rudder and engine speed variations.

4.3   Fuzzy set approach

Most navigators used to interpret and plan their course on confined waterways as a concatenation of lanes and circle-shaped bends. Hence, the planning model is designed to supply the track-following model with a desired track in terms of straight lines and arcs of a circle. Apart from that, the planning model yields a fuzzy definition of the navigator's perception of safe maneuvering zones in order to enable the track following model to assess the safety consequences of deviations from the ideal track. Additionally, the safety zone's definition is also used by the planning model itself, since track planning is impossible without an appropriate representation of safety.

Supervisory and manual control behavior, the execution of the travel plan, is simulated by the track-following submodel. Every second, future states are predicted. The predicted states are evaluated by relating them to the perception of safety as determined by the track-planning submodel and by assessing the measure to which future states diverge from the desired states as defined by the planned track. If necessary, a control action is carried out. Contrary to the case for the control-theoretic model, the control actions taken by the fuzzy set navigator model comprise only rudder commands, and not engine settings. Constant engine speed setting is assumed during a whole run.

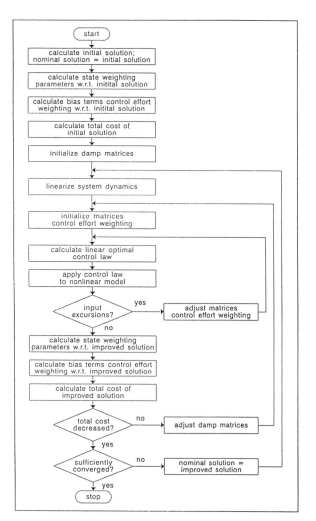

Figure 7.   Track planning algorithm

5 Validation of the navigator models

Validation and evaluation of the models has been accomplished by comparing the outcomes of a large number of ship bridge simulator experiments with the corresponding outcomes of the two navigator models. It was considered essential to be able to judge the performance of the navigator models in inland navigation situations as well as in confined sea navigation situations. Another important issue to be studied was the suitability of the models for dealing with different types of ships. Further, the ability of the models to cope realistically with environmental influences, wind and current in particular, had to be investigated. The experimental program is discussed in some detail in Section 5.1. In Section 5.2 some examples of experimental results are shown.

5.1 EXPERIMENTAL SETUP

For the experiments, two fairways, both located in the Port of Rotterdam area, were selected. For the sea navigation experiments, the approach to Mississippi Harbor was simulated. The inland navigation experiments were performed on the Hartel Canal, the route from the village of Geervliet westward to the Rozenburg lock. For the first fairway, a moderate wind and a heavy wind condition were selected. For the inland navigation experiments, apart from a simple reference condition, a heavy wind condition was defined, as well as one with a relatively strong current from behind.

For both the sea navigation trials and the inland navigation trials, a relatively small and relatively bulky ship type were used. As a small sea vessel, a loaded 210 m long container carrier was employed. The heavier deep sea vessel was a loaded 290 m long bulk carrier. For the inland experiments a relatively small loaded tanker was used, and, being the more difficult vessel to handle, an empty six-barge push tow unit.

Fortunately, it was possible to employ only highly experienced subjects for the experiments. Four harbor pilots cooperated on the sea navigation trials, each of them performing four runs per condition. Two inland tanker captains carried out the inland navigation experiments with the tanker, both conducting a share of five runs per condition. Two push tow captains contributed to the experiments with the push tow convoys, each also performing five runs per condition. As a result, a total number of 124 trials were carried out, divided into 10 experiments. Additionally, the same experimental program was carried out with both the control-theoretic simulation model and its fuzzy-set counterpart.

5.2   SOME RESULTS

In this section, some results of one of the sea navigation experiments are presented as an example. For this purpose, the experiment with the bulk carrier under moderate wind conditions was selected. Figure 8 shows the results of one of the experimental 16 trials performed by a pilot on the ship bridge simulator. To the left, the track of the ship is plotted, as well as the hull of the ship, twice a minute. To the right, a time plot of the engine setting and the rudder setting is given. This example demonstrates the technique mariners use to execute a maneuver such as this. Major course changes are initiated with a generous rudder shift, which is maintained until the ship has started to respond noticeably. Keeping the rate of turn constant at a certain level then, is achieved by setting a somewhat smaller rudder angle, until the turn is compensated for by applying a rudder angle in the opposite direction. Course keeping on more or less straight tracks is realized by relatively frequent application of relatively small rudder adjustments. Further, a minor engine speed variation is executed when starting the first turn. A more significant engine speed adjustment is applied to help counteract the second turn.

Results of a trial of the same experiment, carried out with the control-theoretic navigator model, are shown in Figure 9. Obviously, an optimal control strategy would result in perfectly smooth tracks and control signals if it were not for a shaping filter that modifies the calculated optimal control settings. However, large deviations between the optimal trajectory and its realization are not allowed, as a consequence of which the control signals in Figure 9 are clearly a discrete version of an otherwise almost identical pair of smooth signals. Unfortunately, it may be clear that the character of the control signals generated by the pilots is anything but represented by the optimal control navigator model. On the other hand, the ability of the model to cope with wind and current disturbances is quite remarkable and enables the model to stay well within the limits that are set to the validity of a linear control approach. Furthermore, in spite of the unusual rudder control signal, the resulting track seems realistic enough. Another interesting point is the way in which the model manages to combine engine speed variations and rudder control, navigating the ship apparently effortlessly to its destination.

A similar trial, performed by the fuzzy-set navigator model, is presented in Figure 10. It may be clear that here the character of the rudder signal comes much closer to reality than is the case with the control-theoretic model. This trial involves sailing a bend as well as lane keeping. As indicated earlier, the model does not feature engine speed control, which, apparently, does not prevent it from coping adequately with environmental

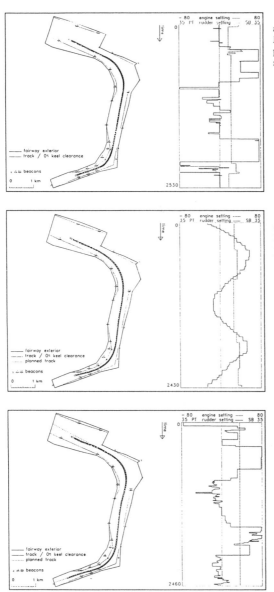

Figure 8.   Simulation of an approach with a bulk carrier into Mississippi harbour (southward), as performed by a harbour pilot on the ship bridge simulator.

Figure 9.   Simulation of an approach with a bulk carrier into Mississippi harbour (southward), as performed with the control theoretic navigator model.

Figure 10.   Simulation of an approach with a bulk carrier into Mississippi harbour (southward), as performed with the fuzzy set navigator model.

disturbances and realizing an acceptable track.

In order to be able to judge the model performance with respect to track variation and control effort aspects for all trials that belong to an experiment, a second output format was also used, as indicated in Figures 11 to 13. To the left in the figures, the total fairway space that was occupied by any ship, the swept path, is indicated by dashed lines. An impression of the distribution of all tracks that relate to a given experiment is given by drawing the individual tracks.

Further, the results of analyzing the corresponding rudder setting signals are shown to the right. The curved solid line represents the distance accomplished by the ship, running

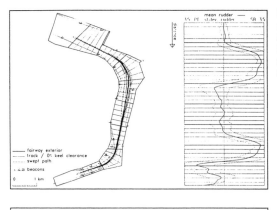

Figure 11. Analysis of all trials of the experiment concerning the approach with a bulk carrier to Mississippi harbour (southward), as performed by harbour pilots on the ship bridge simulator.

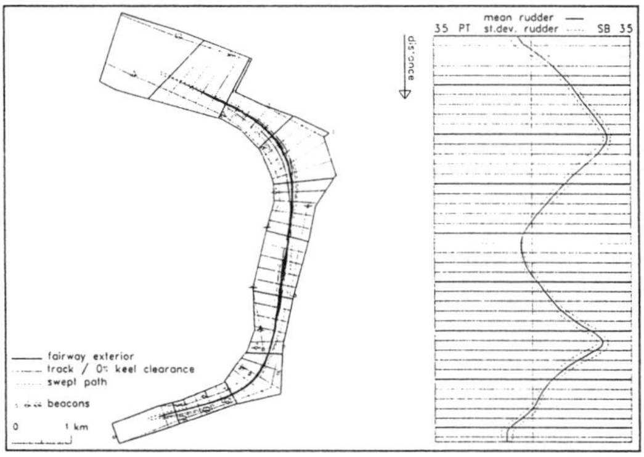

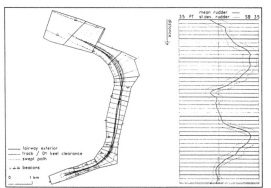

Figure 12. Analysis of all trials of the experiment concerning the approach with a bulk carrier to Mississippi harbour (southward), as performed with the control theoretic navigator model.

Figure 13. Analysis of all trials of the experiment concerning the approach with a bulk carrier to Mississippi harbour (southward), as performed with the fuzzy set navigator model.

from top to bottom, the average mean value of the rudder settings, $m_{av}(s)$, the definition of which is clarified in Figure 14. In this figure the definition of the distance-dependent average standard deviation of the rudder setting, $s_{av}(s)$, is also given. The dashed lines in the right part of Figures 11 to 13 indicate the band of $s_{av}(s)$ around $m_{av}(s)$. In this way, $m_{av}(s)$ gives an idea of the tactical decisions that underlie the maneuver, whereas $s_{av}(s)$ serves as an indicator of the control effort spent all along the maneuver. Thus, as argued by Schuffel (1986), the complexity of the various parts of the maneuver can be judged, as well as the potential risk incurred.

The relationship between the left and the right parts of the analysis output is indicated

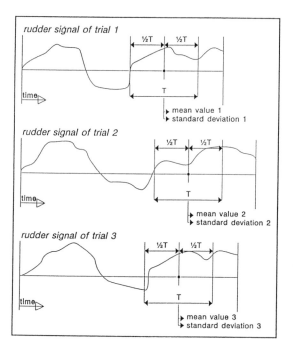

Figure 14. Definition of average mean value of rudder settings

by drawing cross lines at a nominal spacing of 200 m, with the lines in the track plot corresponding one to one to those in the control signal diagram.

Figures 11 to 13 show the analysis plots that correspond to the experiments to which the trials in Figures 8 to 10 belong, based on all 16 runs making up each experiment. Here, a value of 50 m was used for the nominal distance between the cross sections in which the analysis steps were performed, and the time window T was 90 seconds.

As for the control-theoretic model output, a tentative conclusion can again be drawn from Figures 11 and 12 that fairly realistic tracks are produced. At the same time, the corresponding rudder signal analysis plots appear to be perfectly ruthless when it comes to distinguishing between the performance of this mathematical concept and the behavior of a real navigator. By comparing Figure 13 with Figure 11, it can be concluded for the fuzzy set model that the track itself, as well as the apparent variation in track shapes, closely approximates reality. The conclusion that was drawn in the foregoing, namely, that on the whole the navigator's behavior with respect to rudder control is much better represented by the fuzzy-set model than by the control-theoretic model, is strongly confirmed here.

6 Discussion

After more than 20 years of research in the field of modeling the human operator controlling a vessel, it is time to step back and reflect on the outcomes of this research program. In Sections 6.1 and 6.2, the technical soundness of the concepts and the usefulness of the models with respect to the original user requirements were discussed. Generally, the execution of a research program yields a number of byproducts, which sometimes even develop into its principal products. Alternatively, new applications for the intended concepts may open up, putting the value of an entire project into a new perspective. The navigator project has been no different, as argued in Section 6.3. In Section 6.4, an attempt was made to evaluate the entire research program and to summarize the interrelation between the helmsman project and the navigator project.

6.1   OUTCOMES OF THE HELMSMAN PROJECT

Two entirely different models have originated from the helmsman project, a smart and simple linear one, and a somewhat more complicated nonlinear one. As was hoped before the project was started, the models more or less complement each other, as may be clear from the following. If calm water is assumed, and if only the overall performance of the system, i.e., the heading signal, needs to be realistic, the linear model is quite adequate, and thus is preferable because of its simplicity. A realistic rudder setting output, in addition to realistic heading characteristics, can only be produced by the nonlinear model. The original, relatively simple version of the nonlinear model is appropriate if no significant wave influences have to be taken into account. Otherwise, the extended version has to be used.

Consequently, both models could be used as part of a navigation simulation facility, with the choice between the two depending straightforwardly on the experimental conditions. Thus, fairway dimensioning and fairway safety issues can be investigated. Originally the models were only intended to simulate a helmsman effecting a predefined heading signal. However, by coupling either helmsman model to a ship model and a navigator model, the simulation of situations in which a navigator gives heading commands instead of rudder commands also can be achieved.

Additionally, it was concluded that the nonlinear helmsman model could be used for the following purposes:

- *Assessing the handling qualities of a vessel.* In fact, the parameters of the decision-making element of the nonlinear model are divided into a group of parameters that are related to the *method* of steering (i.e. the magnitudes and durations of the rudder angles to be applied), and a group of parameters that are related to the steering *precision* (setting the thresholds to be passed for a control action to be taken). It was found that, after fitting the model characteristics to that of a given helmsman, the first group of parameters could be used as an indicator for the helmsman's opinion on the handling qualities of the ship. The second group of parameters may serve as an indicator of the helmsman's perception of the accuracy of the information supplied.

- *Assessing the benefits of additional navigation information.* As discussed in Section 3.2, it is possible to quantify the influence of additional instruments, such as a predictive display, on the helmsman's behavior in general and on the steering performance in particular by means of the nonlinear model.

- *Investigations in the field of control of slowly responding systems.* It was found that the use of the internal model concept should not necessarily be restricted to the realm of modeling the behavior of the helmsman. On the contrary, it was concluded that this principle could be very useful for a variety of research projects concerning slowly responding systems. Many developments since that time, one of which is the navigator project that was to follow the helmsman project, have strongly confirmed this assumption.

6.2   OUTCOMES OF THE NAVIGATOR PROJECT

On the basis of numerous experiments it was concluded that both approaches, the control- theoretic approach and the fuzzy-set one, have resulted in a practical model that yields realistic ship tracks, given single ship situations. There are some important differences between the two models, however, concerning either the functionality of the models as they now exist and possibilities for future enhancement of their applicability. In the following, this point is examined in more detail, discriminating between seven major issues.

- *Performance*. Neither of the models appears to pose serious problems with respect to the necessary computing time. Generally, both models run about twice as fast as real time on a 50 MHz 486 PC.

- *Basic philosophy*. The basic philosophy underlying the control-theoretic approach is that explicit modeling of the navigator's behavior is achieved only on the level of the navigator's control objectives. The resulting control actions are derived from this, applying a plausible mechanism that does not necessarily ensure perfect resemblance of the model output to the individual actions of a real navigator. The performance of the system to be controlled, however, may just as well be quite realistic. However, the fuzzy-set model is a descriptive human operator model, featuring a more explicit representation of individual decisions made by the navigator. Consequently, the output of the fuzzy-set navigator model is much more realistic, which is essential when, for instance, safety studies are carried out that focus on control effort assessment.

- *Track planning strategy*. The two models differ substantially in the way in which track planning is accomplished. The control-theoretic model determines the optimal track on the basis of weighted minimization of risk, control effort, and terminal error. On the one hand, the result is a highly generic method, which, in principle, enables the model to cope even with such complex navigation tasks as docking. On the other hand, this technique excludes the possibility of tracks that consist only of lanes and arcs of circle to emerge from the planning process. It can be concluded from the experiments, however, that generally the planned tracks of the control theoretic model look quite realistic. Only if unconditional representation of the real planning process, as performed by, for instance, a harbor pilot, is imposed, may this method be considered to lack realism. In those, probably rare, instances, the fuzzy-set model would be superior. However, the cost of this is some loss of universality, because for certain complex maneuvers it could be too much of a simplification to consider the planned track as a concatenation of a limited number of lanes and arcs of circle.

- *Engine speed control*. Originally it was decided that at this stage it is not yet imperative to include engine speed settings in the navigator models. Eventually, however, this should be implemented. For the control-theoretic model this has been achieved, since engine speed control is inevitably inherent in the control-theoretic implementation. However, extension of the fuzzy set model to a navigator model that includes engine speed control as well as rudder control is much more complicated. The point is that it is not sufficient to modify only the track-following submodel. Engine speed setting at a given moment should not only be concerned with the particular maneuver that is then performed. Much more important is the general plan about where to be at what instant, and what speed to possess then, thus anticipating the maneuvers to come. As a consequence, the fuzzy-set model can only be made to feature engine speed control at the cost of considerable modification of the entire model.

- *Time-varying environment definition*. If the experiment involves navigating tidal waters and if the trial covers significantly more than, say, a quarter of an hour, a time-dependent definition of the fairway geometry and current and wind influences has to be employed. This is something that requires no basic extensions of the control-theoretic model, since the linear optimal control law is principally time varying, as are the corresponding weighting parameters. However, in the planning process of the fuzzy-set model there is no direct relationship between time and the resulting planned position path. Although an attempt could be made to compensate for this, the success of such a modification is doubtful.

- *Traffic simulation.* Eventually, it should be possible to cross the single-ship barrier and to develop a simulation system that includes more than one ship, featuring realistic interaction between the individual navigators. In its simplest form, only one ship would be equipped with a navigator model, and the other ships follow predefined tracks. This so-called floating island concept is fairly simple to implement for the control-theoretic approach. It means that the risk definition has to be based on time-varying effective fairway boundaries, which, as discussed earlier, is not a serious complication. If, however, all ships are supposed to react with one another, the situation becomes more complicated. In that case, the state space has to be extended to include the state variables of all individual ships in order to realize an overall optimum. For the fuzzy-set model it will be more difficult to extend it to include floating island traffic simulation. As stated earlier, the planned track lacks a direct relationship with time, which is essential for considering a moving object as part of the physical environment. In fact, full interaction would be even easier to realize, as this field has already been extensively explored, given the work that has been devoted worldwide to collision avoidance systems.

- *Other applications.* Apart from looking at the suitability of the various concepts for modeling the behavior of the navigator, it is interesting to consider whether there are entirely new uses that have been developed for the navigator models. The most obvious, and at the same time most daring, one is the use of a navigator model for automatic guidance of a real ship. Neglecting, for the time being, such important issues as legislative restrictions and economic relevance, it is felt that automatic navigation under some conditions need not be far off, given the results of this project. In principle, automatic navigation in single-ship situations could be performed directly by coupling a navigator model to standard positioning and course-measuring equipment, and supplying adequate chart, wind, and current data. If encounters with other vessels have to be dealt with, a radar-based recognition and prediction system must be available. In addition, major extensions to the navigator model would have to be accomplished, analogous to what was discussed previously.

The suitability of either of the navigator models for automatic navigation purposes can be determined on the basis of the following considerations. To start with, the conclusions of the discussion on traffic simulation are particularly relevant here. Secondly, in the case of automatic navigation, it is not enough to employ methods that cover most, but not all, possible situations. As a consequence, particularly based on the second consideration, it is concluded that for automatic navigation the control-theoretic model is probably the best option. This conclusion is supported by the fact that in the experiments the control-theoretic model managed to steer the vessels extremely accurately along the planned tracks, even though

- There were deliberately considerable mismatches between the hydrodynamic model of the real vessel and that in the navigator model as an internal model of the ship's behavior.

- Considerable environmental disturbances were acting on the real ship.

In order to illustrate this, Figure 15 shows the result of a trial with the push tow unit, steered by the optimal control model, subject to a heavy southwest wind.

In conclusion, it can be stated that the fuzzy-set model is the better model when it comes to realistic simulations of the way in which a real navigator accomplishes rudder

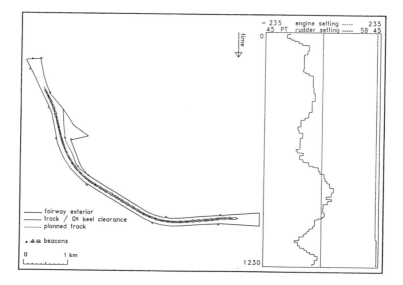

Figure 15.   Trial with the push tow unit, steered by the optimal control model, subject to a heavy southwest wind.

control. The control-theoretic approach is probably superior for development into a system that includes complex planning behavior with significant engine speed variations and complex collision avoidance maneuvers. If, within the framework of such a project, realistic rudder control is essential, the idea of combining the two concepts is suggested. Apart from this, it could be recommended to probe the development of a sophisticated navigation system based on the control-theoretic navigator model that features not only automatic track following but also automatic track definition.

6.3   SPIN-OFFS

Within the framework of the navigator project, two interesting byproducts have been created: COLUMBUS, which aids navigation evaluation utility, and the Complete Compatibility Reconstruction filter (CCR filter). Both concepts have been developed into valuable stand-alone products, which, apart from their origin, serve to endorse the feasibility and usefulness of the navigator-model concepts that have been proposed.

Competent and objective evaluation of aids to navigation layouts can be accomplished using the COLUMBUS program, an interactive excerpt from the state estimation component of the navigator model. Experience has shown that this program offers a useful fairway design facility. In addition, it has provided extra ground for the assumption that the Kalman filter approach to state estimation modeling will yield a simple, flexible, and realistically performing submodel.

The other byproduct is the CCR filter, which is now an indispensable research instrument used by the Transportation and Traffic Engineering Research Department for highly accurate velocity component reconstruction, for precisely measuring and assessing maneuverability characteristics of vessels. Its efficient and secure operation is another demonstration of the reliability of the control-theoretic track-planning algorithm, from which the CCR filter inherited its basic design (see Section 4.2).

6.4   IN RETROSPECT

Around 1970, ship owners and authorities faced substantial scale enlargement, resulting in the need to master new skills and to control unknown risk levels. It was during this period that the helmsman project was begun, with its main objective being the development of a research tool to assess the handling qualities of a given vessel. From the discussion in Section 6.1, it should be clear that the helmsman project has been successful in achieving this objective. Additionally, the helmsman project has yielded useful components for navigation simulation experiments in general.

Eight years after the completion of the helmsman project the navigator project was seriously undertaken to cover the remaining aspects of the combination of human operator modeling and navigation simulation. The resulting models are primarily meant to describe navigation under more or less risky conditions. This explains why the models are being used mainly in the rudder control mode, rather than in the heading control mode. However, if heading control mode, or switching between both modes, is to be described, either helmsman model could be coupled with either navigator model, taking into account the particular fields of application for the various models, as discussed in the preceding sections.

7 Conclusions

Starting with the development of an instrument to evaluate ship handling characteristics, and concentrating only on the implementation of a heading command, the entire research program has evolved to where fully automatic navigation appears to be within reach. Obviously, such a development would require a lot of research, but it is certainly considered feasible from a technical point of view. The main restrictions will be of an organizational or a legislative nature.

In the meantime, the various products resulting from this research program can be used and, in fact, are being used. Further model extensions, such as equipping the navigator model with the possibility of coping with dual-or multi-ship situations, are being considered, and completely new offspring of the program have developed.

References

Lee, C. C. (1990). Fuzzy logic in control systems: fuzzy logic controller: Part I and II. *IEEE Transactions on Systems, Man and Cybernetics* SMC-20: 404-435.

Papenhuijzen, R. and Stassen, H.G. (1987). On the modeling of the behavior of a Navigator. *Proceedings of the 8th Ship Control Systems Symposium,* The Hague, 2.238-2.254.

Papenhuijzen, R. (1988). On the modeling of the behavior of a navigator; background and proposal for a control theoretic approach. In J. Patrick and K. D. Duncan (Eds.) *Training, Human Decision Making and Control.* Amsterdam, Elsevier (North-Holland):189-203.

Papenhuijzen, R. and Stassen, H.G., (1989). On the modeling of planning and supervisory behavior of the Navigator. *Proceedings of the 4th IFAC/IFIP/IFORS/IEA Conference on Man-Machine Systems,* Xi'an, China, 19-24.

Papenhuijzen, R. and Stassen, H.G. (1992). Fuzzy set theory for modeling the navigator's behavior. *Proceedings of the 5th IFAC/IFIP/IFORS/IEA Conference on Man-Machine Systems,* The Hague, 2.1.4.1-2.1.4.7.

Papenhuijzen, R. (1994). Towards a human operator model of the navigator. Ph.D. thesis, Delft University of Technology, Delft.

Schuffel, H. (1986). Human control of ships in tracking tasks. Ph.D. thesis, TNO-IZF, Soesterberg.

Stuurman, A. M. (1969). *Modeling the Helmsman: A Study to Define a Mathematical Model Describing the Behavior of a Helmsman Steering a Ship along a Straight Course.* Report no. 4701, TNO Institute of Mechanical Engineering, Delft.

Veldhuyzen, W. (1976). Ship maneuvering under human control; analysis of the helmsman's control behavior. Ph.D. thesis, Delft University of Technology, Delft.

Chapter 16

Development of Human-Machine Interfaces for Manually Controlled Space Manipulators

Paul Breedveld

On the space station Alpha, the 10 m-long space manipulator ERA will be used for a variety of activities. The movements of the ERA will be monitored with cameras and manually controlled by an astronaut. For the astronaut, it is difficult to estimate the three-dimensional movements of the manipulator from the two-dimensional camera pictures. Furthermore, the large freedom of movement, slow dynamics, and flexibilities in the arm make controlling the manipulator difficult. When the robot is controlled from Earth, the time delays make control even more difficult. Research is being done at the Delft University of Technology on the development of a human-machine interface for the manipulator in order to find solutions for these problems. This chapter discusses the results of this research. Various displays that assist the astronaut in accurately positioning the hand of the manipulator for grasping an object are described. Control methods that transform the movements of the control device to movements on the television screen are also discussed, and the results of some human-machine experiments with these control methods are given. Next, some predictive displays that can be used to reduce the effects of the flexibilities and the slow dynamics, and to compensate for the time delays are described. Finally, a display that makes it easy to see a collision coming when the hand of the manipulator is moved along a track is presented.

1 Introduction

1.1 ERA SPACE MANIPULATOR

Around the end of this century, the first segments of a new space station will be put into orbit around the Earth. This space station, called the International Space Station Alpha, is being developed and constructed in the United States, Russia, western Europe, Canada, and Japan. The European contribution to the space station consists of a manned module, called the Columbus Attached Pressurized Module, and a *space manipulator* that will be used on the Russian part of the Alpha, MIR-2 (Van Woerkom et al., 1994). The space manipulator is called ERA, which is short for European Robot Arm. The robot is shown in Figure 1. The ERA is a 10 m long anthropomorphic space manipulator with 6 degrees of freedom (DOFs). The joints of the robot are driven by electric brushless DC motors, and its links are made of lightweight carbon fibre. In spite of the fact that this is a very stiff material, the long and slender upper arm and forearm of the ERA are flexible. Since the power of the electric motors is limited, however, the arm can only move slowly and its vibrations will be small.

The figure also shows the two hands or *end-effectors* of the manipulator. The ERA uses the first hand to connect itself to the space station and the second one to grasp objects. Due to its symmetrical construction, the robot is able to walk over the space station: The two end-effectors alternately grasp a new connection point, and the manipulator swings to another position.

It is intended for ERA to be launched in the year 1999 by a Russian Zenith Cargo Vehicle. Initially, it will be used by Russian astronauts for assembling the MIR-2. When the assembly phase is finished, the manipulator will be used for servicing and inspection tasks and for the displacement of Orbital Replaceable Units (ORUs). These are containers

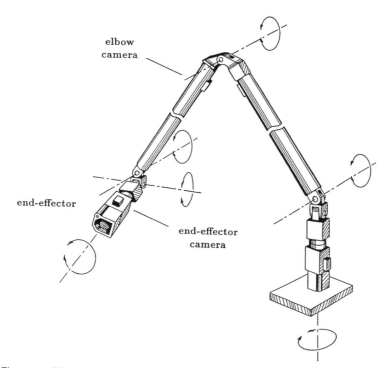

elbow
camera

end-effector

end-effector
camera

Figure 1.   The ERA space manipulator

with objects that must be moved from one part of the space station to another. During these activities, the robot will be controlled from inside the MIR-2 by a human operator. In future applications, however, space manipulators like the ERA will also be used on unmanned space vehicles, and controlled from Earth. In this scenario, time delays due to data transmission through space are present that make controlling the robot more difficult.

Due to a lack of direct vision on the MIR-2, *camera pictures* will be used to monitor the movements of the ERA. The cameras are located on the manipulator itself as well as on the space station. Examples of cameras located on the manipulator itself are the *end-effector camera* which will be used for inspection and for fine positioning the end-effector, e.g., for grasping an ORU and the *elbow camera* which will be used for moving the end-effector along a track. Whenever possible, the ERA will be controlled by *supervisory control*. In that case, the operator gives commands such as "put that there" to the manipulator, and the manipulator performs these tasks automatically. The operator uses the camera pictures to *monitor* the movements of the manipulator. If anything goes wrong, the operator intervenes and switches to *manual control*. In that case, the operator uses a *control device* to move the manipulator into the desired direction.

1.2 RESEARCH AT THE DELFT UNIVERSITY OF TECHNOLOGY

In the case of manual control, the human operator is faced with three problems that make control difficult:

1. It is difficult for a human to estimate the three-dimensional movements of the manipulator from flat, two-dimensional camera pictures.

2 . The ERA is a multivariate system with a large freedom of movement. In general, it is difficult for a human to control such a system manually.

3 . Slow dynamics, flexibility, and time delays make controlling the ERA even more difficult.

To find solutions for these problems, a project was started in 1986 at the Delft University of Technology in the Netherlands, Faculty of Mechanical Engineering and Marine Technology, Laboratory for Measurement and Control, in which the human-machine aspects of remotely controlled space manipulators were investigated. The outcomes of this project can be found in Bos et al. (1995). In 1991 the project was followed by a second project that focuses on the design of a human-machine interface for a manually controlled space manipulator such as the ERA. Although both projects focus on the space industry, their results are also suitable for other related manual control applications, such as endoscopic surgery, undersea robotics, and robotics in the nuclear industry.

In the first project, the hardware facility consisted of a VAX II GPX workstation, which was used to simulate and animate the movements of the manipulator in real-time. The control device consisted of two joysticks, both with one rotational and two translational DOFs. At that time, the graphical workstation was not powerful enough to animate complex three-dimensional spatial pictures in real time. Therefore, the major part of the research was carried out in two dimensions.

The second project was carried out completely in three dimensions. It was initially intended to construct a scale model of the ERA that moved on frictionless air bearings on a horizontal surface plate, to exclude the effects of gravity. Using a control device and some cameras, it was necessary to be able to control this scale model by hand. However, since the scale model could only move in a two-dimensional horizontal plane, the facility appeared to be unsuitable for investigating the first problem mentioned earlier. Therefore, it was decided not to build the scale model, but to also use a graphical workstation, a fast Silicon Graphics Indigo II Extreme workstation, to simulate and animate the movements of the ERA in real time.

In the second project, a *Spaceball controller* was used as control device. This is a 6 DOF force-operated control device that consists of a sphere on a base. The sphere, or Spaceball, can be slightly translated and rotated in three perpendicular directions. The forces and torques that the operator imposes on the Spaceball are transformed into the desired translational and rotational velocities of the end-effector. These signals are inputs for the automatic controller of the manipulator that makes the end-effector move in the desired direction. Instead of velocity control, end-effector *position control* could have been used instead. In Kim et al.(1987), however, it was shown that position control is more suitable for *fast* manipulators with a *small* workspace, while velocity control is more suitable for *slow* space manipulators with a *large* workspace, like the ERA. *Joint control* did not seem interesting to investigate since this means of control is much more difficult for a human operator. Furthermore, many aspects of this means of control had already been investigated in the first project (Bos, 1988).

The research in the second project is phased as follows:

A. *Activities in an operating point*
    1.   Manual control of an ideal manipulator
    2.   Manual control of a flexible manipulator
B. *Activities along a track*

In Stage A, which is focused on positioning the end-effector, e.g., for grasping an ORU, the demanded accuracy is high, but the risk of an unexpected collision between the manipulator and the environment is small, since the movements of the manipulator are small. In Stage B, which is focused on moving the end-effector along a track, e.g., for displacing an ORU or for inspection tasks, the accuracy demanded is small, but the risk of a collision is large. In Phase A1, the first and second problem mentioned earlier are being investigated, and the manipulator dynamics have not been implemented yet. In Phase A2, the third problem is being investigated, and the dynamics of the manipulator are modeled in an accurate real-time simulation model.

2   Displays for positioning the end-effector

2.1 END-EFFECTOR CAMERA PICTURE

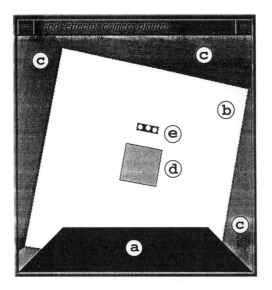

Figure 2.   The end-effector camera picture: (a) end-effector, (b) back plane,
(c) side planes, (d) target, (e) vision target.

When an object, e.g., an ORU, must be grasped, the end-effector camera is used for fine positioning the end-effector. Figure 2 shows a stylized impression of the end-effector camera picture when the distance between the end-effector front and the object to be grasped is decreased to 1 m. The end-effector camera is placed *on top* of the end-effector and shows a view of the environment. The end-effector front is visible at the bottom of the picture. The environment is simplified to five planes. The *back plane* represents the object to be grasped. The four *side planes* represent the environment around the object. On the back plane, a *target* is present. This is a connection point on which the end-effector front must be accurately positioned to grasp the object. Above each connection point on the MIR-2, a *vision target* is present. This object consists of a black base plate with three white discs, with the middle one placed at an elevation. It is used by a *proximity sensor*, which is a computer program that uses image processing techniques to calculate the distance and orientation of the end-effector with respect to the target from the camera picture. This distance and orientation is described by six numbers, as shown in Figure 3. In the case of supervisory control, the numbers can be used to automatically position the end-effector on the target. The slow dynamics and the flexibility of the manipulator, however, make automatic positioning difficult. In some cases, e.g., if an object is present between the end-effector and the target, or if the automatic controller is damaged, automatic positioning might even be impossible. Therefore, it is also possible to position the end-effector by hand as well.

Translations [mm]		Rotations [degrees]	
Up	-153	Yaw	12
Right	- 87	Pitch	-12
Forward	-237	Roll	-21

Figure 3.   The position and orientation numbers

2.2 OVERLAYS ON THE CAMERA PICTURE

In the last stage of the positioning task, when the distance between the end-effector and the target is decreased to only a few centimeters, only the vision target is left visible in the camera picture. It is very difficult for the operator to estimate the position and orientation of the end-effector from the size and locations of the three white discs. To assist the operator, the proximity information can be used to superimpose a *graphical overlay* on the picture that makes it more easy to position the end-effector. This overlay could, e.g., consist of the position and orientation numbers that are calculated by the proximity sensor. When all numbers are made equal to zero, the end-effector is positioned against the back plane, and the positioning task is completed.

With this method of presenting the information, however, the operator has to constantly translate the six numbers into spatial movements in the camera picture. This is a tiring activity that increases the operator's mental load. The risk exists that in order to decrease the mental load, the task will be simplified to just setting the numbers to zero one by one, without paying attention to the picture anymore. This method of controlling is not only quite clumsy but also decreases the operator's involvement with the real situation, which can be dangerous in abnormal circumstances.

In fact, there are two basic ways in which the presentation of the proximity sensor information can be improved:

1 . The way in which the six numbers are displayed is improved, but they are still displayed *individually*.

2. The six numbers are not displayed individually, but are *integrated* in a graphical overlay that makes it possible to estimate the distance and orientation of the end-effector at a single glance.

In the first project, the first, classical method was followed to present the proximity sensor information in the camera picture. This resulted in the end-effector camera overlay in Figure 4 Bos et al.(1995). At the left and right side of the overlay, the position and orientation misfits are displayed individually with six analogue dial indicators. The display was animated in real time on a VAX II GPX workstation and was used in a number of human-machine experiments. However, the operator still had to translate six individual quantities to spatial movements of the end-effector, and the above mentioned risk was still present. Due to the limited computing and graphical power of the workstation, however, it was not really possible to animate more complex graphical end-effector camera overlays in real time.

In the second project, it was decided to follow the second method mentioned earlier, which seemed to lead to a more direct and natural means of presenting spatial information. In this project, the proximity sensor information is integrated in a novel graphical overlay with *three-dimensional graphical objects*, which is constructed in such a way that the graphical objects seem to be part of the *environment*. Due to this, the overlay *increases* the operator's involvement with the real situation. The working principle of this overlay, called the *pyramid display*, is briefly described in the next section. In Breedveld (1995a), the development of this display is described in detail, and some further improvements are given.

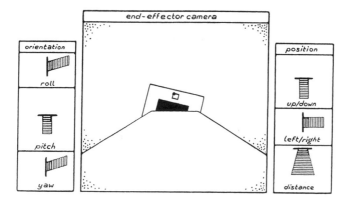

Figure 4.   The end-effector camera overlay of the first project

2.3 PYRAMID DISPLAY

Figure 5a, b, and c show the pyramid display, at a distance of 50 cm, 5 cm, and 0.5 cm from the target, respectively. To improve the sensation of depth, the proximity sensor information is used to project rectangular grids on large unicolored planes in the environment. Besides the grids, the proximity sensor information is also used to animate three graphical objects in the environment: an *insert box*, a *frosted glass*, and a *pyramid*. The insert box and the pyramid are fixed to the back plane, at the vision target, which is therefore not visible. The frosted glass is a transparent square plane fixed to the front of

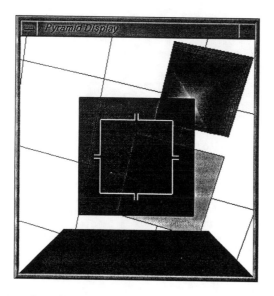

Figure 5a.   Pyramid display at distance of 50 cm from target

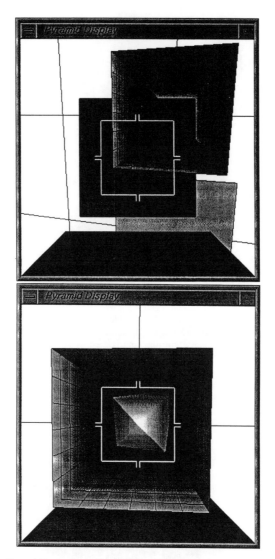

Figure 5b and 5c.   Pyramid display at distances of 5 cm (b, above) and
0.5 cm (c, below) from the target.

the end-effector, similar to the sight of a gun. In the first phase of the task (see Fig. 5a) the operator uses the frosted glass as a viewfinder and moves it towards the pyramid. In the second phase of the task (see Fig. 5b and Fig. 6) the frosted glass intersects the insert box. Then, the rectangular grid in the box is used as a distance indicator, and the rotational misfit of the end-effector is estimated from the orientation of the box. In the third phase of the task (see Fig. 5c) the frosted glass intersects the pyramid. From the size and the shape of this intersection, the small translational and rotational misfits that are left can easily be estimated, and collisions against the back plane are easily avoided. When the intersection finally coincides with the square on the frosted glass, the end-effector is positioned against the target correctly and the positioning task is completed.

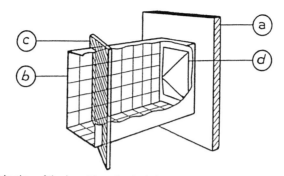

Figure 6.   Side view of the insert box, frosted glass, and pyramid: (a) back
plane (b) insert-box, partially cut away (c) frosted glass (d) pyramid.

Note that in the pyramid display a graphical overlay is superimposed on a *real camera picture*. Instead of this, the proximity measurement can also be used to animate a *virtual view* of the end-effector and the environment that does not correspond with a real camera picture. However, since there is always the possibility that something is wrong, the operator cannot fully rely on the animation, and real camera pictures must always be present for verification. In the case of a graphical overlay, the camera picture and animation are combined into one. This makes verification easy. In the case of a virtual view, however, the operator will have to spread the attention between two or more pictures. This is less handy and less safe. Therefore, in the second project priority was given to the development of graphical overlays.

3   Control methods for positioning the end-effector

For the two-dimensional investigations in the first project, a joystick with one rotational and two translational DOFs was used to control the velocity of the end-effector. Since the DOFs of the joystick exactly matched the DOFs of the two-dimensional end-effector, it turned out to be a convenient control device. For the three-dimensional investigations in this project, two such joysticks were used. For this case, however, the joysticks were less suitable, since together they had two rotational and four translational DOFs, while the end-effector has three rotational and three translational DOFs. Thus, the operator had to translate at least one translational DOF to a rotational DOF to control the end-effector. In the second project, which is being carried out completely in three dimensions, it was therefore decided to use a more dedicated Spaceball controller with three translational and three rotational DOFs. This control device is suitable for controlling the end-effector in three dimensions.

In the second project, several *control methods* have been developed that transform the movements of the Spaceball into desired movements of the end-effector (Buiël & Breedveld, 1995a). The two most important methods are shown in Figure 7. Here the Spaceball controller, target, vision target, and end-effector with camera are shown. With both methods, the movements of the hand of the human operator correspond to the movements of the end-effector in the camera picture. This gives the operator the feeling of holding the end-effector in the hand. A Spaceball controller makes such intuitive control possible, since it can be translated and rotated in three perpendicular directions.

With *forward end-effector control* the operator has to push the Spaceball forward to move the end-effector toward the target, while with *downward end-effector control* the operator has to push the Spaceball downward to move the end-effector toward the target. With forward control the end-effector camera picture is considered to be a forward

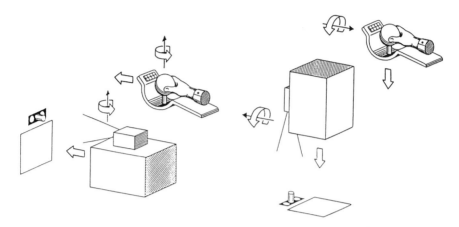

Figure 7.   Forward (left) and downward (right) end-effector control

view, while with downward control the end-effector camera picture is considered to be a downward view. In accordance with the suggestion of a downward view, the monitor should be rotated until the screen lies horizontally in front of the operator. For practical reasons, however, this was not done. Therefore, with downward control, the movements of the Spaceball and the movements on the screen are rotated over 90 degrees with respect to each other.

The control methods were evaluated with a human-machine experiment (Buiël & Breedveld 1994, 1995a). During this experiment six inexperienced subjects had to position the end-effector several times by using the pyramid display and the two control methods. The initial positions and orientations of the end-effector were varied randomly, and the dynamics of the manipulator were not implemented; with the result that the end-effector responded exactly to the desired velocity control commands of the subjects. For each control method, the session took about 1 hour.

Three of the subjects practiced forward end-effector control first. The others practiced downward end-effector control first. At the end of the session, a verbal interview was given, in which the subject was asked to give his or her opinion of the two control methods.

One subject had no preference between the two control methods, while the other five all preferred downward control to forward control, in spite of the rotation over 90 degrees. This surprising preference had the following basis:

- To most subjects, the boxlike environment in the animated camera picture gave the impression of a downward view.
- Many subjects related the horizontal tabletop with the vertical screen plane, and thus instinctively rotated their control actions over 90 degrees. The relation between these two planes is also present in Windows-based software, where the horizontal movements of the mouse on the tabletop are transformed into vertical movements of the cursor on the screen.
- The ergonomic design of the Spaceball controller makes it more suitable for downward control than for forward control. Figure 8a shows the way in which almost all subjects hold the Spaceball in the case of downward control. The grip is comfortable, and it is easy to push the Spaceball down to move the end-effector toward the target. In the case of forward control, however, the grip of Figure 8a is

less handy, since it is quite difficult to only push the Spaceball forward without also pushing it downward a little bit. Pushing the Spaceball exactly forward would be easier if the Spaceball could be gripped as in Figure 8b. However, the shape of the console makes it impossible to place the hand behind the Spaceball. Therefore, most subjects hold the Spaceball as in Figure 8c and use their fingertips to push the Spaceball forward. This grip is less comfortable than the grip used for downward control.

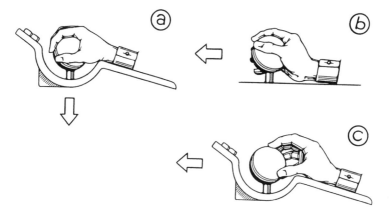

Figure 8.   Different ways to grip the spaceball. (a) grip with downward control (b) desired grip with forward control (c) real grip with forward control.

At the end of the experiment, the subjects were still in the learning phase. In another human-machine experiment with the downward end-effector control method (Buiël & Breedveld, 1995b), learning time and final task performance were measured. During this experiment, six other inexperienced subjects used the pyramid display to position the end-effector several times on the target with a desired accuracy of 3 mm and 1.5 degrees. They were asked to perform the task as fast as possible without colliding against the back plane. Again, the dynamics of the manipulator were not implemented, and the end-effector responded exactly to the desired velocity control commands. The maximum velocity of the end-effector was 0.1 m/s, and the initial positions and orientations of the end-effector were varied randomly in such a way that the distance to the target was always equal to 1 m. This made the minimum time required to complete the task equal to 10 s. After a relatively short training period of about 3.5 hrs, the subjects were able to perform the task with surprisingly fast completion times between 17 and 34 s. The pyramid display and control method proved to be very handy for positioning the end-effector.

4   Predictive displays

4.1   REAL-TIME SIMULATION MODELS OF THE ERA

In the experiments described earlier, the dynamics of the manipulator had not yet been implemented. To investigate the effects of slow dynamics, flexibility, and time delays on both projects, the dynamics of the manipulator were modeled in a real-time simulation model. Due to the limited computing power in the first project, ERA links were assumed to be stiff, and only the effects of the slow dynamics were taken into account. In the second project, the flexibility and time delays are also modeled. Some *friction* is also present in ERA joints. The automatic controller of the ERA is not able to neutralize the effects of this friction completely. When the end-effector is positioned to grasp an object, the friction and flexibility cause a poorly damped vibration that makes it difficult for the human operator to position the end-effector accurately. The real-time model of the second project therefore also includes a novel friction model that simulates the effects of the friction very accurately and takes less computing time (Diepenbroek, 1994).

4.2 DISPLAYS TO REDUCE THE EFFECTS OF SLOW DYNAMICS AND FLEXIBILITY

Since the power of the electric motors in ERA's joints is limited, the manipulator can only move slowly; it takes some time for the end-effector to accelerate or stop. Consequently, it takes a few seconds for the operator to see how the end-effector reacts to the velocity command, and it is not clear at what position the end-effector will come to a standstill when the control device is released. In the first project, the effects of these slow dynamics were investigated with some human-machine experiments. These experiments showed that these effects make controlling the manipulator very difficult.

Since the automatic controller of the ERA is already optimized, it is not really possible to reduce the effects of the slow dynamics further at the control side of the human-machine interface. At the display side of the human-machine interface, however, the effects of the slow dynamics can be reduced by using a *predictive display* (Sheridan, 1992, pp. 214-227). In the past, such displays have been employed on ships and submarines, and as optical landing aids for aircraft pilots. In the case of the ERA, a predictive display is a graphical overlay on the camera picture that shows the direction in which the end-effector will move when the operator gives a control command, and at what position it will come to a standstill when the control device is released.

Figure 9 shows the predictive display that was developed in the first project for the two-dimensional situation. The figure shows a two-dimensional picture of the manipulator. The end-effector had to be moved along the square objects and positioned horizontally in the "hole" at the bottom right of the picture. In the predictive display, the control signals of the operator are used in a computer program, called the *predictor*, that predicts the movements of the end-effector. The output of the predictor is used to animate the prediction in Figure 9. The predictor uses a simplified model of the manipulator to calculate the prediction. Human-machine experiments proved the prediction to be a handy aid that made it much easier to control the manipulator (Bos, 1991, pp. 88-121).

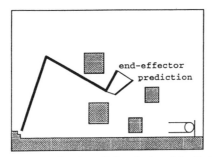

Figure 9. Two-dimensional predictive display of the first project

In the first project, the concept of a predictive display was also tried out in three dimensions. On a VAX II GPX workstation, a three-dimensional version of Figure 9 was animated. The animation showed a side view of the manipulator in a simple environment consisting of a horizontal surface plane with some square objects that had to be avoided. The movements of the end-effector were predicted in a manner analogous to Figure 9. To improve the spatial observability of the picture, a rectangular grid was projected onto the surface plane, and vertical lines were drawn from the grid to the end-effector and the prediction to help one visualize the height of these two objects. The prediction turned out to be very helpful in the three-dimensional case as well, but it also turned out that further research was needed to improve the visualization of the spatial information.

Due to the good results achieved with predictive displays in the first project, a predictive display for the end-effector camera picture has been developed in the second project that makes it easier to position the end-effector by hand. In the predictive displays of the first project, a prediction of the movements of the *end-effector* is animated in a side view of the manipulator and the environment. In the end-effector camera picture itself, however, a prediction of the end-effector would look very confusing. Therefore, in the end-effector camera picture, a prediction of the movements of the *environment* is animated. Figure 10 shows a picture of the resulting predictive display. Since the pyramid display proved to be very handy for positioning the end-effector, its concept is also used in this display. The spatial information is presented with three-dimensional graphical objects that are projected in the real and predicted environment, and the intersections of these objects with the frosted glass are used to visualize the misfits (Breedveld, 1995b). The predictive display turned out to be quite handy. In the future, a human-machine experiment will be performed with the display to prove its suitability for positioning the end-effector.

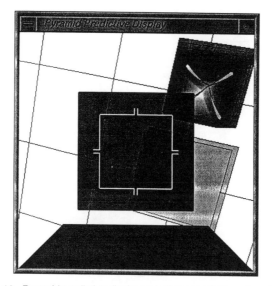

Figure 10. Pyramid predictive display at a distance of 50 cm from the target

4.3 DISPLAYS TO COMPENSATE FOR THE TIME DELAYS

The left-hand side of Figure 11 shows the manual control setup if a space manipulator like the ERA were to be controlled from Earth. The operator looks at the movements of the manipulator on a television screen. With the control device, velocity commands are generated that are transmitted into space and arrive at the manipulator with a time delay D_t (of up to about 3 s) (Sheridan, 1992, p. 212). The delay is *variable*, due to changes in the distance and the number of satellites (one or two) between the manipulator and Earth, and due to the processing time within the satellites. The automatic controller of the manipulator transforms the delayed control signals into movements of the arm that are monitored with a camera. The video signal is transmitted back to Earth, and the camera picture appears on the television screen, again with a time delay D_t. In the control loop, the overall time delay of $2D_t$ (of up to about 6 s) makes controlling the manipulator difficult. To reduce the risk of instability, the operator is forced to use a *move and wait* strategy when controlling the manipulator. With this cautious, time-consuming method of control, the operator generates a small incremental control signal, then stops and waits for a few seconds to see what happens, then gives another small incremental control signal, etc.

The right-hand side of Figure 11 shows that a predictive display can also be used to compensate for the time delays. In the predictive display, the same predictor can be used as in the case of slow dynamics and time delays. In this case, however, the predictor is located on Earth. Due to this, the prediction can be controlled without time delays, and the above-mentioned problems are no longer present. The control signals from the human operator are transmitted into space, and the real robot in the camera picture follows the prediction. If control signals are no longer generated, the robot in the camera picture coincides with the prediction after at least $2D_t$. If desired, the delayed camera picture can be used to tune the predictor in order to make the prediction more accurate (Breedveld, 1992, 1995a).

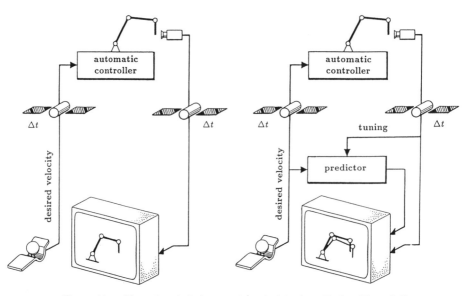

Figure 11.   Manual control of a space manipulator from Earth, without (left) and with a predictive display (right).

4.4 METHOD TO COMPENSATE FOR VARIATIONS IN TIME DELAY

If the manipulator is controlled from Earth, the control and video signals are transmitted digitally as a discrete series of samples with a constant time interval, e.g., of 0.05 s. Due to the variable time delay, the time interval at the arrival will not be constant and the signals will be deformed. This makes the prediction very inaccurate. In Breedveld (1992, 1995a), a method is described by which the deformation in the signals can be compensated. This method uses two *buffers*; the first one is near the manipulator and the second one near the television screen. At arrival, the samples are put into the buffer, which is emptied on-line, again with the constant time interval of 0.05 s. Since the time interval between the numbers is again constant, the signals are not deformed. The time interval at the input of the buffer varies, while the interval at the output is constant. Therefore, the size of the buffer is variable. Since the buffer should never become completely empty, the constant time delay between the moment of transmitting and the moment of emptying the buffer must be somewhat larger than the maximum value of D_t. Due to this, the time delay increases a little bit. If a predictive display is used, however, it is not expected that a small increase in the time delay will make controlling the manipulator more difficult.

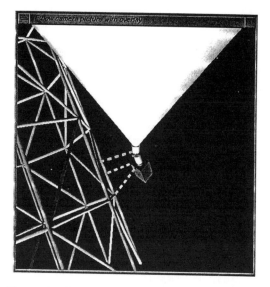

Figure 12.   The elbow camera overlay

5   Activities along a track

The preceding sections focus on Stage A of the second project, in which displays and control methods that make it more easy to position the end-effector with a high accuracy are developed. In Stage B of the second project, which focuses on moving the end-effector along a track, as noted earlier, the accuracy demanded is small, but the risk of an unexpected collision between the manipulator and the environment is large. In activities along a track, the forearm and end-effector of the ERA are most subject to collisions. These parts of the arm are both visible in the elbow camera picture, which will be used for monitoring the movements of the arm during these activities. However, it is quite difficult for a human operator to estimate three-dimensional distances from this two-dimensional picture and to see a collision coming.

It is intended that the complete geometry of the Space Station Alpha will be stored in a database in the onboard computer of the space station. Together with measured information about the configuration of the arm, this database can be used to superimpose a graphical overlay on the elbow camera picture that makes it more easy for the operator to estimate the distance between the forearm and end-effector and the different parts of the space station, and to see a collision coming. Figure 12 shows the elbow camera overlay that has been developed in the second project. The white triangular-shaped object in the upper part of the picture is the forearm of the robot. The end-effector is visible in the middle of the picture. The left-hand side of the figure shows a segment of the MIR-2. In the overlay, the distance from the end-effector and forearm to the three closest tubes of the space frame structure is visualized with three dashed lines. By using these lines, it is quite easy to move the end-effector and forearm through the space frame without colliding against it. Such a complex maneuver of the arm is needed, e.g., if an object must be inspected that is installed inside the space frame. Without assistance of the overlay, it is almost impossible to perform the maneuver without collisions.

Acknowledgments

The research described in this chapter is supported by the Dutch Technology Foundation STW in cooperation with Fokker Space & Systems B.V., the European Space Agency ESA, and the National Aerospace Laboratory NLR.

References

Bos, J.F.T. (1988). Pilot experiments on the manual control of a space manipulator in the single joint control modes. Report N-293, Delft University of Technology, Faculty of Mechanical Engineering and Marine Technology, Laboratory for Measurement and Control, Delft, the Netherlands.

Bos, J.F.T. (1991). Man-machine aspects of remotely controlled space manipulators, PhD Thesis, Delft University of Technology, Faculty of Mechanical Engineering and Marine Technology, Laboratory for Measurement and Control, Delft, the Netherlands.

Bos, J.F.T., Stassen H.G., Van Lunteren A. (1995). Aiding the operator in the manual control of a space manipulator. *Control Engineering Practice,* 3: No 2, 223- 230; also *Proc. 5th IFAC Symp. on Analysis, Design and Evaluation of Man-Machine Systems*, June 9-11, 1992, the Hague, Oxford, the Netherlands, Pergamon Press, pp.215-220.

Breedveld, P. (1992). Controlling a space manipulator from earth with predictive display techniques. *Proc. 11th European Annual Conference on Human Decision Making and Manual Control,* Nov. 17-19, Valenciennes, France, paper 27.

Breedveld, P. (1995a). The development of a Man-machine interface for telemanipulator positioning tasks. *Proceedings. 6th IFAC Symp. on Analysis, Design and Evaluation of Man-Machine Systems*, June 27-29, MIT, Cambridge, Massachusetts.

Breedveld, P. (1995b). The development of a predictive display for space manipulator positioning tasks. *Proeedings. 14th European Annual Conference on Human Decision Making and Manual Control,* June 14-16, Delft University of Technology, Delft, the Netherlands, Session 7-1, pp. 1-8.

Buiël, E.F.T. (1994). A laboratory evaluation of four controller configurations for a teleoperated space manipulator. Report N-472, Delft University of Technology, Fac. of Mechanical Engineering and Marine Technology, Laboratory for Measurement and Control, Delft, the Netherlands.

Buiël, E.F.T. and Breedveld P. (1995a). A laboratory evaluation of four control methods for space manipulator positioning tasks. *Preprints 6th IFAC Symposium on Analysis, Design and Evaluation of Man-Machine Systems*, June 27-29, MIT, Cambridge Massachusetts.

Buiël, E.F.T. and Breedveld P. (1995b). A laboratory evaluation of two graphical displays for space manipulator positioning tasks. *Proceedings 14th Annual Conference on Human Decision Making and Manual Control*, June 14 - 16, Delft University of Technology, Delft, the Netherlands, Session 7-2, pp. 1-8.

Diepenbroek, A.Y. (1994) Development of a real-time simulation model, including friction, of the ERA space manipulator (in Dutch). Report A-704, Delft University of Technology, Faculty of Mechanical Engineering and Marine Technology, Laboratory for Measurement and Control, Delft, the Netherlands.

Kim, W.S.,Tendick, F, Ellis, S.R, Stark, L. W. (1987). A comparison of position and rate control for telemanipulations with consideration of manipulator system dynamics, *IEEE Journal of Robotics and Automation,* RA-3, 426-436.

Sheridan T.B. (1992). *Telerobotics, Automation and Human Supervisory Control*, Cambridge, MA: MIT Press.

Van Woerkom, P.T.L.M, et al. (1994). Developing algorithms for efficient simulation of flexible space manipulator operations, *Proceedings 45th Congress of the International Astronautical Federation*, Oct. 9-14, Jerusalem, Israel, pp. 1-13.

Chapter 17

Visualization in Tele-Manipulator Control

Gerda J.F. Smets

In this paper an overview of the Delft Virtual Window System (DVWS) is given, referring to definitions of concepts from robotics and perception psychology that are crucial in order to explain the working principle of the system, as well as the system's performance measurements. The second section refers to the theoretical background. It addresses the question of what is the relevant visual input in optic flow to allow for depth perception and, therefore, for telemanipulation and for enhanced tele-presence? The third section discusses tele-presence applications using the DVWS.

The paper emphasizes the critical human-machine interaction research problems from the perspective of the Gibsonian theory of perception. It is parochial in that it deals with research and design engineering problems tackled at our laboratory. Our research effort concerns the role of (observer-generated) movement in perception and its application in technology.

1   Introduction

The question of how we come to see a world containing objects that have stable sizes and shapes and positions is of practical technological relevance for the design of workstations that can carry out physical tasks safely in a hostile environment, e.g., a nuclear hot cell. However, this question is also of high theoretical interest, and it has dominated the psychology of perception. We want to show how the answers to this question offered by perception psychology, and especially the answer given by the Gibsonian theory of perception might be of help in optimizing the design of teleoperators. We will limit ourselves to the visual aspects, television. The whole world has been exposed to television, and the technology involved is inexpensive and highly developed. However, there remain some problems that, while not critical in ordinary television programming, are critical for teleoperation. Foremost among these is depth perception, which continues to be a major reason why direct manipulation performance is not matched by telemanipulation (Sheridan, 1989; Stassen & Smets, 1995). Analogous to color televisions we will speak of depth television when referring to television allowing for telemanipulation. In this chapter we present a new working principle for such a system, referring to its performance measurements, theoretical background, and applications.

Sheridan (1989, 1992) gives an excellent overview of the history of tele-operating and tele-presence, including a survey of current applications. Vertut & Coiffet (1984) provide a review of the mechanical technology that still relates to recent developments and current problems. As background for the psychology of perception, we refer to Bruce and Green (1990), since they consider perceptual theory from the point of view of its relevance for daily life as well as its scientific merit.

2   Definitions and overview of the DVWS

2.1   DEFINITIONS

The concept of teleoperation is discussed along with the related concepts of robotics and tele-presence. The definitions used are adapted from Sheridan (1989). *Robotics* is the science and art of performing, by means of an automatic apparatus or device, functions ordinarily ascribed to human beings, or operating with what appears to be almost human intelligence. A *teleoperator* includes at the minimum artificial sensors, arms and hands, a vehicle for carrying these, and communication channels to and from the human operator. The term *teleoperation* refers to direct and continuous human control of the teleoperator. Teleoperation problems can be divided into four broad categories:

1. Tele-sensing, including vision, resolved force, touch, kinesthesia, proprioception, and proximity
2. Tele-actuating, combining motor actuation with sensing and decision making
3. Computer aiding in human supervision of a teleoperator
4. Meta-analysis of the human-computer teleoperator-task interaction

This chapter is limited to tele-sensing and, more specifically, to depth television, from a meta-analytical point of view. *Telepresence*, then, is the ideal of sensing sufficient information about the teleoperator and task environment, and communicating this to the human operator in a sufficiently natural way that the operator feels physically present at the remote site.

When talking of depth television we are talking of television allowing for telemanipulation and providing tele-presence. The television screen forms a virtual window such that what lies behind it forms a rigid whole with the world before it, just like a window in daily life. We call this system the Delft Virtual Window System (DVWS).

The present work is embedded in the search to account for the perceptual consequences of active movement generated by an observer as opposed to merely passive motions presented to a passive onlooker. Therefore, we define the concepts of optic array, optic flow, motion, and movement along with the related concepts of movement parallax and motion parallax, proprioception, exteroception, and exproprioception. A good lexical discussion about those concepts is given by Owen (1990). Experimental data illustrating the relevance of this distinction are discussed in the second section of this chapter.

In the Gibsonian approach to perception, the concept of reciprocity between individual and environment is fundamental. Perceiving is an act of an individual in an environment, not an activity in the nervous system (Mace, 1977; Gibson, 1979, pp. 239 - 240; Owen, 1990). The visual system has a dual role, being at once both *exteroceptive* (giving information about extrinsic events in the environment) and *proprioceptive* (giving information about the actions and reactions of the individual) (Gibson, 1966). Following Lee (1976) we also mention exproprioceptive information about the position of the body or parts of it relative to the environment.

Gibson (1966) defined the ambient optic array as the arrangement of variations of the light intensities (structured light) available at an observation point, a point in space where an observer can sample light. For each observation point a unique optic array exists.

The optic flow is the optic array at a moving observation point. Variations in the observation point can be generated by the observer or can be imposed on the observer. Imposed displacements occur when members of the body are moved, when the head is accelerated or turned, when the whole individual is passively transported and the eyes are stimulated by motion perspective, or when the observation point is passively moved as when looking at motion pictures. Following Cutting (1986) we use the terms *motion* for imposed displacements, resulting in a purely exteroceptive experience, and *movement* for observer-generated displacements, occurring when they behave or perform with any of the motor systems of their body, giving potential information for proprioception and exteroception as well as exproprioception. In accordance with the distinction between motion and movement, we differentiate between motion parallax and movement parallax. This difference is comparable to Sedgwick's (1986) distinction between relative and absolute motion parallax. *Parallax* refers to the fact that when two objects are at different distances from a moving observation point, the objects seem to shift relative to each other. We speak of *movement parallax* when the movement of the observation point is self-generated, and of *motion parallax* when this motion is imposed.

Suppose that you are walking in the countryside and looking at the landscape. Let us also suppose that your gaze direction is perpendicular to your moving direction, that your direction of movement is from right to left, and that you are gazing at the fixation point. Under these conditions all of the objects closer to you than the fixation point would appear to move in a direction opposite to your movement. On the other hand, objects that are farther away will appear to move in the same direction you are moving. Not only the direction but also the speed of movement varies with the object's proximity to your fixation point and your movement speed and direction. This is movement parallax. Motion parallax contains the part used by a cinematographer. He uses the apparent movements around the fixation point as registered by camera movements, without them being linked to the spectators' movements. With movement parallax, potential information for the absolute distance of the object from the observer would be available if the observer were able to register both the translatory component of the eye's movement relative to the object and the visual direction of the object. With motion parallax there is no potential information for absolute distance, the distance between ourselves and the perceived object. Both movement and motion parallax offer potential information for *shape perception,* a distance order within a limited cluster of coherent features (defined by Koenderink, 1990). With the help of those definitions, an overview of the DVWS is given, with reference to performance measurements.

2.2   OVERVIEW AND PERFORMANCE MEASUREMENTS OF THE DVWS

Figure 1. The Delft Virtual Window System operates by sensing the observer's head position and moving the camera in the remote site accordingly. The screen then no longer acts as a screen but rather as a window, wherein what lies behind it forms a rigid whole (and therefore a virtual reality) with what is in front of it. The sensor at the observer's head is only symbolic. Using existing technology, head position can be registered without a sensor attached to the head.

If perception and movement are tightly interlocked, as is the case in movement parallax, then exproprioceptive information must be provided. Therefore we contend that depth television, where the screen is perceived as a window with a virtual reality behind it, can be created by continuously updating the display output to match the observer's movements in front of the screen. This is accomplished by sensing the observer's head position and moving the camera in the remote site accordingly (Fig. 1). Patents on the working principle have been obtained for Europe and the United States (Smets, et al., 1988, 1990). The original version of the system used a simple video camera and an observer watching with one eye only. Other versions used stereo and computer-generated images, but the monocular version already provides a compelling and reliable spatial impression, contrary to what most textbooks on perception and the majority of existing spatial display systems (reviewed in Overbeeke & Stratmann, 1988) suggest, in which space perception is often equated with the use of two eyes.

Exproprioception is lost when the visual input is considered without the observer's movements, as is the case with ordinary television programming. In such a case the visual input cannot be scaled relative to the observer's body movements. This means that the perception of a rigid world, allowing for adequate spatial behavior and, hence, telepresence, is impossible. Shape perception, however, as we all know, remains possible.

Experiments with the DVWS (Smets, et al., 1987; Overbeeke, et al., 1987; Overbeeke & Stratmann, 1988) indicate a perceptual advantage of the active observer, whose head movements steer the camera, over the passive onlooker, whose movements are not coupled to the display output (Fig. 2). Both active and passive observers receive identical output on their monitor screens, yet their perceptual input differs. The former receives movement parallax information, the latter motion parallax (Fig. 3). The task consists of aligning wedges by means of a remote control.

Figure 2. Performance measurements of the DVWS: A sketch of the experimental design. Two observers receive the same display output, yet the head movements of the left-hand observer steer the camera. His or her visual input contains movement parallax information, whereas the observer on the right disposes of motion parallax information. Although the displays are the same, the perceptual input differs (see Fig. 3). This proves to have a significant effect on spatial performances.

As shown in Figure 2, the experimental setup consists of a frame with a central rod and three wedges. The camera and the rotation of the camera (as steered by the active observer) are focused on the central rod. Yet another wedge is mounted in front of the monitor. The lower rear wedge and the upper frontal wedges are reference wedges. They can be placed in one of three positions (0, -2.5, and +2.5 cm). The upper rear wedge and

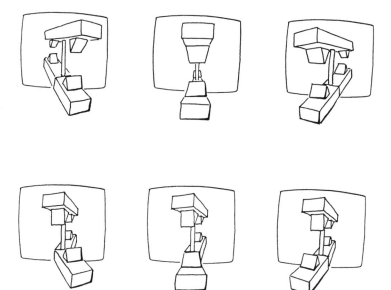

Figure 3. Views of the DVWS display. Like a hologram, the DVWS implements a stable link between the visible layouts of the remote (or virtual) and real environments. If, for example, the observer whose movements steer the camera places his finger "visually on" a remote object, it will appear there even when he moves his head (top). Passive observers, whose movements do not drive the camera, do not experience this overlay (bottom).

the wedge mounted in front of the screen are the aligning wedges. They are randomly placed within a range of 15 cm forward or backward from the 0 position of the corresponding reference wedges.

Both observers were required to align the wedges "in" or "in front of" the screen. In half of the conditions the active observer aligned the wedge "in front of" the screen (condition 1), while the passive observer at the same time was aligning the wedges "in" the screen (condition 2). In the other trials the reverse was done: The active observer aligned the wedges "in" the screen (condition 3), whereas the passive one aligned those "in front of" (condition 4). A control condition was added to assert the maximum level of reliability of depth estimates for a comparable task in a real-life condition (Fig. 4).

Figure 4. Control condition with real wedges.

Given reports in the literature of a learning effect of depth estimation based on motion parallax (Ferris, 1972), each subject went through a training phase in the real-life setup. The learning phase was terminated once the deviation of the alignment wedge from the reference wedge did not increase more than 5% in three successive trials. The reliability of the depth estimates in the real-life condition was then measured in 10 alignment trials per subject, without feedback. When this was over, the subject was passed on to either the active or the passive condition according to a preset random order.

Forty subjects took part in the experiment, 10 in each condition, each of whom carried out 10 depth estimates. All 40 were tested in the control condition. Were more subjects necessary to obtain significance, the working principle under investigation would appear not to lead to useful teleoperation applications. For the same reason, we used a somewhat insensitive test: We calculated the general mean and variance of depth estimates for each condition in the experimental design without eliminating the variance among subjects. Since we wanted to estimate the reliability of depth estimates, the variances of the depth estimations are more important than their means: An unreliable depth estimation cannot be corrected, whereas a systematic over- or underestimation of depth can. Therefore F-tests were used.

Telemanipulations of the active operator almost match direct manipulation performance: The average over- or underestimations are consistently small and their variance does not differ significantly from the variance of errors in real life condition ($F = 1.24$, with 10 subjects in each condition). The tele-performance of the passive onlooker, on the contrary, is not consistent, although the average error size remains small. The variance of this error size is significantly greater than that of the active observer ($F = 3.25$, $p < 0.01$, 10 subjects in each condition) and than obtained in the real-life setup ($F = 2.25$, $p < 0.01$, 10 subjects in each condition). Data indicate that the active condition, where movement parallax information was provided on the display, even allows for things to apparently leap out of the screen. In this case perceived depth is somewhat less compelling, however, than in the real life setup. There is a systematic underestimation in aligning a wedge virtually leaping out of the screen with a real wedge mounted in front of the display, yet the variance remains relatively small. Movement parallax allows for telemanipulation, and motion parallax does not. How can one account for these data?

3   Perceptual theory and tele-presence

In order to describe the importance of motion and movement for telemanipulation, we begin by contrasting two positions, which we call traditional and Gibsonian. We describe how this came about and what the consequences are for telemanipulation and telepresence. By the traditional approach, we mean psychologists and computational theorists of the tradition who describe the visual input in terms of point intensities of light at the retina. By the Gibsonian approach to perception, we mean those theorists who argue that the input for vision can be described in terms of the structure of light. The theory is rooted in the research of James Gibson and his followers. We regard the two approaches as complementary.

3.1   TRADITIONAL APPROACH

Often the eye is thought of as a camera acting to focus light onto a mosaic of retinal receptors. At any instance of time, therefore, one can conceive the pattern of excitation of retinal receptors as a picture curved around the back of an eyeball (Bruce & Green, 1990, p. 142). Though curved, the image is two-dimensional, and yet our perception is of a three-dimensional world. How might depth be recovered? Without invoking learning or memory, pictorial cues and binocular disparity are the most popular answers, with motion parallax being a good third. The movement of the observer is thought to be irrelevant for visual perception. In fact, experiments about visual perception, even those about parallax effects, are often designed so that the subject cannot move with his or her head being fixed by a chin rest, for example.

As a result, the adherents of this approach create tele-presence by reproducing a high resolution stereoscopic image of the tele-environment. As far as motion is concerned, it is restricted to presenting parallax shifts on the stereo display as a consequence of performances in the tele-environment. Movement of the human operator is important for controlling the mobility of the effectors of the workstation, but it is not considered important for visual perception as such. The human operator provides largely symbolic commands (concatenations of typed symbols or specialized key presses) to the computer. However, some fraction of these commands still are analogous (body control movements isomorphic to the space time force continuum of the physical task) since "they are difficult for the operator to put into symbols" (Sheridan, 1989, p. 488). Yet this analogy is not considered important to enhance either tele-performance or telepresence.

A rather unpleasant human factor with which these teleoperator designers must contend is simulator sickness, a euphemism for heavy nausea if there is a conflict between the operator's perception of motion (shifts on the stereo display of the head-mounted system) and his or her own movements (which is strengthened by the stereo display often producing pseudo-parallax and pseudo-convergence).

3.2   GIBSONIAN APPROACH

In the traditional approach, shape and depth perception are expressible in essentially static and pictorial terms. However, they can also be defined, in accordance with Gibson, as the detection of formless invariants over time. For a moving observer, the retinal projection undergoes a continuous serial transformation on the retina, due to movement parallax, that is unique for this movement pattern in this environment. We believe that movement parallax is sufficient to create (tele-)presence, by rigidly linking the observer's proprioceptive experiences to his or her exteroception, providing for exproprioceptive information.

Reduction of movement parallax to motion parallax

Movement parallax causing space perception was first described by von Helmholtz (Gibson et al., 1959). He postulated that a difference in angular velocity between the projections of objects on the retina is an indication of the perception of distances between objects and between the observer and an object. From that time on, typical experiments were designed so that the subject could not move, thus reducing movement parallax to motion parallax. This was done partly for reasons of convenience (it is sometimes easier to simulate movement of the display than to allow the head to move) and partly to eliminate questions about the role of proprioceptive information concerning head movements (Sedgwick, 1986, p. 45).

At the same time, a distinction has been drawn between the role of movement parallax in absolute and relative distance estimation, with the emphasis being placed on estimation of relative distances. It was shown that motion parallax provides information about the relative distances of points or surfaces around the subject, but it cannot specify absolute distances or surfaces. A human or animal may be able to solve this problem by scaling its visual input through head or body movement of a fixed velocity (Bruce and Green, 1990, pp. 259-260), thus using movement parallax instead of motion parallax. It has been proven that animals do this.

Head movements by human beings might be useful to their visual perception as well. Rogers & Graham (1979) and Graham & Rogers (1982) built a setup in which movements of the subject's head caused shifts on a screen (Fig. 5) corresponding to the projection of different wave patterns (square wave, sine wave, triangular wave, or sawtooth wave) on the screen. When the subject kept his head still, he saw a random dot pattern. However, if he moved he had a convincing and unambiguous shape impression of the different patterns, seen from different angles. The phenomenon only occurred when the pattern on the screen was shifted according to the movements of the observer. Perceived depth was more pronounced than in an analogous experiment where motion

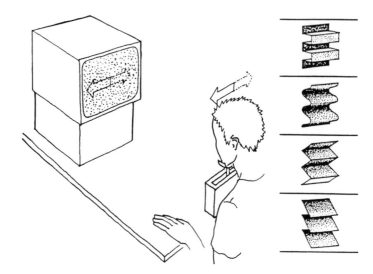

Figure 5. Experimental setup of the Rogers and Graham (1979) experiment (for explanation see text).

parallax was used. In this experiment the observer's head remained fixed and the oscilloscope moved back and forth, with internal depth in the display specified by motion gradients tied to the movement of the oscilloscope. This demonstrates the effectiveness of movement parallax alone in producing vivid perceptions of surface layout in depth and suggests that previous failures to find such pronounced effects have been due to limitations in experimental displays used to simulate movement parallax, rather than to limitations in the visual system's ability to make use of it (Sedgwick, 1986, pp. 21-46).

Evidence for the relevance of movement, and therefore also movement parallax, for visual perception has also come from studies of the development of perceptual skills. The applicability of the principle of active control for perception and, hence, movement parallax is witnessed by the successful experiments in sensory substitution, such as Bach-y-Rita's (1972) tactile visual substitution system and Bower's sonar visual substitution system (1977, 1978; Aitken & Bower, 1982). Stassen (personal communication) mentions that thalidomide babies, born without arms, did not show adequate spatial behaviour when they started walking. Yet they did when they received elementary (rigid) arm prostheses from their first months of life. The fact that they encountered obstacles with their arm prostheses seemed to be of help in developing their depth perception.

Importance of the task

Whether motion is sufficient for adequate spatial behavior, or whether movement is required at all, depends on the visual task. For some perceptual tasks shifts in the optic array can be reduced to shifts on the retina. This applies to those tasks that do not imply relative shifts in the optic array, or, in perceptual terms, where two-dimensional form perception is at stake with neither shape nor depth perception being necessary. For this kind of task the movement of the observer is irrelevant. Think of the recognition of a bird high in the sky, or the recognition of a landmark far below when you look out of an airplane, or a graphic task . This type of task can easily be simulated on a computer display, with standard animation software and video processor hardware. The screen is perceived as a flat screen.

Yet, most perceptual behavior, e.g., all manipulation tasks, implies shifts in the optic array. Different parts in the optic array shift relative to each other as well as to the observer. Those kinds of tasks are concerned with shape and depth perception, or, in other words, with the perception of a rigid world. For them, the visual input cannot be simulated in the same way as for the former tasks, since here the exploratory movements of the observer causing the serial transformation in optic flow are essential. Simulating a virtual world in which the visual input forms a rigid whole with the perception-action system of the observer, and which in turn allows for telemanipulation, requires the movements of the observer to be coupled to the shifts in the optic array. This can be realized on an ordinary television screen if the video camera in the tele-environment is steered by the head movements of the observer. This creates a monocular virtual window display: The television screen does not look like a flat display any more, but like a window with a virtual reality behind it, forming a rigid whole with what is in front of it, allowing for reliable and convincing depth and shape experience. There is no need for stereo images. As long as movement (head movements of the operator) and motion (shifts on the display) are consistent, there is no simulator sickness.

The optic flow of those tasks cannot be simulated on a two-dimensional display without the movements of the observer and the motion on the screen being coupled, as described earlier. In that case simulation of flow in the ambient optic array is confounded with and reduced to the simulation of retinal flow. Flow on the retina or in the image does not necessarily copy flow in the array. How failure to distinguish flow in the optic array from flow on the retina can result in confusion is clearly stated by Owen (1990, pp. 44-47). A second implication when studying the second kind of task is that we think it might be necessary to extend the models of optic flow to include the movements of the observer, thus pointing to a kinetic concept of optic flow rather than a merely kinematic one. This is explained in Smets (1995) and Stappers (1992).

4   Applications: New perceptual aids using telepresence

A history of applications of teleoperation extension of human sensing and manipulating capability by coupling to (remote) artificial sensors and actuators is given by Sheridan (1989). This review of relevant historical developments shows that teleoperation is primarily used for manipulation in hazardous or inaccessible environments (e.g., in outer space, undersea, or nuclear "hot laboratories") or to compensate for motor deficiencies (e.g., for the disabled or, more generally, when forces have to be exerted that are far too large for a human operator). Until now teleoperation has not been used to overcome visual (or, more general, perceptual) deficiencies. Through the recent attainment of the sense of tele-presence in telemanipulation, however, we are convinced that this is a promising, yet virgin, territory for applications. Two applications illustrate how we faced this challenge. We will demonstrate the advantage of simple human-supervised teleoperation for visual tasks, using the DVWS working principle.

Needless to say, although we will not elaborate on them, those designs are subjected to extended performance measures and assessment techniques. We are conducting studies considering the specific mission requirements of both industrial design engineering projects described later. This is being done in collaboration with industry. At the moment there are essentially no accepted standards for asserting the accuracy, repeatability, linearity, etc. of tele-operating systems in general.

4.1   HELPING THE VISUALLY IMPAIRED

Leifer's (1983) review of the use of tele-robotics for the disabled illustrates that most applications are designed for motor disabilities. One (famous) exception is the development of a tele-operated electromechanical guide dog for the blind (Tachi, 1981). However, we think that this is a promising field for implementation. We will illustrate this with two examples.

4.2   THREE DIMENSIONAL ENDOSCOPY

For medical and industrial applications, a 3D endoscope based on the DVWS will support tele-exploration and telemanipulation tasks. In medical endoscopy, depth perception is crucial, because the surgeon might otherwise damage tissue. When the head movements of the surgeon are coupled to the movements of the tip of the endoscope around the fixation point, he or she is able to make correct depth estimates and to distinguish front-back reversals. The same applies for industrial endoscopy. The image of the interior of an engine is ambiguous. For example, is a visible crack really a crack in the surface or just a droplet of oil on the surface? A DVWS endoscope might be the answer here.

4.3   LUGGAGE INSPECTION

An especially exciting challenge is to develop a luggage inspection teleoperator. For security reasons, the inspection of personal and freight luggage becomes more and more important. Part of this control consists of inspecting luggage with an X-ray camera. Luggage conveyed through such security devices can only be inspected from one viewpoint, that of the camera. Nevertheless, the safety attendant often would like to look in and around every hole and corner to solve perceptual ambiguities. A pistol, for instance, can be easily confounded with a lot of other objects when only looked at from one point of view. If the X-ray camera is steered by the head movements of the attendant, perceptual ambiguities disappear and depth perception is provided for. The working principle of this luggage inspection device, which can be used for freight as well as for personal luggage, is shown in Figure 6. The security attendant can do the inspection task nearby or at a safe distance. The advantages of this security system as compared with the existing security systems have been praised by Amsterdam airport's security service. Product development is under way.

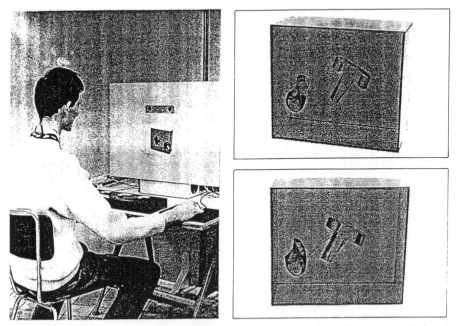

Figure 6. Experimental setup for DVWS X-ray luggage inspection. The observer's head movements are measured (by the Dynasight above the screen) and steer the retrieval of images from the computer memory. Are the two objects in the box connected by a thin wire, or not? On the right side of the figure two possible views are shown. This illustrates that, when moving, one can make out that they are not.

5 Conclusions

Research with and on teleoperators persuades us that the principles of the Gibsonian theory of perception offer a powerful engineering paradigm and also demonstrates that this applied research can help to clarify the nature of tele-presence itself. Since tele-presence is commonly claimed to be important for direct telemanipulation, a theory of presence/tele-presence is sorely needed (Sheridan, 1992). We think that the ecological theory of perception might be of help here, by explaining the difference between motion and movement parallax in terms of their perceptual meaning, i.e. their importance in allowing for adequate spatial behavior.

6 Acknowledgments

Work mentioned in this paper was supported by almost the entire staff of the laboratory of Form Theory at the Department of Industrial Design Engineering at Delft University of Technology. This is a multidisciplinary team consisting of, among others, industrial design engineers, psychologists, and physicists working together to build an interdisciplinary approach to perception. I specifically want to thank C.J. Overbeeke, P.J. Stappers, O.A. van Nierop, R. Wormgoor, A.M. Willemen, G.W.H.A. Mansveld, H. Subroto, A.C.M. Blankendaal, J.P. Claessen, and C. Kooman.

References

Aitken, S. and Bower, T.G.R. (1982). The use of the sonicguide in infancy. *Journal of Visual Impairment and Blindness*, 91-100.

Bach-y-Rita, P. (1972) *Brain mechanisms in sensory substitution*. New York: Academic Press.

Bower (1977). Blind babies see with their ears. *New Scientist*, 73:255-257.

Bower (1978). Perceptual development: Object and space, In E.C. Carterette and M.P. Friedman (Eds.) *Handbook of Perception*, Vol. VII, Perceptual Coding. New York: Academic Press.

Bruce, V. and Green, P. (1990) *Visual Perception: Physiology, Psychology, and Ecology*. London: Lawrence Erlbaum Associates.

Cutting J.E. (1986). *Perception with an Eye for Motion*. Cambridge, MA: MIT.

Ferris, S.H. (1972). Motion parallax and absolute distance. *Journal of Experimental Psychology*, 95: 258-263.

Gibson, J.J. (1966). *The Senses Considered as Perceptual Systems*. Boston,: Houghton-Mifflin.

Gibson, J.J. (1979). *The Ecological Approach to Visual Perception*. Hillsdale, NJ: Lawrence Erlbaum.

Gibson, E.J., Gibson, J.J., Smith, O.W., and Flock, H. (1959). Motion parallax as a determinant of perceived depth. *Journal of Experimental Psychology* 58: 40-51.

Graham, M., and Rogers, B. (1982). Simultaneous and successive contrast effects of depth from motion-parallax and stereoscopic information. *Perception*, 11: 247-262.

Koenderink (1990. Some theoretical aspects of optic flow. In: R. Warren and A.H. Wertheim (Eds.) *Perception and Control of Self-Motion*. Hillsdale, NJ: Erlbaum, pp. 53-68.

Lee, D.N. (1976). A theory of visual control of braking based on information about time-to-collision. *Perception*, 5, 437-459.

Leifer (1983). Interactive robotic manipulation for the disabled. *Proceedings of the 26th IEEE Computer Society International Conference COMPCON*, San Francisco, CA, pp. 46-49.

Mace, W.M. (1977). James J. Gibson's strategy for perceiving: Ask not what what's inside your head, but what your head's inside of. In R. Shaw and J. Bransford (Eds.), *Perceiving, Acting, and Knowing*. Hillsdale, NJ: Erlbaum, 43-65.

Overbeeke, C.J., and Stratmann, M.H. (1988). Space through movement. Ph.D. thesis, Delft University of Technology.

Overbeeke C.J., G.J.F. Smets, and M.H. Stratmann (1987). Depth on a flat screen II. *Perceptual & Motor Skills*, 65: 120.

Owen, D.H. (1990). Lexicon of terms for the perception and control of self-motion and orientation. In R. Warren and A.H. Wertheim (Eds.) (1979). *Perception and Control of Self-Motion*. Hillsdale, NJ: Erlbaum, pp. 33-50.

Rogers, B. and Graham, M., Motion parallax as an independent cue for depth perception. *Perception*, 6: 125- 134.

Sedgwick, H.A. (1986). Space perception. In K.H. Boff, L. Kaufman, and J.P. Thomas (Eds.). *Handbook of Perception and Human Performance* Vol. I. New York, NY: Wiley.

Sheridan, T.B. (1989). Telerobotics. *Automatica*, 25: 4, 487-507.

Sheridan, T.B. (1992). *Telerobotics, Automation and Human Supervisory Control.* Cambridge, MA: MIT Press.

Smets, G.J.F. (1995). Designing for telepresence. In J.M. Flach, P.A. Hancock, J.K. Caird and K.J. Vicente (Eds.), *An Ecological Approach to Human Machine Systems* II: Local Applications. Hillsdale, NJ: Erlbaum, pp. 182-207.

Smets, G.J.F., Overbeeke, C.J., and Stratmann, M.H. (1987). Depth on a flat screen. *Perceptual & MotorSkills* 64: 1023-1024.

Smets, G.J.F., Stratmann, M.H., and Overbeeke, C.J. (1988, 1990). Method of causing an observer to get a three-dimensional impression from a two-dimensional representation. U.S. Patent 4757, 380 (1988) and European Patent 0189233 (1990).

Stappers, P.J. (1992). Scaling the visual consequences of active head movements. Ph.D. thesis, Delft University of Technology.

Stassen, H.G. and Smets, G.J.F. (1995). Telemanipulation and telepresence: A contribution to the design and evaluation of man machine systems. *Proceedings of the IFAC MMS*, MIT: Cambridge, MA, June 27-29.

Tachi, S. (1981) Guide dog robot, feasibility experiments with MELDOG Mark 3. *Proceedings of the 11th International Symposium on Industrial Robots*, pp. 95-102.

Vertut, J. and Coiffet, P. (1984). *Robot Technology. Vol. 3a: Teleoperations and Robotics, Evolution and Development.* Englewoods Cliffs, NJ: Prentice Hall.

Chapter 18

Minimally Invasive Surgery: Human-Machine Aspects and Engineering Approaches

Kees A. Grimbergen

Minimally invasive surgery is a recent development in surgery with important consequences for the functioning of the surgeon. In this technique the view of the surgeon is based on a laparoscope, a camera presenting a two-dimensional image of the operating field. The tactile observations and perception of forces are obtained via special tools devised for this type of surgery. A model of the surgeon in conventional open surgery and in minimally invasive surgery is generated to explain the consequences of the differences between the techniques.

Engineering approaches to improve minimally invasive techniques are discussed. A method of analyzing surgical procedures and their technological aspects is described. Its use in introducing changes in the technique is addressed. The evaluation method is based on the standardized evaluation of time synchronous multiple video recordings of surgical procedures. The application of the analysis is illustrated in the case of advanced surgical procedures showing the evolution of the technique during the learning curve. The demands for a solution with active tools based on force reflection are treated, and the prospects for telemanipulation surgery are addressed.

Finally, an outline of the project on minimally invasive surgery based on cooperative work between the Delft University of Technology and the Academic Medical Center of the University of Amsterdam is given. The involvement of clinicians, engineers, and industry in this multidisciplinary project is discussed.

1 Introduction

Important developments have been observed in surgery recently, and minimally invasive surgery or keyhole surgery is probably the most outstanding (Satava, 1993). This type of surgery is based on the use of a limited number of feedthroughs, so-called trocars, applied, for instance, through the abdominal wall. Surgical tools are inserted through these trocars, which have a diameter of 5 - 12 mm. In order to enable the surgeon to observe his or her actions, a small camera is introduced through one of the trocars. The video camera with its fiber optics is called an *endoscope* (internal scope) or *laparoscope* (lap scope). Owing to the prominent role of the camera, the method is also referred to as *endoscopic* or *laparoscopic surgery*. The abdominal cavity is inflated, using carbon dioxide, or mechanically expanded so that a working space is created in which to operate the surgical tools.

Minimally invasive surgery in principle can be applied to any area of the body, including the abdomen (laparoscopy), chest (thoroscopy), joints (arthroscopy), gastrointestinal tract (e.g., coloscopy of the colon), and blood vessels (angioscopy). As we are mainly concerned with the abdominal applications of the technique, most of the material presented herein will pertain to laparoscopy.

The minimally invasive surgical technique may have great benefits for the patient. Injuries to the abdominal wall are much less than with conventional surgery, in which extensive incisions produce larger scars that are detrimental to the speed of recovery after the operation. Also, in principle, the risk of infections is smaller because of a virtually closed system. Finally, there is the tendency to operate with higher accuracy due to magnification by the camera.

All these advantages for the patient have stimulated the development of this type of surgery, which has only been in use since 1989, when it really took off after the introduction of the Charge Coupled Device semiconductor chip (CCD) camera (Classen et al., 1987). It was primarily the success of the laparoscopic removal of the gallbladder that contributed to its increased use. That operation has now become generally accepted and has reduced hospital stays from 8 to 2 days. The potential advantage of fast recovery of the patient has led to a strong increase in interest in recent years, and new applications are reported continually (Boutin et al., 1991; Dallemagne et al.; 1991, Jansen, 1994; White & White, 1989).

Minimally invasive surgery, however, is also characterized by the surgeon's lack of direct mechanical contact with the tissue to be operated upon and by the lack of a direct view of the area of operation; the surgeon has to operate using a video image for guidance. There are severe disadvantages for the surgeon involved with the application of this complicated technique (Herfarth et al., 1994). Another consequence of the complexity of the procedure is that the quality and safety of surgery depend even more on the skills of the individual surgeon than with conventional surgery. It can nevertheless be expected that this technique will continue to supplant conventional operations in the abdominal and thoracic cavities. With these new applications, however, the disadvantages of the technique are much more prominent than with relatively simple gallbladder removal. The minimally invasive approach will have to be improved and the disadvantages be reduced to extend the range of its surgical applications.

In the following discussion, the human factors involved in open surgery and minimally invasive surgery are first treated. Then two engineering approaches to improve the technique of minimally invasive surgical procedures are presented. Finally, the multidisciplinary approach resulting from the cooperation between the Mechanical Engineering Department of the Delft University of Technology and the Faculty of Medicine of the University of Amsterdam is discussed.

2   Surgeon's model

2.1   OPEN SURGERY

In conventional, or open, surgery the procedure is characterized by a surgeon who invades the body with an incision through healthy tissue, which is made large enough to expose the organ to be operated on to the surgeon's eyes and hands.

Perception. In open surgery, a direct three-dimensional view of the operating field is offered based on the binocular vision of the surgeon. Sometimes this view is impaired by a location deep in a cavity of the body or by the small dimensions of the organs involved. Then assisting provisions, such as special illumination or microscopes for microsurgery, are needed. There is, however, great freedom to approach the organ in a versatile manner for optimal perception. This same freedom is present for tactile observations by the surgeon. The surgeon is free to handle the organs with the hands restricted only by gloves. In this way the surgeon can judge the condition of the tissues and feel the normalities and abnormalities in strength or consistency of tissues based on experience. Recapitulating, the surgeon can use the appropriate senses, most importantly the vision and tactile senses, in a nearly optimal way in accordance with daily use of these senses.

Tools. During the open procedure, the surgeon can use the whole realm of handheld surgical tools, with the lancet being the most important representative. Handheld tools are ergonomically well developed and provide mechanical information through force feedback of the interaction between the tissue and the tool. This is especially true when the surgeon uses his or her own hands as a tool, for instance, for stretching or separating tissue. Complex motions with the tools can be performed due to the fully exposed operating field.

Open Surgery

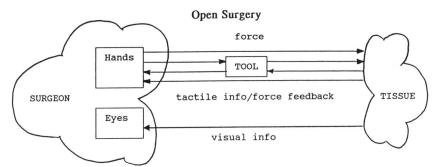

Figure 1.   Model of a surgeon in open surgery. The tissue is exposed for visual and tactile observations by the surgeon. Direct force and force feedback via the tool are used.

Open surgery can thus be modelled by a feedback loop containing the surgeon with his or her senses, hands, and tools (Fig. 1). It can be noted that in open surgery the surgeon is working autonomously with only some additional handheld tools and perception aids.

2.2 MINIMALLY INVASIVE SURGERY

Minimally invasive surgery is characterized by the use of multiple small incisions producing minimal damage to healthy tissue of the patient. As a direct consequence, the visibility and accessibility of the operating field are restricted and organs are not freely exposed to the eyes and hands of the surgeon. The sensing and actuator capabilities of the surgeon are thus severely restricted.

Perception. In minimally invasive surgery, perception of the operating field is provided by a camera, an endoscope with illumination giving a two-dimensional view on a television screen in the operating theater. Although magnification may enhance vision, the field of view and the restricted directing freedom in one of the trocars limit perception capabilities. The properties of the video signal, such as limited resolution, contrast, and color fidelity, present a further reduction in the vision of the surgeon (Tendick et al., 1993). Additionally, in practice, the image quality is often impaired by contamination of the endoscope lens and by smoke produced by diathermy cutting. So the view on the operating field is significantly reduced by an image of limited quality, a limited field of view, and a lack of disparity in depth perception.

The reduction of perception also pertains to the tactile sense. In minimally invasive surgery, long tools of limited cross section are used and the surgeon is no longer capable of feeling by touching the organs with his or her hands. The internal friction caused by the elongated construction of the tools is considerable. Information on the consistency of the tissue as passed through the instrument to the hands of the surgeon is therefore severely reduced. Another disturbing factor is the force needed to position the relatively heavy instruments with the unsupported arm and hand.

Tools. The surgical treatment is performed by special endoscopic instruments with only four degrees of freedom because they have to pivot about the point of incision through the skin (Jansen & Cuesta, 1993; Melzer, 1992). This results in a significant reduction in manipulation capabilities relative to the human arm, and hands with its redundant seven degrees of freedom, which make alternatives for the same pose of the hand possible (Anderson & Romfh, 1980). Due to the aforementioned internal friction of the tools and the relatively high positioning forces needed, mechanical information on the interaction between the tissue and the tool by force feedback is also diminished. The capacity of the hand of the surgeon to serve as a versatile tool is also lost. This all leads to a manipulation capability in minimally invasive surgery that is strongly restricted compared with conventional surgery due to a decreased freedom of motion and the increased mass and internal friction of the endoscopic tools.

The procedure of minimally invasive surgery can thus be modeled by a feedback loop containing the surgeon and including senses, hands, and tools, but now handicapped by limited vision and tactile perception, and by tools of lower mechanical quality (Figure 2). It can be observed that now in the feedback loop some additional constraining mechanisms are in series with the senses and hands of the surgeon, hampering the work as a craftsman. In addition, the surgeon is now dependent on another person handling the video camera (an assistant surgeon). This, together with the use of elongated tools, impairs the hand-eye coordination that is so important for the delicacy of surgical actions.

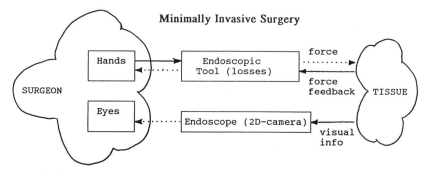

Figure 2.   Model of a surgeon in minimally invasive surgery. Visual and tactile observation of the tissue are limited by the 2D image of the endoscope and by the friction of the endoscopic tools. Indirect force and force feedback are used.

2.3 COMPARING OPEN AND MINIMALLY INVASIVE SURGERY

When comparing minimally invasive surgery with conventional open surgery using both models, it is noted that both the sensing and actuator capabilities of the surgeon are reduced by the minimally invasive technique. Consequently the surgeon is not helped by this technique, and there is hardly any advantage indicated for the operation procedure itself. It is only the patient who reaps a benefit by reduced damage to healthy tissues and a resulting fast recovery. It is important in the evolution of minimally invasive surgery that not only the drawbacks of the technique be reduced but also that the technique be essentially improved to present some advantages to the surgeon and the surgical procedure as well to really become an attractive alternative for any surgeon.

3   Engineering approaches

3.1 RECORDING AND ANALYSIS OF SURGERY

Surgical procedures are complicated processes involving an operating team with several members working together. The role of the techniques used is one complicated aspect of the operation. In general, the surgeon has gained experience from many operations in which skills were adoped and there was learning to cope with the limitations presented by the minimally invasive procedure in the various patients treated (Herfarth et al., 1994). During this learning period the surgeon had in mind the patient outcome, safety, and prevention of complications. The surgeon was not trying to improve the technology itself, which is the task of an engineer. The engineer, however, must be aware of the most important limitations of the technique. Regular interaction with clinical practice is imperative, and we thus designed a method to record the surgical procedure in all its aspects on a regular basis (Claus et al., 1995). These recordings are evaluated in an effective and objective way to support the interaction between the surgeon and the engineer.

A portable instrument setup has been designed that can record four video images synchronously. These might be one or two endoscope images, a general view of the operation room, and possibly images of X-ray or ultrasonic examinations. After the operation, the video images are analyzed with the aid of a computer (spreadsheet) program. In this analysis, the surgical actions are counted and evaluated as objectively as possible by defining a thesaurus, a limited number of standard terms and scores from which the analyst has to choose. This may be done not only for the actions of the surgeon but also for the assisting surgeon, surgical assistant, and all others present. The recorded data can subsequently be evaluated, producing a chart of the various actions involved and the time spent on them (Fig. 3). The use of the tools involved or the judgment of the complexity of the actions can also be analyzed (Claus, 1993). This type of analysis is performed for every intervention. One might also compare different interventions in this way (see, e.g., Fig. 3), because standard terms are used for the various actions and tools. Strategies to improve surgical technique can be selected on the basis of these analyses, e.g., different allocation of tasks and use of different tools or simple facilities that are developed in house (Claus, 1993). After introduction of a modification in the surgical technique, its results can be quantified using the same analysis method.

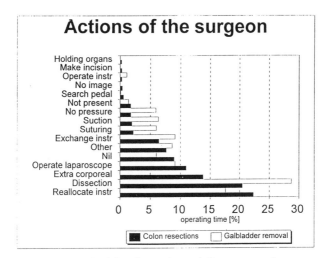

Figure 3. The average time required for the actions of the surgeon for seven advanced laparoscopic procedures (colon resections). For comparison the data from a laparoscopic gallbladder removal are shown as well. The relative time spent on the various actions appears to be roughly the same for these very dissimilar procedures.

This approach of recording and evaluating surgical procedures has several important benefits:

• The recordings of one operation can be analyzed over and over again for the analysis of its various aspects.
• The analysis makes it possible to have documented discussions between the clinician

and engineer.

- The surgical technique is continuously improved by alterations based on the common findings of the surgeon and engineer. This also makes it attractive for the surgeon to spend time on the interaction with the engineer.

- During the operation the engineer has enough time to take notes or make additional observations.

- Using the standard terms, comparisons between open surgery and minimally invasive versions of the same surgical treatment can be made. These comparisons will show the merits and limitations of both approaches.

The method of recording and analysis of the surgical procedures is seen as a basic tool leading to interaction between the surgeon and the engineer in the development of new techniques.

3.2 TELEMANIPULATION SURGERY

There are a number of limitations in visual and tactile perception introduced by the endoscope and endoscopic tools in minimally invasive surgery. The force feedback from the interaction of the tools with the tissue is attenuated by the endoscopic tool, and direct tactile information cannot be obtained. An improvement would be to provide the tools with sensors, making force feedback and tactile information available to the surgeon. By using an actuator in the tool, both the force and force feedback as well as visual information would be transferred by electrical means and telemanipulation would become feasible.

In telemanipulation, tools are operated and controlled using the force feedback information from the sensors of the tool (Anderson & Spong, 1989; Ham et al., 1994; Sheridan, 1992; Jongkind, 1993). The action is not automated but the operator controls the action from a distance. The technique of minimally invasive surgery has many similarities to telemanipulation. The operating field cannot be sensed directly, only by means of an electronic camera, and the surgical workspace cannot be accessed by the surgeon without endoscopic tools, while manipulation and tactile sensations are limited (see Fig. 2). So the operating field is effectively remote. Therefore, remote manipulation solutions from space, under sea, and nuclear plants as an approach to improve minimally invasive surgery have been considered.

For telemanipulation, active tools are needed with force feedback instead of the present passive endoscopic tools. Telemanipulation may have a number of advantages, dependent on the properties of the actuator and sensors. The accuracy and sensitivity can in principle be augmented if the sensors are capable of presenting accurate information with the needed sensitivity. Improved tactile information could also be presented analogous to the magnifying property of the endoscope. In telemanipulation, a number of limitations in force or location could be controlled, thereby increasing safety. Tactile sensors and sensors to assess tissue slip would be beneficial for optimal handling of tissues, because it would be possible to keep the forces of interaction to a minimum.

Although potential advantages exist, there are a number of barriers to be overcome before useful telemanipulation tools can be developed:

- The actuators and sensors, as known from space, undersea, and nuclear applications, have to be miniaturized to pass the trocars.

- The tools with the actuators and sensors should be able to withstand body fluids and sterilization procedures.

- The tools should have a high inherent safety, which may not be fully dependent on the proper functioning of one of the sensors or actuators.

Promising actuator solutions are based on hydraulics, because of the safety and compatibility with the body environment of water, which permit the actuator mechanics and electronics to remain at a distance. Glass fiber optics seem to provide a suitable technology for the development of rugged and biocompatible sensors because most of the critical mechanics and electronics can be kept outside the body. Research and development are under way in these directions but are still in their formative stages.

4   Delft - Amsterdam project

Improving the technique of minimally invasive surgery is an excellent example of a multidisciplinary problem. The surgical procedure comes first, and any solution must fulfill all the demands of a safe and effective treatment. In minimally invasive surgery, however, the surgeon is exceptionally dependent on technology both for perception and for tools.

In a project with the aim of improving this technique, it is of paramount importance to have an intensive and regular interaction between surgeons and technical specialists. The surgeon is responsible for the well-being of the patient and is not specialized in the capabilities and limitations of modern technology. The engineer, however, cannot bear responsibility for the patient and lacks the complete insight in the medical background and restrictions imposed on technical solutions.

Clinical cooperation with the surgeon Tom Jansen working in the St. Antonius Hospital in Nieuwegein, and since 1994 in the Kennemer Gasthuis in Haarlem has been essential in this respect in our work. The analysis of surgery and the engineering interaction with the surgical field is conducted by the Man-Machine Systems group of the Department of Mechanical Engineering of the Delft University of Technology in cooperation with the Department of Medical Physics of the University of Amsterdam.

Within this project on improvements in minimally invasive surgery a second group is using telemanipulation as a starting point for innovation. The development of the telemanipulation tools was initiated by the Control Group of the Electrical Engineering Department of the Delft University of Technology, based on its experience with force reflection systems and telemanipulation in the nuclear power plant environment (Ham et al., 1994; Jongkind, 1993).

Using these two lines of approach, an extensive cooperation has been initiated with groups from Delft and from the Faculty of Medicine of the University of Amsterdam. In the clinical field, work is done with the departments of Surgery and Experimental Surgery of the University of Amsterdam. There is a trend toward using the recording and analysis technique for evaluating surgery with more clinical groups. In the field of perception, the Department of Industrial Design of Delft University of Technology is working on depth perception and trying to introduce its Delft Virtual Window approach (Smets et al., 1988) into endoscopy. In the field of robotics, the group working on the European Robotic Arm (ERA) within the Man-Machine Systems group of the Department of Mechanical Engineering is involved with its experience with human factors in displays and control tasks. In the field of sensors, there is cooperation between the Electrical Engineering Department of the Delft University of Technology and the sensor group of TNO (Dutch organization for Applied Scientific Research) in Delft. Our project is directed toward improving the technique of minimally invasive surgery, and eventually these developments should lead to commercially available products. Industry has thus been involved from the beginning. There is cooperation with one company interested in new surgical developments and display technology (Philips Medical Systems) and with another company working in the field of robotics and telemanipulation (Vermaat Technics) (Ham et al., 1994). This summary shows that a multidisciplinary project such as this can lead to extensive cooperation and interaction with a large group of researchers from clinical groups, engineering groups, and industry. The circumstances, relations, and expertise present in Delft and Amsterdam seem to be very favorable for such a joint effort.

5 Discussion

Minimally invasive surgery is characterized by a decreased capability of the surgeon to observe the tissue to be operated upon. The endoscope limits the vision of the surgeon, while tactile information is lost and the interaction of the surgical tool with the tissue is obscured. Improvements in the minimally invasive technique will be directed at alleviating these restrictions. The most important limitations in the minimally invasive approach must be uncovered by analyzing the role of these restrictions in clinical practice. To this end a recording and analysis method has been designed that is used as a basic tool to define developments necessary to improve the quality and speed of the operation. The same analytic method can also be used for defining improvements in conventional surgery or to make comparisons between conventional surgery and the minimally invasive approach. It is already clear that the development of new procedures and tools for endoscopy can also be beneficial for the advancement of conventional surgery.

In telemanipulation, the transfer of visual information by electrical means as used in minimally invasive surgery is extended to the tactile and force feedback information from the tissue and the application of force to the tissue. In principle, this could improve the accuracy of the surgery. With suitable actuators and sensors, applications such as microsurgery could be performed with a high sensitivity by reducing the scale of movement and amplifying the force feedback involved. The quality of telemanipulation depends totally on the properties of the sensors and actuator, which must be designed to be small and resistant to sterilization procedures.

The reduction of cost to an acceptable level is a major issue in the development of new technical solutions. Especially in the telemanipulation approach, the question is whether sophisticated and expensive telemanipulation tools will lead to a surgical procedure that is superior to conventional manual surgery. If these tools are able to reproduce the capabilities of the human hand but do not have additional capabilities, then a very high price is paid for the minimally invasive approach. If these tools, however, are more accurate or if scaling to smaller dimensions or smaller forces is possible, then these tools will make new surgical approaches feasible and the cost involved could be worthwhile.

References

Anderson, R. and Romfh, R. (1980). *Technique in the Use of Surgical Tools*. New York: Appleton-Century-Crofts.

Anderson, R. and Spong, M. (1989). Bilateral control of teleoperators with time delay. *IEEE Transactions on Automatic Control*, Vol. 34: pp. 494-501.

Boutin, C., Viallat, J. and Aelony, Y. (1991). *Practical Thoracoscopy*. New York: Springer-Verlag.

Classen, M., Knyrim, K., Seidlitz, H. and Hagenmuller, F. (1987). Electronic endoscopy, the latest technology. *Endoscopy*, 19: 118-123.

Claus, G.P. (1993). Analysis of laparoscopic surgery. Improving the technique (in Dutch). Master Thesis, Delft University of Technology, Report No. 679.

Claus, G.P., Sjoerdsma, W., Jansen, A. and Grimbergen, C. A. (1995). Quantitative standardized analysis of advanced laparoscopic surgical procedures. *Endoscopic Surgery and Allied Technologies* 3: 210-213.

Dallemagne B., Weerts, J.M., Jehaes, C., Markiewicz S. and Lombard R. (1991). Laparoscopic Nissen Fundoplication: preliminary report. *Surgical Laparoscopy Endosospy*, Vol. 1: 138-143.

Ham, A.C. van der, Honderd, G. and Jongkind, W. (1994). A bi-lateral position-force controller with gain scheduling for telemanipulation. IEEE/SPRANN94, Lille, France, pp. 282-285.

Herfarth, C., Schumpelick, V. and Siewert, J. R. (1994). Pitfalls of minimally invasive surgery. *Surgical Endoscopy* 8: 847.

Jansen, A. and Cuesta M. A. (1993). Basic and advanced instruments needed for developments in minimally invasive surgery. In M.A , Cuesta., A.G., Nagy (e.d.): *Minimally Invasive Surgery in Gastrointestinal Cancer*, Edinburgh Churchill, Livingstone: pp. 15-25.

Jansen, A. (1994). Laparoscopic-assisted colon resection. *Annales Chirurgiae et Gynaecologiae*, 83: 86-91.

Jongkind, W. (1993). Dextrous gripping in a hazardous environment. Ph.D. thesis, Electrical Engineering Faculty, Delft University of Technology.

Melzer, A. (1992). Instruments for endoscopic surgery. In A. Cuschieri, G. Buess, J. Périssat (Edso*): Operative Manual of Endoscopic Surgery*. Berlin, Springer Verlag: pp. 14-36.

Satava, R.M. (1993). Surgery 2001, A technologic framework for the future. *Surgical Endoscopy*, 7: 111-113.

Schultz, L., Graber, L., Pietrafitta, J. and Hickok, D. (1990). Laser laparoscopic herniorrhaphy: A clinical trial preliminary results. *Journal of Laparoendoscopic Surgery* 1: 41

Sheridan, T. B. (1992), *Telerobotics, Automation and Human Supervisory Control*. Cambridge MA, MIT Press.

Smets, G.J.F., Stratmann, M.H. and Overbeeke, C.J. (1988). Method causing an observer to get a three-dimensional impression from a two-dimensional representation. US Patent, no. 4,757,380.

Tendick, F. (1993). Visual manual tracking strategies in humans and robots. Ph.D. thesis. University of California, Berkeley.

Tendick, F., Jennings, R.W., Tharp, G. and Stark, L. (1993). Sensing and manipulation problems in endoscopic surgery: Experiment , analysis, and observation. *Presence*, 2: 66-81.

White, G. and White, R. (1989). *Angioscopy: Vascular and Coronary Applications*. Chicago: *Medical Year Book*.

Chapter 19

Continuous-Discrete Control-Theoretic Models for Human-Machine Performance

William Levison
Sheldon Baron

1   Introduction and background

Complex technological systems, such as aircraft, power plants, and weapons systems, almost always require humans to monitor, control, and/or supervise their operations. These "engineered" systems are designed and developed using a variety of methods for analysis, testing, and evaluation that rely heavily on mathematical models that describe the inanimate systems involved and the environment in which the system operates. It is essential to account for the human component in the design/development process for these combined person-machine systems, and it is highly desirable to do so in a manner similar to that in which the rest of the system is treated. These needs have led to the development of a large number of models of relevant human performance. We refer to these models as *human performance models* (HPMs).

The theory and techniques of control systems design, analysis, and evaluation have served as a basis for developing one class of HPMs commonly referred to as *control theoretic models*. The technological systems that motivate the development of these HPMs are dynamic in nature and are described by differential (or difference) equations. In continuous manual control problems an alternative representation for the system that is often used is a frequency-response or transfer function (i.e., a frequency-domain representation).

Although control-theoretic models of the human operator treat many psychological phenomena, e.g., psychomotor performance, perception, human information processing, and even cognitive behavior, the aims of the models and the methods employed to describe and/or predict human performance are quite distinct from those of mathematical and/or experimental psychology. In particular, the focus of these models is on prediction of total, i.e., combined person-machine performance. This leads inevitably to engineering characterizations of human operation and performance that are significantly different from , but relate to, those of traditional psychology.

The earliest control-theoretic models of note dealt principally with problems in which a human operator manually controlled a dynamic system (e.g., the motion of a gun or an aircraft) to track a single input (e.g., a target) in a continuous manner so as to maintain small tracking errors. These tracking problems often required very rapid and precise responses. To perform them well, human operators had to be highly practiced and had to virtually "automate" their response to given situations (e.g., specific system dynamics). In such cases, the models for the operator could be relatively simple input-output models. The HPMs that emerged grew out of a servo-mechanism paradigm and the operator was represented by a transfer function or, later, by a describing function and a remnant. Model parameters were those appropriate to such a description (i.e., gain, phase, delay, and time constants). By far, the most successful approaches involved frequency domain representations of the human (McRuer et al., 1965). Verbal rules were used to transform empirical results into mechanisms for selecting the forms and parameters of transfer functions as a function of task parameters.

A major increase in complexity occurred when interest turned to multivariable problems (ones in which there are multiple outputs to be controlled by one or more controls). Because the human operator of such systems cannot observe or process simultaneously all of the outputs to be controlled, he or she must attend selectively to individual variables. This complication led to a need for a more complex model for the human controller's observation process as well as a requirement for modeling the decision processes associated with goal-oriented selective attention. Furthermore, more complex forms of control structure were needed to model multivariable control. Finally, where visual sampling of individual displays was required, a discrete aspect to the control task was introduced that was not present in single-variable manual control problems. Under certain assumptions, however, it was still possible to treat important and complex problems of this type using frequency-domain methods from continuous control theory. This was accomplished by integrating an information-theoretic approach to the visual sampling process developed by Senders (1964) within a multiloop describing function description of the overall control task. This ingenious extension proved to be useful for analysis of certain display problems, but it was ultimately limited by the constraining assumptions necessary to apply frequency-domain techniques and the overall complexity of the approach.

The need to analyze increasingly complex control problems was fueled by technological advances both in the systems being controlled and in the means for controlling them. Perhaps the most important influence was the widespread introduction of digital computers into the control process itself. These computers provided more versatile and powerful means for automating very difficult control tasks or for assisting humans in accomplishing them. Their introduction changed dramatically the fundamental nature of human control and operation of complex systems from one that primarily involved continuous control to a control and monitoring task that included both continuous and discrete tasks. Over time, the discrete portions became more and more prominent as the level of automation increased and the human's role became more and more supervisory in nature.

In parallel with (and partially stimulated by) the changes associated with the introduction of digital computers, there occurred a change in direction, and many advances in the mathematical theory of control and the computational approaches to control problems. The principal developments centered around the widespread use of time-domain and state-space representations for multivariable systems and from advances in stochastic optimal control and estimation theory. These and related developments, which in the 1960s and 1970s were collectively referred to as *modern control theory*, became the basis for a new and different approach to modeling human performance. The HPM that emerged from these developments was called the Optimal Control Model (OCM) (Baron & Kleinman, 1969; Kleinman et al., 1971). It has proven to be a successful tool for predicting human and system performance in complex manual control tasks; more importantly, and of more interest here, is the fact that it has also provided a springboard for modeling other tasks and multitask situations.

In the next section of this chapter we discuss a recent application of an OCM-based approach to modeling car driving. This application has been receiving increased attention recently and illustrates the approach in a combined continuous-discrete task. The concluding section of the chapter provides a brief discussion of how the OCM approach was used as the basis for modeling human performance in tasks that are more and more supervisory in nature (i.e., they involve primarily monitoring, decision making, and discrete control rather than continuous control) and a brief prognosis for the future of this approach.

2   Integrated driver/vehicle model (DVM)

In this section we discuss a recent model of driver-vehicle systems, based on OCM foundations, that attempts to account for driver steering performance as well as the allocation of attention between the primary control (steering) task and other monitoring and control tasks associated with advanced automotive information systems. This model was developed in support of an experimental study to develop evaluation methods and human factors guidelines for in-vehicle information systems (Green et al., 1993). A more detailed summary of this modeling effort is provided by Levison (1993) and Levison & Cramer (1995).

Given descriptions of the tasks to be performed and of the driver's information-processing limitations, the model predicts a variety of performance measures for typical scenarios. Representative measures include lane deviations, control use, and monitoring times for a variety of in-vehicle systems, and various measures of driver attention, such as eye fixation times and scan frequencies, task-to-task transitions, and statistics relating to task interruptions.

The primary intended uses of this model are to aid in the design of manned simulation experiments and to help extrapolate experimental results to conditions not yet tested. Pre-experiment model analysis is of particular value where, because of the expense or limited access to resources, it is critical to have the experimental program well defined before starting a set of simulation or on-road studies. By exploring a range of potential experimental variables (typically, much wider than would be practical to explore in properly controlled experiments) one can use the model as an all-digital simulator to predict which choices of parameter values will yield results that are sensitive to experimental variation and which candidate experiments will tend not to show significant effects. Armed with these results, one can presumably make better choices as to which candidate experimental variables to explore, the range of values to be explored, optimal settings for other independent variables, and so on.

The model presented here, which is called the *Integrated Driver Model* (IDM), is an integration of two previously existing computerized models referred to below as the *procedural model* and the *driver/vehicle model*. The procedural model represents the driver of the vehicle in terms of perceptual, neuromotor, and cognitive responses (Corker et al., 1990). Submodels may include visual scanning and detection, auditory perceptual processing, neuromotor reaction time, and choice and decision in the selection of activities. The procedural model deals primarily with in-vehicle auxiliary tasks (i.e., tasks other than continuous vehicle control) and with the task-selection and attentional-allocation procedures.

The driver/vehicle model predicts closed-loop continuous control behavior. This model, which is currently used to predict lateral path (steering) control, is based on the "optimal control model (OCM)" for manually controlled systems (Levison, 1989). The structure and predictive value of the OCM has been verified via extensive application to laboratory and operational manual control tasks, and the OCM has been applied successfully to the design of manned simulation studies (Levison, 1985). The driver/vehicle model is currently implemented to simulate a constant-speed steering task.

The resulting integrated model allows one to predict continuous steering performance as visual attention is intermittently diverted from the roadway to one or more monitoring locations associated with the auxiliary in-vehicle tasks. The model also allows the driver to attend visually to the roadway while simultaneously processing auditory information. Attention-switching and task selection are made on the basis of time-varying priorities that consider, at each decision point, the penalties for tasks not performed. Presentation of auxiliary tasks is controlled in part through dependencies on the state of the driving environment as predicted by the model and in part through "scripting" (i.e., state-independent time-based occurrence of events defined prior to the model run).

Figure 1 contains a diagram of the integrated driver model showing the principal functional elements of the model and the major communications paths.

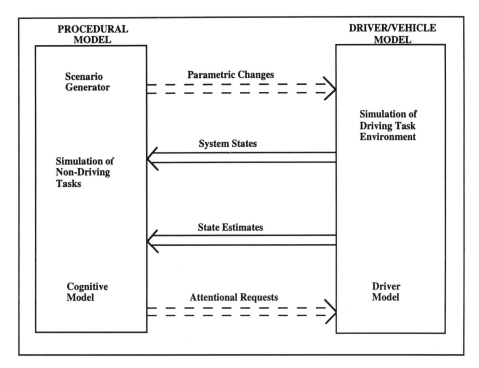

Figure 1. Overview of the integrated driver model

3. Driver / vehicle model

The major assumptions underlying the driver/vehicle model are as follows:

1. The operator is sufficiently well trained and motivated to perform in a near-optimal manner subject to system goals and limitations.

2. The driver constructs an internalized representation (mental model) of the driving environment in which all dynamic response processes are represented by linear equations of motion.

3. Performance objectives can be represented by a quadratic performance index (e.g., minimize a weighted sum of mean-squared lane deviation and mean-squared control activity).

4. Driver limitations can be represented as response-bandwidth limitations, time delay, and wide-band "noise" processes to account for information-processing limitations.

In order to obtain a model solution, the user must provide information sufficient to describe the task environment, the performance goals, and the operator's response and information-processing limitations. Because the model is a simulation model, timing parameters must also be specified. The kinds of inputs required for the driver/vehicle models are listed in Figure 2.

Description of Driving Environment
Vehicle response dynamics
Perceptual variables
Command and disturbance inputs
Performance requirements
Initial conditions
Driver Characteristics
Mental model of the task environment
Information-processing limitations (S/N)
Perceptual limitations (thresholds)
Time delay
Motor lag
Simulation Parameters
Simulation update interval
Data recording interval

Figure 2. Inputs to the driver/vehicle model

The flow of information within the driver/vehicle model component is shown in Figure 3. For applications in which the vehicle is maintained at near-constant speed and undergoes relatively low lateral accelerations, the model components enclosed in boxes are implemented as linear dynamic processes for which the behavior of the "system states" is described by a set of linear differential equations. The "vehicle response behavior" element contains a description of the dynamical response of the automobile, the kinematic equations that relate turn rate and speed to lateral displacement, and any dynamic response elements that might be needed to model external disturbances.

The "cue generation" element accepts the system states and external command inputs to generate the set of perceptual cues assumed to be utilized by the driver. This element contains a linearized approximation that relates the perspective real-world scene cues to system states and command inputs. (For a constant-speed steering task, typical perceptual cues are lane error, drift rate, heading relative to the road, turn rate relative to the road, and road curvature.) These perceptual cues are then corrupted by wide-bandwidth "observation noise" and delay, where the observation noise reflects both a signal-to-noise type of information-processing limitation as well as perceptual "threshold" limitations.

The driver's adaptive response behavior is represented by the optimal estimator and predictor, the optimal control laws, and the response lag, with an additional "motor noise" corrupting the motor response. The "mental model" noted earlier is a component of the optimal estimator. The estimator and predictor construct a least-squared error estimate of the current system state, and the (linear) optimal controller generates the optimal control response operating on these state estimates. The motor noise serves to provide some uncertainty concerning the response of the vehicle to the driver's inputs, and the response lag may be thought of as reflecting a penalty for generating a high-bandwidth control response.

The form and quantification of the estimator, predictor, and controller are determined by the specific problem formulation according to well-developed mathematical rules for optimal control and estimation (Kleinman et al., 1970, 1971). Model outputs consist of quantities similar to those measurable in a manned simulation (e.g., time histories for all important system variables), as well as quantities that cannot be directly measured (e.g., the driver's estimate of the value of any system variable).

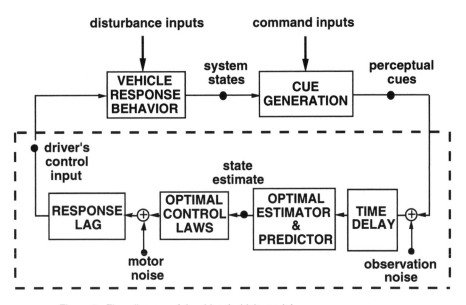

Figure 3. Flow diagram of the driver/vehicle model

The driver's assumed mental model of the driving environment is a key feature of the driver/vehicle model. Typically, we assume the driver to be sufficiently well trained in the specific driving task to allow the mental model to replicate the model of the physical environment. However, we can explore the consequences of the driver's misperception of the external world by making the mental model different from the world model in terms of parameter values and/or structure.

When the driver is required to share attention between the vehicle control task and one or more auxiliary tasks (e.g., look at the rearview mirror, tune the radio), performance of the control task will in general degrade. The effects of such interference are accounted for in one of two ways. For intervals in which visual attention is directed away from the roadway cues to some other visual input, the mathematical "driver" receives no perceptual inputs relevant to vehicle control, and the model continues for a short time to generate control inputs based on the internal model only.

We assume the driver can attend simultaneously to vehicle control and to auxiliary tasks requiring speaking or listening. In this case the driver is assumed to continue to fixate visually on roadway cues, but central-processing resources are now shared between the two tasks. The effects of less than full cognitive attention to the driving task is modeled by degrading the driver's signal-to-noise ratio, in effect, by increasing the observation noise level (Levison, 1979). Either type of attention-sharing tends to decrease the portion of the driver's control response that contributes to effective control and to increase the stochastic component of the driver's control, with the net effect of degrading vehicle control performance.

3.1 PROCEDURAL MODEL

In addition to acting as the supervisory element of the integrated model, the procedural model simulates the in-vehicle auxiliary tasks and performs task selection. We consider first the task selection algorithm and then discuss the overall logic of the procedural model.

3.2 TASK SELECTION ALGORITHM

Task selection is based on assumptions that are generally consistent with the multiple-resource theories of Wickens & Liu (1988). Specifically, we assume that:

1. If two or more tasks require different visual fixation points, only one such task may be performed at any given instant.

2. If two or more tasks require listening or speaking, only one such task may be performed at a given instant.

3. If one task requires visual inputs and another requires auditory inputs, they may be performed concurrently (with presumably some performance degradation) if they require different "processing codes" (i.e., one requires spatial processing and the other verbal processing).

4. Task selection is based on the perceived relative importance of competing tasks and is computed by minimizing the expected net penalty of tasks not performed.

5. If an auditory and a visual task are performed concurrently, cognitive attention is allocated according to the penalty functions.

6. When a task is first attended to, or first reattended to following attention to another task, attention must remain on this task for some minimum "commit time", after which the driver is free to allocate attention as described earlier.

Note that the steering task (which requires attention to the road) is always competing for attention. The logic for selecting a task when multiple tasks compete for attention is diagrammed in Figure 4.

3.3 MODEL INPUTS AND OUTPUTS

The kinds of inputs must be specified for the procedural model are listed in Figure 5.

An auxiliary task may consist of one simple activity (e.g., glance at the rearview mirror) or a sequence of activities, such as use of an in-car telephone. An elemental activity may require visual attention (eyes), or both visual and manual (eyes and hand). Two categories of parameters need to be specified for each activity: parameters relating to performance of the task, and parameters that determine the relative importance or "urgency" of the task. Performance is usually defined by one or more time parameters, which may include (1) times to move eyes and hands, if necessary, in preparation for the task; (2) time to complete the task, and (3) minimum commit time following initial (re-)attention to the task. Some tasks are described by a simple task completion time. Other tasks are defined by a rate of progress, with the driver allowed to interrupt after some commit time and later continue the task.

For tasks that consist of sequences of activities, sequencing rules need to be implemented, as well as rules for which sequences of tasks must be performed as a unit before the driver is allowed to select another task. (For example, the driver may be assumed to dial the entire area code before deciding whether to look back at the road or to continue dialing.)

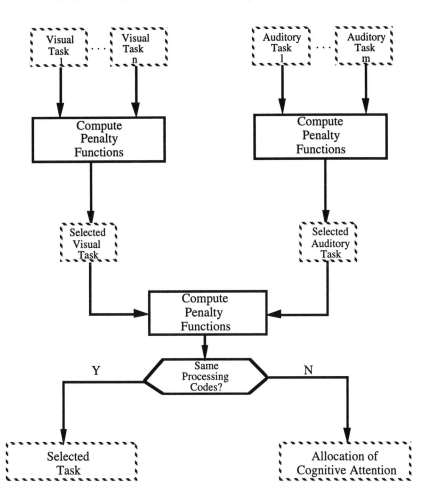

Figure 4. Task selection logic

Description of "Activities"		
Models of performance versus time		
Penalty functions (penalty for not performing task)		
Script of Events		
Times at which activities are spawned		
Simulation Parameters		
Simulation update interval		

Figure 5. Inputs to the procedural model

Penalty functions for in-vehicle tasks may be specified in terms of a single number, or as a number that (typically) increases with time, up to some limit, until the task is completed. A different kind of penalty function is used for the driving (steering) task; namely, the predicted probability of exceeding a lane boundary within a "prediction time" that consists of the time required to perform the in-vehicle task segment plus an assumed time to recover control of vehicle path upon reattending to the road. This computation is based on the driver's current estimate of lane deviation, drift rate, and heading, and is similar to the "time-to-line-crossing" metric proposed by Godthelp et al. (1984).

The output file produced by the procedural model includes time histories of the driver's visual fixation point, the position of his "free" (nonsteering) hand, and measures of performance for each in-vehicle task in progress (e.g., number of words read so far from the visual monitor, time elapsed since initiation of the task.) As with a manned simulation experiment, post-trial analysis of model outputs can be performed to yield a variety of performance statistics, such as means and standard deviations for all continuous variables relevant to the steering task (including variables internal to the driver), statistics relating to the duration of a given in-vehicle tasks, and statistics on dwell times and intervals of inattention.

3.4 SIMULATION CYCLE

After the model has been initialized, the "simulation cycle" is executed once per update interval until some stopping criterion has been reached (typically, a stopping time specified at the start of the run). The cycle begins with a check on which new tasks, if any, are to be added to the "active" list (the set of tasks now competing for the driver's attention). New tasks may be spawned according to the time-based script, or because of completion of an antecedent subtask.

If the task currently attended to is locked up, the driver must continue to attend to that task. If the task is not locked up, the task selection algorithm described earlier is executed to determine the task to be next attended (which may be the same task). Active tasks are updated, and simulation variables needed for postsimulation analysis are recorded in the output file. Application of this model to combined driving and in-car telephone use is illustrated in Levison (1993) and Levison & Cramer (1995).

3.5 MODEL CALIBRATION AND VALIDATION

To the extent that the driving task(s) of interest may differ in important respects from driving tasks modeled previously, a certain amount of initial empirical data are desired to calibrate the model for the baseline experimental condition(s). Calibration data may be needed for the driving task, the auxiliary in-vehicle tasks, or both, depending on the amount of pre-existing data relevant for model calibration. The driver/vehicle model was first calibrated against data obtained in a simplified laboratory simulation of a driving task as described in Levison (1993) and Levison & Cramer (1995), and then against on-road data obtained in a study by Green et al. (1993). Details of these calibrations can be found in Levison & Cramer (1995).

The model was calibrated against on-road data by adjusting independent parameters to provide a best joint match to lane-deviation and steering wheel deflection statistics for segments where the vehicle speed was approximately 96 km/h (60 mi/h). To test the predictive usefulness of this calibration, the model parameters obtained in the calibration against 60-mph data were then used to predict the scores corresponding to a forward speed of 64 km/h (40 mi/h). Measured and predicted SD scores are compared in Table 1.

All predicted SD scores were within 5% of the corresponding experimental measures. The model correctly predicted the trend of the path error scores with nominal forward velocity (about a 30% higher score for the higher speed). This

Table 1. Comparison of measured and predicted SD scores

Speed	Variable	Data	Model
96 km/h (60 mi/h)	path error (ft)	.653	.674
96 km/h (60 mi/h)	wheel (deg)	.860	.904
64 km/h (40 mi/h)	path-error (ft)	.535	.522
64 km/h (40 mi/h)	wheel (deg)	.908	.877

relatively close match between model and experiment provides at least a partial validation of the model structure.

3.6 QUALITATIVE VALIDATION OF THE TASK-SHARING MODEL

The foregoing model analysis focused on calibration of the "drive-only" situation. This section explores the task- and attention-sharing aspects of the model. Specifically a sensitivity analysis is performed to demonstrate that variations in model parameters predict qualitatively expected performance trends, and also to support an experimental finding of unexpected performance trends. Two sensitivity analyses were performed. The first explored predicted model trends as a function of selected task- and driver-related parameters, and the second tested the ability of the model to reproduce some non-intuitive trends reported by Noy (1990). Both analyses employed a simple simulated driving task-lane maintenance on a straight road in the presence of external disturbances (e.g., wind and road surface effects).

3.7 SENSITIVITY TO SELECTED PARAMETERS

The scenario posed for the first analysis was that the driver shared visual attention between the forward scene cues (i.e., the information needed for lane maintenance) and "sightseeing" (i.e., a nonspecific task competing for visual attention). The "penalty" for not sightseeing was set at a constant value of unity, whereas the criterion level of concern for exceeding the lane boundary was either fixed at 1% or varied as an independent variable of the analysis. (A criterion error of 1% means that the level of concern for a probability of exceedence of 1% equals the level of concern for not sightseeing. See Levison & Cramer (1995) for details concerning the algorithm for task sharing.)

Two independent parameters of the analysis were varied: (1) the probability of lane exceedence acceptable to the driver (i.e., the relative importance of the driving and monitoring tasks), and (2) the root mean square (rms) disturbance level (an index of task difficulty). The simulated trial time was 60 s. Each model run used the same random number sequence for stochastic variables (external disturbance and operator "noise" processes) to eliminate performance differences due to factors other than the model parameters of interest.

Predicted scanning behavior and lane-keeping performance are given in Table 2. The first data column contains the independent variable of the analysis: error criterion in terms of acceptable probability of lane exceedence for set "a" and rms disturbance amplitude for set "b." The remaining two columns contain the dwell fraction, indicating the fraction of time that the driver attends to the roadway cues, and the within-trial standard deviation (SD) of the path error computed for the simulated 60 s trial.

The effects of manipulating the error criterion on performance trends were as expected. As the criterion was relaxed, the fraction of attention to roadway cues decreased and lane-keeping performance generally degraded. Because task difficulty was held constant for this analysis, manipulation of the task-selection model parameters influenced the trade-off between attending to the driving cues and driving performance.

Manipulation of task difficulty, however, did not allow such a tradeoff. Table 1b shows that as task difficulty was increased, the "driver" performed less well while attending more to the roadway cues. This analytic result is qualitatively consistent with data reported in the literature (Noy, 1990).

4 Summary of Noy's results

Noy (1990) reports the results of a series of experiments exploring multitask behavior in a driving simulator. His results are particularly useful for providing a qualitative test of the integrated driver model in that they include measures of task performance as well as a usable set of eye-movement statistics.

Table 2. Results of the sensitivity analysis

Independent Variable	Mean Dwell Fraction	Path Error SD (ft)
(a) Effects of Error Criterion		
1%	0.34	0.86
2%	0.25	1.18
5%	0.19	1.10
(b) Effects of rms disturbance amplitude (feet)		
5.0	0.13	0.85
10.0	0.34	0.86
20.0	0.84	1.15

Noy's subjects were required to maintain lane position and headway while driving a simulated truck along a roadway having randomly-sequenced straight and curved segments. In some of the experiments, the subjects were required to perform a concurrent visual search task in which the subjects looked for a short line in a set of otherwise uniform long lines.

Experimental variables included (1) difficulty of the driving task as determined by the curvature of the road segments, (2) difficulty of thesearch task as determined by the number of lines presented, and (3) relative importance of the auxiliary task, which the experimenters attempted to manipulate via instructions to the subjects. Driving was continuous, and the search task was self-paced; i.e., once the subject indicated presence or absence of a deviant line, a new screen was presented.

Driving performance measures included standard deviation of the lane position, time to line-crossing (a measure derived from vehicle position, velocity, and relative turn rate), and headway. Auxiliary task performance was defined in terms of the time required to respond following the onset of the presentation. Typically, subjects required a number of scans to complete the search task.

Principal findings of the Noy study included:

• Driving errors were larger under dual-task than under single-task conditions.
• Manipulating the difficulty of the driving task had a modest effect on driving performance, a larger effect on scan behavior, but little effect on secondary task performance.

- Manipulating the difficulty of the auxiliary task had the opposite effect: There were only small effects on driving performance and scanning, but relatively large effects on auxiliary task performance.

- When the concurrent-task driving performance data were aggregated according to where the subject was looking (i.e., at the road or at the auxiliary display), analysis showed that driving performance was, on average, slightly better when the subjects were looking at the auxiliary display than when looking at the road.

This last result at first appears counter-intuitive. Noy explains it by hypothesizing that, since visual attention was under the control of the subjects (rather than forced by the experiment), the subjects would look away from the road when they felt that the vehicle state was satisfactory and would look back at the road when errors were starting to build up. This hypothesis is entirely consistent with the philosophy underlying the decision module of the integrated driver model. Implied (but not tested by Noy) is that the reverse trend would be found if the driver were to adopt a scanning strategy (say, periodic scan) that did not consider instantaneous vehicle state.

Model runs were formulated and analyzed to determine whether or not the model would replicate two experimental trends: (1) the degradation in driving performance when a concurrent auxiliary monitoring task is required, and (2) the better performance found when the driver was not looking at the road, given the driver's ability to control his scanning behavior. To provide an additional check on the reasonableness of the model predictions, the model was run under the assumptions of periodic scanning behavior.

The driving task explored in the model was the same as that used in the preceding sensitivity analysis. Analysis allowing the "driver" to share attention according to the task-sharing algorithm implemented in the IDM was first performed. Dwell times for the periodic-scanning trials were subsequently selected to provide a close match to the average times predicted for the driver-controlled scanning condition. The simulation update rate for the task-selection model was 0.2 s.

Because the driver/vehicle model uses random noise sequences to generate the external disturbance process as well as various sources of driver response randomness, model predictions obtained in a single trial are somewhat dependent on the specific random number sequences. In order to assess the reliability of any performance trends that might be seen, four 60-s trials were run with different random sequences, and means and standard errors for each performance variable of interest were computed.

Average predicted path error standard deviation (SD) scores are shown in Table 3 for the three attention-sharing conditions explored with the model. Because predicted mean errors were relatively small compared with SD scores, the SD scores may also be considered to be predictions of rms path error. Three scores are shown for multitask conditions: the overall score computed from the full data set, the "on-road" score computed from error predictions corresponding to times when the driver is predicted to be looking at the road, and the "off-road" score computed from data correlated with attention away from the road.

Table 3. Average path error SD (ft)

Attention-sharing Mode	All Data	On-Road	Off-Road
Driver-controlled	1.03 (.07)	1.20 (.11)	0.94 (.06)
Periodic	0.98 (.06)	0.96 (.06)	0.99 (.06)
Full attention	0.67 (.04)	---- ----	

244 *William Levison and Sheldon Baron*

Table 3 shows the following predicted trends: (1) Driving performance is worse when attention is shared between roadway cues and some unrelated monitoring task, compared with driving alone; (2) when attention is controlled by the driver, performance is better on the average when the driver is looking away from the road than when attending to driving-related cues; and (3) when scanning is forced to be periodic, performance is slightly better when attending to the road. The mean and standard error scores shown in this table suggest that the first two trends, but not the third, would have been statistically significant had these been experimental data.

4.1 COMPARISON OF PREDICTED AND MEASURED PERFORMANCE TRENDS

Table 4 compares the performance trends predicted by the model sensitivity analysis with the experimental performance trends reported by Noy. Trends are shown for driving performance and scanning behavior. No trends are shown for auxiliary task performance, because this was not a factor in the model analysis. Table 4 shows that where model and experimental results are available for comparison, the model mimicked all but one of the experimental trends. This discrepancy, along with other aspects of the model results, is discussed in the following.

The major discrepancy between model analysis and experiment concerns the effects of manipulating the relative importance of the auxiliary task. The model yielded the expected result: Attention to the driving task decreased and driving performance degraded as the relative importance of the auxiliary task increased. Noy's experiment, however, showed no significant effect on scanning behavior or driving performance. To some extent, this result may have been due to experimental technique, because drivers were generally encouraged to maintain good driving performance in the presence of a side task. Noy suggests, however, that drivers have learned how much attention is generally required by the driving task in a given situation, and that they are reluctant to reduce their attention below the required amount. This assumption can be accounted for in the IDM by assuming a fixed penalty function for the nondriving task, independent of the specifics of the auxiliary task.

5 Modeling tasks of a more supervisory nature

As noted earlier, the OCM structure includes separate perception, information processing, and control modules (see Fig. 3). This structure, as well as the particular nature of the information processing module and the normative assumptions underlying the approach, has allowed the model to be applied to a wide variety of monitoring, decision-making and discrete control tasks. The major modification to the OCM for these applications is that the portions of the model pertaining to continuous control are augmented, or deleted and replaced, by appropriate rule-based or decision-theoretic elements. In this way, the overall framework can be extended to include or model separately a wide variety of discrete tasks and behaviors, such as: simple visual detection tasks (Levison & Tanner, 1971); dynamic task sequencing decisions (Pattipati, et al. 1979); monitoring and discrete control of remotely piloted vehicles (RPVs) (Muralidharan & Baron, 1980); and the multicrew, multitask procedural and control activities and interactions required to execute the approach and landing of a jet aircraft (Baron et al., 1980). The evolution of these and other related modelling efforts and the particular changes and additions associated with them are chronicled and summarized in several references, notably Baron (1984) and Baron & Corker (1989). Detailed discussion of each model can be found in the original references given in these two summarizing papers.

Table 4. Predicted and measured performance trends

Perf	Performance Trends	
	Model	Experiment
Multitask vs single-task driving performance	Multitask worse	Multitask worse
Increasing difficulty of the driving task	More attention to driving task	More attention to driving task
	Driving performance degraded	Driving performance degraded
Increasing relative importance of the auxiliary task	Less attention to driving task	No significant influence on attention or
	Driving performance degraded	driving performance
Driver-controlled attention	Better performance when not attending to road	Better performance when not attending to road
Periodic attention	Better performance when attending to road	

The progression of models mentioned in the preceding paragraph mirrors the changing roles of humans in the operation of complex systems. Except for highway driving, few situations of current interest have a substantial continuous control component (except, perhaps, in backup mode). As we have seen in Section 2, even in the case of automobile driving, human performance and overall safety issues are more and more concerned with the diversions of attention associated with the introduction of other equipment in the vehicle (e.g., cellular phones, navigation equipment) than they are with the particulars of the control task. Increases in task complexity and the accompanying advances in automation have caused researchers and human factors engineers to focus on the cognitive aspects of tasks rather than on the physical and psychomotor aspects. People are interacting more and more with computers rather than with the actual system being controlled or monitored. In this changed and evolving environment, what can be said concerning the future of the control-theoretic models discussed in this chapter?

Although prognostication is always dangerous, it seems clear that control-theoretic models of the type described herein will not play a dominant role in modeling supervisory control and, moreover, will fade out over time. This will probably happen even for classes of supervisory control problems where fruitful application of the approach is possible. This outcome is almost assured by the lack of research funding directed at these techniques as well as by the need to understand a kind of mathematical and computational complexity that goes beyond the normal training and interests of the cognitive scientists who are in the forefront of addressing these problems. Nevertheless, some of the newer modeling approaches are likely to be influenced by the control-theoretic approach, and some specific aspects of control-theoretic modeling may appear in future models. Indeed, we believe that the fundamental underpinnings of the control theoretic approach to modeling the human operator, such as the focus on closed-loop modeling of the human and the system and the use of explicit performance objectives and normative models, will prove to have an enduring influence on the field.

References

Baron, S. (1984). A control theoretic approach to modeling human supervisory control of dynamic systems. In W.B. Rouse, (Ed.), *Advances in Man-Machine Systems Research, 1.* Greenwich, CT: JAI Press Inc., pp. 1-47.

Baron, S. and Kleinman, D. L. (1968). The human as an optional controller and information processor. *IEEE Transactions on Man-Machine Systems*, 10: 9-16.

Baron, S., Zacharias, G., Muralidharan, R.,and Lancraft, R. (1980). PROCRU: A model for analyzing flight crew procedures in approach to landing. *Proceedings of Eight IFAC Work Congress*, Tokyo, Japan, 15: 71-76.

Corker, K.M., Cramer, N.L. and Henry, E.H. (1990). *Methodology for Evaluation of Automation Impacts on Tactical Command and Control (C^2) Systems: Final report.* Report no. 7242. Cambridge, MA: BBN Systems and Technologies.

Godthelp, H., Milgram, P. and Blaauw, G. J. (1984). The development of a time-related measure to describe driving strategy. *Human Factors 26*: 257-268.

Green, P. (1993). *Human Factors of In-Vehicle Driver Information Systems: An Executive Summary*, Technical Report UMTRI-93-18. Ann Arbor, MI: The University of Michigan Transportation Research Institute (in preparation).

Green, P., Hoekstra, E., Williams, M., Wen, C. and George, K. (1993). *One-the-Road Testing of an Integrated Driver information System.* Technical Report UMTRI-93-32. Ann Arbor, MI: University of Michigan Transportation Research Institute.

Kleinman, D. L., Baron, S. and Levison, W. H. (1971). A control-theoretic approach to manned-vehicle systems analysis, *IEEE Transactions. on Automatic Control 16*: 824-833.

Kleinman, D. L., Baron, S. and Levison, W. H. (1970). An optimal control model of human response Part I: Theory and validation. *Automatica, 6*: 257-369.

Levison, W. H. (1993). A simulation model for the driver's use of in-vehicle information systems. *Transportation Research Record 1403*: 7-13.

Levison, W. H. (1989). The optimal control model for manual controlled systems. In G. R. McMillan et al., (Eds.) *Applications of Human Performance Models to System Design*, pp. 185-198, New York, Plenum Press.

Levison, W. H. (1985) *Application of the Optimal Control Model to the Design of Flight Simulation Experiments*. SAE Paper no. 851903. Warrandale, PA: Society of Automotive Engineers.

Levison, W. H. (1983). *Development of a Model for Human Operating Learning in Continuous Estimation and Control Tasks.* Report no. AFAMRL-TR-83-088. Wright-Patterson Air Force Base, OH: Air Force Aerospace Medical Research Laboratory.

Levison, W. H. (1979). A model for mental workload in tasks requiring continuous information processing. In N. Moray, (Ed.) *Mental Workload: Its Theory and Measurement.* New York: Plenum Press.

Levison, W. H. and Cramer, N. L. (1995). *Description of the Integrated Driver Model.* Report No. FHWA-RD-94-092. McLean, VA: Federal Highway Administration.

Levison, W. H. and Tanner, R. B. (1971). A control theory model for human decision making. *NASA* CR-1953.

McRuer, D., Graham, D., Krendel, E. and Reisener, W. (1965). *Human pilot dynamics in compensatory systems. Theory, Models, and Experiments with Controlled Element and Forcing Function Variations.* Technical Report 65-15. Dayton, OH: Wright-Patterson Air Force Base.

Noy, Y. I. (1990). *Attention and Performance While Driving With Auxiliary In-Vehicle Displays.* (Publication no. TP 10727 (E). Ottawa, Canada: Transport Canada.

Pattipati, K. R., Ephrath, A. R. and Kleinman, D. L. (1979). *Analysis of Human Decision Making in Multi-Task Environments.* Technical report EECS TR-79-15, University of Connecticut.

Senders, J. W. (1964). The human operator as a monitor and controller of multi-degree-of-freedom systems. *IEEE Transactions of Human Factors in Electronics,* 5: 1-5.

Wickens, C. D. and Liu, Y. (1988). Codes and modalities in multiple resources: A success and a qualification. *Human Factors,* 30: 599-616.

Zwahlen, H. T. and Balasubramanian, K. N. (1974). A theoretical and experimental investigation of automobile path deviations when driver steers with no visual input. *Transportation Research Record 520*: 25-37.

Section C

HUMAN CONTROL OF LARGE COMPLEX SYSTEMS

Chapter 20

Operator Support and Supervisory Control

Peter A. Wieringa and Ron A. van Wijk

1   Introduction

The term *supervisory control* is closely linked to the term *automation*. In his book, *Telerobotics, Automation and Human Supervisory Control*, Sheridan (1992) addresses many technological and socio-technical aspects of the link. Supervisory control describes the situation, created by design, in which the task of a human operator is mainly to monitor the system. At the other end of the spectrum is the fully manually controlled system. For manually controlled technological systems, the operator is clearly "in the loop," which is not at all obvious for the supervisory control situation.

The human behavior associated with the operator tasks also differs between supervisory control and manual control. For supervisory control the behavior is classified, according to Rasmussen's behavioral model (1986), as knowledge-based and cognitive, whereas for manual control the skills dominate behavior (Fig. 1). The reader should realize that supervisory control tasks still include manual actions. In fact, these actions, such as pressing a button or moving a switch, play a very important role in supervisory control because they are the realization of the human-machine communication.

The study of human-machine interaction during supervisory control can be split into human behavioral studies and interface design studies. The human behavioral studies include:

- "Normal" human behavior
- Human capabilities
- Human error
- Training and selection

When adding more detail, one may substitute for "human being" not only a single operator but also a team of operators, field-operators, managers, designers etc.

The studies on interface design have been focused on

- Information presentation of the plant variables
- Task allocation
- Diagnostic support systems
- Computerized procedures
- Maintenance scheduling

Very little is known on how to use the knowledge obtained from these studies in an early state of the design or modification of a plant.

Associated with the change in operator tasks and behavior is the change in operator training (including education and selection), human-machine interface design (including operator support), and operating procedures to control a plant. One may argue that guidelines for designing the TIP items (training, interface, and procedures) are used to certify reliable operator behavior and to reduce unpredictable, erratic control actions. In other words, they are used to condition human behavior. Thus, a good understanding of

Peter A. Wierenga and Ron A. van Wijk

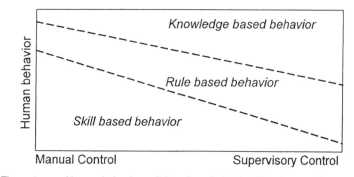

Figure 1.   Human behavior split into three behavioral levels according to
Rasmussen as a function of the degree of automation.

the mechanisms behind human errors is a prerequisite for the design of the TIP items when using new technological developments.

Work the past several years (Stassen, 1995) has focused on

- Cognitive behavior and human error mechanisms
- Task allocation in complex systems
- Design of human-machine interfaces and operator support systems
- Design of simulators for evaluation, experimental validation, and training

It is interesting to note, as Stassen et al. (1995) showed, that Human Reliability Analysis (HRA) and HUman Machine InterFace (HUMIF) designs have been developed in parallel without much interaction. Collaboration between scientists from both fields could lead to interesting new developments.

Fundamental to the design of HUMIFs, task allocations, and simulators is a better understanding of the relation between system dynamics and operator behavior. Two studies will be addressed in this chapter, namely, a study of the relation between system complexity and performance, and a study of the appropriate degree of automation. Several projects in the Delft Laboratory for Measurement and Control include the design of operator support systems. Where appropriate, the latest activities on that subject will be briefly summarized. An overview of the historical development in supervisory control will be given first.

2   Historical overview of the change from manual control to supervisory control

When supervisory control was unknown the operator had to work directly on the machine or process. The term man-machine interface is most obvious for these situations.

Manual control is not abandoned in modern systems. Most transport systems are still manually controlled. Also, teleoperation is a form of manual control that is under development (e.g., in the medical field, in space, and for environmentally difficult or dangerous areas for humans).

Automatic process control was made possible due largely to the introduction of pneumatic controlling devices. The introduction of these new technologies also made it possibile to centralize the process interface into a central control room. In the process industries, the controllers in the early part of the century were located close to the particular equipment (e.g., furnace) or machines (e.g., compressors) involved. The introduction of pneumatic lines opened the possibility of bringing all controls of the entire process plant together in one control room.

The panels in the control room contained all the controllers, relevant indicators, recorders, and alarm displays. In that way the operator had all the information about the process at hand in a parallel fashion. The major advantage of this method of presentation was that the operators could "walk the panel" and have a good overview of the status of the process. The layout hardly changed (it mimicked the layout in the plant itself) so that the operators were always able to find the information they required quickly. Evidently this setup had also disadvantages. The operator was not close to the plant and process equipment anymore, and therefore did lose some important feedback on the functioning of the equipment. The control room operator had to rely on outside operators, and good (radio) communication became paramount. The system itself was very inflexible, so it was difficult to make changes, e.g., to change the control configuration or to introduce extra measurements. Moreover, in troubleshooting, often not enough information was recorded, or when enough records were available it was difficult to derive cause and effect from the rather inaccurate recorder systems.

The first generation of distributed control systems (DCS) was introduced in the late 1970s, also enabled by the fact that intrinsically safe connecting systems in the field (no danger of ignition sources) became available. The prime focus in the DCS was to improve accuracy of control and flexibility in configuration and recording. Only to a lesser extent were the operator interface and human operability addressed.

The operator interface was changed fundamentally as result of the appearance of computer-driven visual display units (VDUs). These, however, presented parallel information in a serial fashion. The operator had to page through the system in order to get an overview of the process state, rather than being able to derive that in one glance, as with the earlier pneumatic panel.

Table 1. Functional attributes of human and machine

Man	Machine
single channel information processing	multi channel information processing
limited capacity for information processing	high capacity for information processing
poor computational skills	excellent computing power
limited reliability	high potential reliability
limited repeatability	high repeatability
high short-term precision	continuous performance
graceful performance	sudden performance degradation (redundancy required)
good long-term memory possible	infinite memory capacity
relatively poor short-term memory	volatile memory
self error-correcting	need for error-correcting system
complex pattern recognition	pattern recognition possible
loose job specification	needs exact job specification
can generalize and make inductive decisions	efficiency requires specialized and "narrow minded" behavior
can handle short-term overloads	robust against overload if so designed

With the increasing information density available, the need for a more fundamental approach in control interface design emerged. Later generations of DCSs did address the operator interface. Information was grouped and served the operator's need for a good mental picture of the plant operating variables. However, much research is still needed to improve the information presentation in DCSs.

3   Task allocation and automation

Capabilities and limitations of humans and automation have been discussed extensively in the literature (Moray (1996), Neboyan & Kossilov 1992). Fitts' fundamental approach was that the assignment of functions to humans who operate complex industrial systems should match their capabilities. The same holds for automation. The Fitts list was first published in 1951 and is still used in the ergonomic literature for function allocation (Kirwan & Ainsworth 1992). Typical functional attributes that appear in such a list are given in Table 1.

The supervisory control tasks that are assigned to human operators can be grouped into five categories (Sheridan and Hennessey, 1984): planning, instructing, monitoring, intervening, and learning. During various process control modes, like normal stop and start operation, normal operation, fault management, and maintenance, the operator has difficulty executing all of these tasks well. Especially during fault management the operator demonstrates rule-based and knowledge-based behavior (according to the levels of behavior defined by Rasmussen (1980).

Like all human beings, operators make errors when using HUMIFs despite good guidelines for the design of the TIP items (extensive selection, training, and well-designed human-machine interfaces and procedures). Operator errors have been discussed extensively in the literature (e.g., Rasmussen, 1986; Reason, 1990). Beside the efforts put into understanding human erroneous behavior, much effort has been spent on diminishing the need for human interaction and on limiting the consequences of human errors (to improve both production and safety).

Associated with automation is the tendency to leave the human operator out of the operating theater. However, for instance, operating rooms and airplane cockpits are still designed to be manned by human operators. The reason is that because of some unique characteristics, the human is considered to be a last resort for rare and unforseeable situations.

Increasing technological performance has led to increased levels of automation. Associated with increased automation and technological performance is increased complexity of systems. It is well known that for optimal control of a system the operator needs a good mental representation of the system and the functional goals of the system's output (Stassen et al., 1990). But there are obvious limits to what a human being can comprehend (the volume of information and knowledge) or can pay attention to (temporal information density). Hence, a proper function analysis and task allocation of the system is required to set the conditions for optimal control of the system.

Despite available guidelines for the design of automated systems, automation clearly has some limitations. For instance, such designs are limited by our lack of knowledge of the cognitive behavior of any human supervisors of the automation.

Research on the limitations of the degree of automation is needed. Wei and Wieringa (1994) used a simulation of a linear, time-invariant system that consisted of 12 subsystems, called cells. Figure 2 shows an example of such a system for the case in which only three cells are defined. The cells were ordered and numbered. The output of each cell with a lower order number could affect a cell with a higher order number. The influence was adjustable by a gain that was kept constant during Wei's experiments. Each cell consisted of a first-order system. The number of inputs and outputs of the distributed system equaled the number of cells. The task of the operator or the automation was to react on requests to change the set-point of any cell in the system. Meanwhile, the

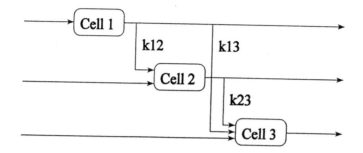

Figure 2. An artificial system showing only three cells that are forwardly connected by gains k12 . . . .k13. The cells are first-order systems with equal parameters.

other cells had to be kept at their last set point value. Each cell could be automatically or manually controlled. This was determined and fixed prior to the experiments.

One can easily accept the idea that with an increasing number of automated cells the degree of automation increases. Wei defined the degree of automation as the ratio between the number of tasks that are automated and the total number of tasks that exist. Only those tasks are considered that can be allocated to either operator or automation. The main task for the Wei experiments was to perform changes at the inputs of the cells. In this respect, all tasks were the same. However, it could be shown that the same task on a cell with a lower order number caused considerably more mental load than a task on a cell with a higher order number. Thus the task complexity was affected not only by the number of manually controlled cells but also by the location of the cells in the network and the associated tasks. Wei & Wieringa (1994) were also able to show that the time interval between set-point requests, which was kept in the order of 30 s on average, significantly influenced both the mental load and operator performance. Sheridan (1992) added to this list the degree of overlap between the tasks. Although Wei's definition clearly has some shortcomings, it provides a good framework to relate experimental work. However, a link to industrial processes is still difficult to make because so far Wei studied a limited number of tasks with small differences.

Many modern systems allow the operator to decide which task to control automatically and which to control manually. Trust in the system's automation plays an important role in deciding when to turn on or off the automation. Furthermore, the operator takes over the system in order to learn, to gain experience, or when the automation is entering an unpredictable state domain. The operator can then give permission for execution of each part successively, thus regaining control. Experience in the field has taught that if automation is too fast, the operators become uncomfortable and request reconfiguration so that automation is split up into parts. Laboratory setups like that described earlier can also help us to understand proper dynamic task allocation (Wei & Wieringa, 1994).

From the above-mentioned description it appears that system complexity and degree of automation are closely linked. Human operator support systems are designed to help the operator, but at the cost of system complexity. The operator now has to understand not only the system itself but also the philosophy of the support system. Moreover, the number of components involved in the automation increases the level of complexity.

The term *complexity* is often used in situations that are not well understood or are difficult to comprehend. Beside this qualitative use of the word, attempts have been made to define quantitatively a system's or task's complexity (Scuricini, 1988; Thelwell, 1994). Nevertheless, very few attempts have been made to quantify complexity. As with the definition of the degree of automation, such measures could be used for comparative evaluations of human- machine interaction and techniques to improve such interaction.

Stassen & Wieringa (1993) reported earlier the use of the same simulation as used by Wei to study the degree of complexity. With such systems it is easy to change the number of components and the gain of the links between the components. In a preliminary study, it was shown that a qualitative relation could be made between the degree of complexity (defined by a function of the number of components, the gain of the links, and the frequency of set-point requests) and the performance. However, a translation to nonlinear and less artificial systems needs to be made so that laboratory experiments can be used in practical situations.

Beside the work done on the relation between the degree of complexity and operator and system performance, it is tempting to investigate means to reduce the perception of complexity by mimicking the appearance of the system to the operator (Andriessen & Wieringa, 1995) such that the operator can get a realistic picture of the totality of how control is affecting the system.

4   Human operator support systems for manual control

As noted earlier, technological developments have not led to automation of all tasks and to abandonment of the human operator. Typical areas where operators will be found for decades to come are situations where it is almost impossible to foresee and measure the disturbances that may act on the system, or where it is too expensive to design or install sensors to measure these disturbances. Two examples are given, both from the dredging industry.

When removing highly polluted mud from harbors, predefined profiles need to be followed. Operators have to position the "grab" under water to high accuracy (±10 cm). The operator has to compensate for different local mud densities and unexpected objects on the bottom. An enormous effort is needed to automate such a process so that for the time being it seems more pragmatic to develop systems to support instead of replace the operator.

Jonkhof (1995) developed a graphical operator support system for the above-mentioned situation based on the findings of the research of Breedveld (see chapter 16). The graphical image Jonkhof developed shows the profile that needs to be obtained. Measurements of the actual position of the crane and grab are used to project the grab above this profile. Simple color bars at the edges of the simulated grab indicate when the grab is within a specified tolerance so that the operator may close the grab.

Another example is also from the dredging industry. Operators controlling the mud sucking process have to closely watch the density of the mud. If this becomes too low, then the hulk of the ship is not efficiently used. If the density becomes too high, the suck-opening risks getting plugged. Information on the mud density is therefore very important. The density is measured by using a radioactive source. Regulations prohibit this source from being placed under water. Hence, the operator does not get the density of mud that enters the pipe but that arriving some 15 s later at the measuring device. Since operators have great difficulty in operating systems with large time delays and time constants (Veldhuyzen & Stassen, 1977), this control of the mud density on board dredging ships is difficult.

Leenders (1995) looked into this problem and developed a method to predict the mud density on the basis of rheological models and hydrostatic measurements along the sucking pipe. The algorithm he developed calculates the density of the mud and updates the information by using the radioactive density measurement device that is still in place. Preliminary results show that his ideas should work. This example shows that the design of operator support systems requires a good understanding of the technical process being controlled.

5   Decision support systems for supervisory control

Human operator support systems are designed to help the operator diagnose plant conditions and to plan actions. Decision support can be advantageous prior to the initiation of an alarm and can help predict possible alarm conditions if no action is taken by the operator. The introduction of digital computer systems creates the possibility for development of advanced features (Kossilov, 1994), such as

- Increasing time for operators to diagnose and plan
- Reducing the complexity of the human-machine interface and the instrumentation
- Automating tasks that are prone to human error

Among the practical applications of operator support systems in the nuclear field (Kossilov, 1994) are

- Fault detection and diagnosis
- Intelligent alarm handling

Many simulators have been used to study operator behavior. One of the most complex low fidelity simulators is the generic nuclear plant (GNP), a model of a pressurized water nuclear reactor power plant developed by the RISØ Institute in Denmark. Sassen (1993) adapted this model and used it to evaluate a knowledge-based fault detection and diagnosis system. The Multi-level Flow Model (MFM) concept (Rasmussen, 1986) was used for the interface between the knowledge-based system and the operator. The performance of operators aided by the diagnostic system and the MFM interface was compared with the performance of operators using no support and only the conventional interface. The experiments showed that the number of false detections, incorrect diagnoses with multiple causes, and incorrect diagnoses for untrained faults in the system was reduced when the operators used the knowledge-based system. However, these operators did not detect and diagnose the faults faster.

6   Alarm management

Alarms are provided to the operator to indicate a change in state of a component or the crossing of some threshold of a process variable. Due to the increasing complexity of systems, one alarm may cause an avalanche of other alarms. Too little has been done to develop a generic strategy for alarm management, coping with the confusion that occurs just following the alarm initiating event. This may be confirmed by listening to complaints by operators, especially about the presence of insignificant alarms. In principle, a serious alarm (high-high alarm) should indicate that if the operator doesn't pay attention to a specific part of the plant and take action, a trip of (a part of) the process will occur within a certain time. From experience the operator knows how much time is available for action. With the introduction of modern control concepts, such as model-based predictive control and expert systems, it should be possible to come up with a prediction of the time window available for action. Miazza et al. (1993) described a computerized alarm system (CASH) under development at the Halden Reactor Project, which consists of three parts:

- Alarm generation
- Alarm structuring
- Alarm presentation

It has been noticed that at a very early stage of a serious alarm the operator needs to be informed about rather abstract goals associated with critical functions.

Information about the state and integrity of the plant tells him whether certain functions are still available. Much further in the fault diagnosis process detailed information on components is needed. Hence, alarms and other information made available to the operator should match the level of abstraction at which the operator is working.

An alarm may be irrelevant in certain situations and crucial for understanding the fault in another situation. It has been reported that in intensive care units of hospitals alarms are false in 80% of the cases, i.e., not related to a change in state of the patient but mostly related to loose connectors or cables. Irrelevant and false alarms distract the operators attention, slow down the diagnoses process, and may cause operating errors (e.g., ignoring alarms) due to an increased mental load.

Pauwels (1995) studied the false alarm generation of the heart rate (HR) signal at an intensive care unit (ICU) in a hospital. In the normal situation HR is obtained by a device using electrocardiogram (ECG) measurement. However, alarms occur when electrodes get loose or when the patient is moving or being moved. Such alarms do not warn the medical personnel about the state of the patient but more about the state of the medical devices.

In most cases the blood pressure (BP) of a patient in an ICU is also measured. Both signals are periodic and include information about HR. Effective use of the redundancy in the periodicity measurement can be made to suppress false HR alarms, and thus to improve alarm generation. Furthermore, ECG measurements are based on electrical contact with the patient, whereas BP measurements need mechanical contact. This difference in measurement principle reduces the likelihood of a common mode for the generation of a false alarm.

Pauwels generated a decision tree to decide whether the HR signal obtained from the ECG should be used to compute the HR or not. In cases where the HR-ECG measure showed a deviation from the normal trend the HR-BP was analyzed. If this signal was also abnormal, there was sufficient reason to believe that the patient's heart function is decreasing, and so an alarm occurred.

Instead of using redundant information, it is worthwhile to investigate the use of global measurements that indicate the patient's state. For this purpose van Rooij (1995) has studied scoring methods such as the Acute Physiology And Chronic Health Evaluation (APACHE) and the Simplified Acute Physiology Score (SAPS). Such measurements may be used to set the alertness level of the medical personnel rather than to reduce false alarms. It is expected that if physiological state indicators show a decrease, medical personnel may respond to an alarm more promptly than in other cases.

In process industries alarm handling suffers the same shortcomings. Simple techniques such as masking alarms from equipment that is not in operation are being applied. When it comes to intelligent alarm filtering, not much progress has been made. In the research field, systems are being developed that calculate resultant alarms from a first alarm situation, and, based on a consequence model, filter those alarms that are predicted. These systems have not been applied, since little is known about the importance of acknowledging the redundant information included in forseen alarms. On the other hand, when a severe process upset takes place, the risk of an alarm avalanche may render the operator totally ineffective (as seemed to occur in the Three Mile Island accident). So, wisdom is to balance both aspects.

7   Support for training

The use of training simulators is still increasing. Simulators for vehicles (air, highway, sea) and industrial plants (oil, chemical, electrical power) are getting more and more realistic and detailed. An example is in maritime operations, where, for instance, simulation of the interaction between the main vessel and assisting tugs is desired. For this problem Bakker (1995) developed a semi-autonomous tugboat simulation. Tugboat and captain are for this case treated as one system as has been done for similar cases by Papenhuijzen (1994) (see also chapter 15). During a simulator run, the tugboat receives commands from the navigator on the bridge of the main vessel. Interpretation of the commands should be done in such way that the tugboat's own safety is also considered. This implies that the tugboat should avoid obstacles such as buoys automatically and without receiving new commands. Bakker found solutions for the automatic detection of obstacles, correction of the planned task, and recovery of the task after the obstacle has been passed. Hence, assistance of tugboats can now be used in large vessel simulators in a realistic way such that, for instance, buoys are avoided automatically without additional commands from the bridge of the main vessel.

In process industries simulators are being used more and more. For new plants, simulators are used to train personnel in plant startup and shutdown as well as abnormal event handling. It is believed that with simulator training, plant startup can be done faster. However, at the moment the value of some existing plant simulators (unlike flight simulators) is being questioned. The experience is that if the simulators for the process industry are not maintained following plant changes, they fall into disuse. This is in contrast to the training simulators used for nuclear power plants.

8   Concluding remarks

Human beings will continue to affect system performance despite increased levels of automation. Financial, temporal, technological, political, social, and psychological reasons limit the degree of automation. Increased levels of automation and complexity require a better understanding of human cognitive behavior and error mechanisms. Essential to the design of operator support systems is a good understanding of the system's behavior, the human operator's capacities and behavior, and the dynamic interaction between the two.

References

Andriessen, J.H.M. and Wieringa, P.A. (1995). Influencing complexity perception by means of the man-machine interface. Abstract, *14th Benelux Meeting on Systems and Control*, p. 15.

Bakker, M. (1993). Een decision-support systeem. MSc, thesis, Delft University of Technology, TUD-WBMT-MR-A-612.

Bakker, N.H. (1995). Model voor het genereren van realistische sleepbootassistentie. MSc thesis, Delft University of Technology, TUD-WBMT-MR-A-617.

Jonkhof, M.N. (1995). Display ontwerp met ruimte informatie voor baggerkraanbesturing. Msc thesis, Delft University of Technology, TUD-WBMT-MR-A-687.

Kirwan, B. and Ainsworth, L.K. (1992). *A Guide to Task Analysis*. London, Taylor and Francis.

Kossilov, A. (1994). Role of computerized operator support systems in nuclear industry. *Journal of Scientific & Industrial Research*, 53: 468-474.

Leenders, B. (1995). Ontwerp en evaluatie verbeterde dichtheidsobserver voor slibbaggeren. MSc thesis, Delft University of Technology, TUD-WBMT-MR-A-689.

Miazza, P., Torralba, B., Kårstad, T., Moum, B. and Follseø, K. (1993). CASH: Computerized Alarm System for HAMMLAB. An outline of required functions. OECD Halden Reactor Project, HWR 362, Halden, Norway,

Moray, N. (1986). Monitoring behavior and supervisory control. K.R. Boff, L. Kaufman, J.P. Thomas (Eds.). *Handbook of Perception and Human Performance.*, Vol II Cognitive Processes and Performance. New York, John Wiley and Sons. 40.1-40.59.

Neboyan, V. and Kossilov, A. (1992). *The Role of Automation and Humans in Nuclear Power Plants*, IAEA-TECDOC-668. Vienna.

Papenhuijzen, R. (1994). Towards a human operator model of the navigator. Ph.D. thesis, Delft University of Technology.

Pauwels, W.G.A. (1995). Vals alarmonderdrukking op de Intensive Care afdeling. Msc thesis, Delft University of Technology, TUD-WBMT-MR-A-718.

Rasmussen, J. (1986). Information processing and Human-Machine Interaction. *System Science and Engineering* 12: 214.

Rasmussen, J., (1980). What can be learned from human error reports? *Changes in Working Life*. K. Duncan, M. Gruneberg, and D. Wallis (Eds). New York, John Wiley & Sons.

Reason, J.. (1990). *Human Error*. Cambridge, UK: Cambridge University Press.

Van Rooji, M.P. (1995). Continue toestandschatting van intensive care unit patienten met behulp van

SAPS II. Msc thesis, Delft University of Technology, TUD-WBMT-MR-A-629.

Sassen, J.M.A. (1993). Design issues of human operator support systems. PhD thesis, Delft University of Technology.

Scuricini, G.B. (1988). Complexity in large technological systems. In *Measures of Complexity*. Lecture notes in physics no. 314. L. Peliti, and A. Vulpiani.(Eds.) Berlin: Springer Verlag.

Stassen, H.G., Johannsen, G. and Moray, N. (1990). Internal representation, internal model, human performance model and mental workload. *Automatica,* 26: 811-820.

Stassen, H.G. and Wieringa, P.A. (1993). On the human perception of complex industrial processes. *Proceedings of the IFAC World Congress,* Sydney, pp. 441-446.

Stassen, H.G., Macwan, A.P. and Wieringa, P.A. (1995). Man-machine system studies and human reliability analysis. *Proceedings of a Workshop Human Reliability Models: Theoretical and Practical Challenges,* Stockholm.

Sheridan, T.B. and Hennesy, R.T. (Eds.) (1984). *Research and Modeling of Supervisory Control Behavior.* NRC, Committee on Human Factors Washington DC: National Academy Press.

Sheridan, T.B. (1992). *Telerobotics, Automation and Human Supervisory Control.,* Cambridge, MA: MIT Press.

Thelwell, P.J. (1994). What defines complexity? *Proceedings of Ergonomics Annual Conference.*

Veldhuyzen, W. and Stassen, H.G. (1977). The internal model concept: An application to modelling human control of large ships. *Human Factors* 19: 367-380.

Wei, Z.G. and Wieringa, P.A. (1994). Task allocation problems in human centered automation systems. *Proeedings of the 13th European Annual Conference on Decision Making and Manual Control,* Finland, pp. 5-14.

Chapter 21

Human-Computer Interfaces for Supervisory Control

Gunnar Johannsen

Supervisory control of industrial processes as well as of transport systems and vehicles requires well-designed, easy-to-use, and transparent human-computer interfaces with real-time capabilities. This chapter describes the functional specifications of such human-machine interfaces. A human-centred design approach with several stages of user participation for determining the functionalities of the human-machine interfaces is explained. The structure and functionalities of the interfaces are briefly outlined, such as the levels of presentation, dialogue, user model, and technical systems model. The specific interface requirements for different human user classes and the aspect of cooperative work with multi-human users are discussed. This chapter is based on experience with the design of human-machine interfaces for supervisory control of chemical plants, cement plants, power plants, and aircraft.

1   Introduction

Human supervisory control means the control by one or several humans of a computer, which, at a lower level, is controlling a dynamic system (Sheridan and Johannsen, 1976). The human-interactive computer often coordinates and controls one or several hierarchical layers of subsystem- and subprocess-interactive computers as parts of the decomposed dynamic system (Sheridan, 1992; Johannsen, 1993). The human supervisory functions in such a human-machine system comprise planning, teaching, monitoring, diagnosing, intervening, and learning. The human-computer interfaces between the human supervisors and the coordinating computer as the higher level of the whole dynamic technical system, called the machine for short, have to support the human supervisory functions as well as the internal representations of any human supervisors (Stassen et al., 1990).

The human-computer interfaces for supervisory control of industrial plants and other dynamic technical systems have now been recognized as essential components for process safety, quality, and efficiency. Computer graphics, knowledge-based systems, and other software technologies offer a wide range of alternative designs of such interfaces, which have made them into sophisticated, intelligent human-machine interfaces. Beyond technological possibilities, it is necessary to consider the cognitive capabilities and the information-processing behavior of the human supervisor interacting with the technical systems, as well as the tasks to be performed within a specific human-machine system (Johannsen, 1992).

Intelligent human-machine interfaces comprise not only presentation and dialogue levels but also explicitly embedded user, technical systems, and task modeling components, as well as knowledge-based explanation and justification facilities (Johannsen, 1995b). An object orientation allows one to deal with graphical and multimedia presentations, as well as with the knowledge sources and information flows on all interface levels in a unified manner.

The design of human-machine interfaces for integrated technical systems has become quite complicated and can no longer be handled in an intuitive fashion. The integrated technical system consists of the dynamic technical process and its computer layers of the automation (S&C) systems and, additionally, of one or several decision support systems (DSS). The human-machine interfaces envisage a better symbiosis between this technical system on the one side and different classes of human users on the other side. Both the main components of the integrated technical system as well as some classes of human users are shown in Figure 1. The human-machine interface with its knowledge-supported levels is located between these human users and the technical system. In general, the classes of human users may not only include operators and operational engineers as supervisors, maintenance personnel, and researchers, but also designers themselves.

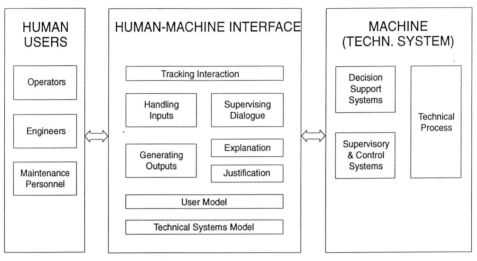

Figure 1.   Functionalities of human-machine interfaces between different human user classes and the subsystems of the machine. (From Johannsen, 1995b, with permission.)

In this chapter, the functional specifications of human-machine interfaces are explained in Section 2 and the whole human-centered design approach in Section 3. The structure and functionalities of the human-machine interfaces are described in Section 4. Additional requirements of human-computer interfaces are discussed for different user classes and cooperative work in Section 5.

2   Functions specification of human-machine interfaces

A specification of Human-machine Interface (HMI) functions has to be derived at an early design stage. This is a very strong design requirement that is still violated much too often in practice (Johannsen et al., 1994). However, there are also fortunately signs of change (e.g., Boutruche & Kärcher, 1995). The functions specification for the human-machine interfaces depends on the goals and the goal structures prescribed for the human-machine system, all kinds of technological and intellectual means for accomplishing these goals, and the tasks to be performed by the human users with the purpose of achieving the goals by appropriate usage of the available means (Johannsen, 1995a); see also Figure 2.

The goals of the human-machine system are mainly (1) productivity goals, (2) safety goals, (3) humanization goals, and (4) environmental compatibility goals. An appropriate goal structure needs to be specified in which these four goal classes are related to each other and further subdivided into several hierarchical levels of subgoals. The productivity goals include economic as well as product and production quality goals. The importance of the safety goals is strongly determined by the application domain. This goal class dominates all others in many large-scale systems and, particularly, in risky systems. The humanization goals comprise team and work organization, job satisfaction, ergonomic compatibility, and cognitive compatibility. Cognitive compatibility includes the subgoals of transparency and human understanding. The environmental compatibility goals refer to the consumption of energy and material resources as well as impacts on soil, water, and air.

The means for achieving the goals can be separated into purely technological means and human-related means. Most issues concerned with the design of the technical process itself are purely technological means. The degree of automation also belongs to this category but represents human-related means as well. Supervision and control (S&C) systems and, even more so, knowledge-based decision support systems, should be designed with a human-centered approach.

Human-related means comprise several categories. Knowledge about the application domain as well as about human users' needs and strategies can be regarded as human-related means. This may be described by structural, functional, causal, and cognitive relationships. Other human-related means are available through different views on the application domain. These views consider the levels of abstraction, the levels of aggregation or detail, parallel versus serial presentation, navigational possibilities, and degrees of coherence. All these means are implicitly inherent in the application domain but often need to be transformed for explicit usage.

The tasks to be performed by the human users depend on the scenarios of use, i.e., situations and contexts of the human-machine system. Different experience and knowledge of the human users, as well as different normal and abnormal situations and states of the dynamic technical system, lead to different subjectively perceived tasks. Further, the prescribed goals and available means for their achievement strongly determine the types of tasks to be performed in a certain scenario.

The functions of any HMI for supervisory control should be defined in such a way that the prescribed goals can be achieved, the available means can be transformed and used appropriately, and the tasks can be correctly perceived and effectively performed. The HMI functions can be specified in the following global categories:

- Supporting the accomplishment of all goal classes (goal functionality)
- Supporting the transformation and usage of all available means (means functionality)
- Supporting appropriate task perception and performance (task functionality)
- Organizing sufficient and timely information transfer from and to the technical system and all its subsystems with respect to goals, means, and tasks (dialogue functionality)
- Visualizing all process and systems states with respect to goals, means, and tasks (presentation functionality)
- Supporting the compatibility and adaptability of the human-machine interaction to the human users (user functionality)
- Supporting human understanding by adequate explanation and justification (explanation functionality)
- Avoiding or compensating for human errors (human error-tolerance functionality).

Several categories of human errors may occur (Reason, 1990) and can partially be handled by appropriate user-model based decision support systems within the technical system (Johannsen, 1992) or within the HMI.

3 Human-centered design approach

It seems most appropriate to combine strict systems engineering life cycle procedures with final-user participation and rapid prototyping for the design of human-machine interfaces (Johannsen, 1995b). Any user-oriented design of interactive software products, such as HMIs, should start with scenario definitions and task analyses in order to have a solid basis for user requirements and systems specifications, particularly for HMI functions specification, as described in the previous section. User requirements, as specified in international standards for software development, should consider the goals-means-tasks relationship and be based on a task-oriented perspective.

The design stages with different forms of user participation are shown in Figure 2. This indicates, as outlined in the previous section, that the functional specification for human-machine interfaces depends on the goals and goal structures prescribed for the system, as well as the technological and intellectual means for accomplishing these goals, and on the tasks to be performed by the human users.

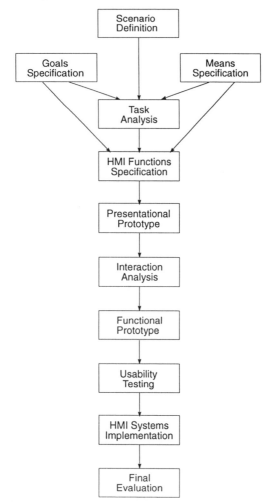

Figure 2. Design stages for human-machine interfaces with different forms of user participation. (From Johannsen, 1995c, with permission.)

The tasks of human users are specified together with final expert users during the scenario definition. The tasks may require cooperation in work situations with many human-machine interfaces (Johannsen, 1995c). In the special case of cooperative work, it is particularly necessary to investigate the work organization and structure. All opportunistic and informal communication channels have also to be discovered. The real information flows need to be clearly understood, more than the formal organization (which may or may not be important). Special expert analyses can be organized, either in the real application field or in a simulator environment. The results are information flow diagrams (see Section 5).

Thorough task analysis, combined with knowledge elicitation techniques, is a strong foundation for the definition of the user requirements and functional specifications of the human-machine interfaces (Johannsen & Alty, 1991; Kirwan & Ainsworth, 1992; Heuer et al., 1993). A methodology for structuring observations in complex field environments and for structuring systems functionalities is based on functional information and knowledge acquisition (Sundström & Salvador, 1995). Another methodology that specifically supports the monitoring tasks in supervisory control systems has been proposed by Thurman & Mitchell (1995).

During later design stages, it is necessary to organize further expert meetings with the participation of different human user classes in order to evaluate intermediate prototype designs of the human-machine interfaces. Cooperative work between the different user classes has a high priority in these evaluations. The same is true for the final evaluation at the end of the human-machine interface design. In Figure 2 this sequence of intermediate prototype designs and final HMI systems implementation, alternating with their corresponding evaluation stages, is outlined. The presentational prototype is more a surface layout followed by an interaction analysis with appropriate user participation, the purpose being to gather more inputs for the functional prototype. Based on the usability testing of the latter, the final design and implementation of the human-machine interface is accomplished followed by its evaluation.

The analysis and evaluation techniques for all these design stages with user participation comprise scenario definition, organizational analysis, task and knowledge analyses (all with unstructured and structured interviews, walk-throughs and talk-throughs in real or simulated situations, etc.), less formal expert analyses, and experimental sessions with usability testing procedures.

4   Structure and functionalities of human-machine interfaces

An appropriate structure of human-machine interfaces for integrated automated systems distinguishes between the presentation level and the dialogue level (e.g., Johannsen, 1992; see Fig. 1). The dialogue level deals with information flows regarding such problems as what information to handle when. The presentation level is concerned with the problems of how to present the information output to the human users and how to transform their inputs. Additionally, knowledge-based support modules with user model and technical-systems model capabilities, as well as explanation and justification functionalities, become more and more attractive submodules of advanced human-machine interfaces (Johannsen & Averbukh, 1993; Fabiano et al., 1993).

As shown in Figure 1, the human-machine interface includes the functionalities of handling inputs and generating outputs, both of which are typical presentation level functionalities. The generating outputs functionality normally contains a lot of graphical and textual dynamic pictures, and is designed by means of special graphical editors. The knowledge contained in the generating outputs functionality is represented in graphical objects and related nongraphical information about the technical system (Fejes et al., 1993; Johannsen, 1990). A lot of research has recently been done in the areas of speech output and multimedia interfaces (Alty & Bergan, 1992; Heuer et al., 1994; Zinser, 1993). In the case of applications of telemanipulation, the combination of proprioceptive feedback and 3D perception of depth is needed for achieving good performance (Stassen & Smets, 1995). Nevertheless, the main mode of presentation is still visualization.

New forms of visual information presentation have been proposed. They consider different levels of abstraction with respect to the goals-means hierarchies of Rasmussen (1986). Examples of such improved human-machine interface presentations for supervisory control are based on ecological interface design and multilevel flow modeling (Vicente et al., 1995; van Paassen, 1995). Further, they can be based on multimodels, including causal nets, goals-means hierarchies and, again, multilevel flow models, in addition to topological structures, as well as on hypermedia structures (Johannsen et al., 1995).

Dialogue functionality, also called supervising dialogue, can be designed as a quite complicated goal-oriented component for handling the dialogue between the human user(s) and all other components and subcomponents of the human-machine interface and, thereby, of the technical system. Among other subfunctionalities, this includes the resolution of possible conflicts between different components with respect to dialogue requests to and from the human users.

Another human-machine interface functionality deals with tracking interaction, i. e., the complete interaction between the human-machine interface and its two sides, namely, the human users and the machine. Knowledge-based subfunctionalities of tracking interaction are responsible for updating the interaction context and for quality control of the whole human-machine interaction.

Both the presentation and the dialogue levels explicitly depend in their functionalities on technical systems or application models, and on user or operator models. The more explicit representation of such models in the human-machine interface leads to more advanced paradigms (Johannsen & Averbukh, 1993).

A technical systems model contains knowledge about the goals, structure, and functions of a particular application. The functionality of the technical systems model internally supports all the other functionalities of the human-machine interface. Frame, rule, and causal network representations can be used for building such a model. Dependent on the intended dynamicity, such knowledge representations may not be sufficient. A combination with proven theories and methodologies implemented with numerical algorithms seems often to be more appropriate.

A user model functionality is needed if a certain adaptability to human user classes or single users is to be achieved (Kobsa & Wahlster, 1989; Sundström, 1991). A more elaborate user model will always include a technical systems model in order to represent the user's view with respect to the technical system. In addition, knowledge of human information processing behavior and cognitive strategies has to be represented in a user model by means of algorithms, rules, and, possibly, active inference mechanisms.

Other functionalities of a knowledge-based human-machine interface are explanation and justification. The explanation functionality informs the human users on request about what the components of the technical system (and possibly also of the human-machine interface) mean and how they function (Cawsey, 1992). On the other hand, the justification functionality explains the reasons why a certain decision of the decision support system is valid.

5   Different user classes and cooperative work

The investigation of human-machine systems was mainly concerned, up to now, with the interaction between a single human and the dynamic technical system, the machine. In the past, the subfield of human-machine interface designs was also pursued with such a preference. Real work situations of many application domains require, however, that several humans of different occupations frequently communicate and collaborate with each other. This communication and collaboration occurs on different levels within task-oriented subareas of the whole organizational hierarchy. Group meetings within subareas across different human user classes are held on the basis of specific task and information needs. Thus, they can better be reflected in information flow diagrams rather than in organizational hierarchies. The complete information flow diagram represents the plantwide or organization-wide communication and collaboration. The interaction and the communication with technical systems is a strong part of the whole organization. Some industrial examples of such technical systems are power plants, chemical plants, cement plants, and discrete manufacturing systems. Other examples of technical systems include transportation systems, public information systems, banking and insurance systems, and many others.

The state of computerization of these systems has often been developed quite far. Thus, they are supervisory control systems. However, the degree of computerization is different on the different organizational or information-flow levels. The need for face-to-face communication can exist on all levels, particularly on those closer to high-level management.

The application domain of the cement industry shows that several persons from different human user classes need to cooperate, partially via the human-machine interfaces, in the plant control room or, also, in office rooms during group meetings. These human user classes are control room operators, field operators, maintenance personnel, operational engineers, instrumentation and control engineers, chemists, laboratory personnel, commissioning engineers, researchers, and managers. As a task analysis showed, the style of cooperation between the different human user classes is very flexible and communicative (Heuer et al., 1993).

Further, the information flow between different human user classes and with the support functionalities of human-machine interfaces for cooperative work between these user classes has been investigated (Johannsen, 1995c). It is assumed that different people from different human user classes work together, share their overlapping skills and knowledge, and interact, at least partially, via human-machine interfaces. In order to prove this assumption, additional expert group meetings with unstructured interviews, walk-throughs, and talk-throughs were performed in the cement plant, mainly in the control room, based on the above-mentioned task analysis.

One of the main results of these expert group meetings is a better understanding of the information flows between different human user classes. The main interactions between these people can schematically be represented by an information flow diagram. This diagram was constructed and discussed during the group meetings with the control room engineer and the process engineer. Such information flow diagrams are independent of any hierarchical organisational structure.

The actual information flows depend on specific tasks and problem solving needs. About 90% of the problems are small ones concerning the ongoing operation of the plant and can be solved online with control room personnel, the production master, and field operators. More complicated problems occur due to major equipment failures or with the experimentation and testing of new equipment or new products. Then, special group meetings between personnel from the control room, production engineering, and process engineering are held, often only for an hour or so. The production master and, sometimes, also field operators may participate. The demand for new equipment and major problems with production quality control will be reported to the manager.

The main objective of the plant is to produce prescribed quantities of different types of cement, sometimes with frequent changes, by strictly controlling product quality, such as fineness, strength, and color. The product quality control is guaranteed in short-term intervals of 2 hours by the laboratory personnel and on a long-term basis by the scientific laboratory. Problems in product quality require group meetings with people from these two laboratories together with personnel from the control room and process engineering.

All these meetings are organized in a flexible manner as soon as possible after the specific problem has occurred. Some of the meetings are held in the control room or they are held in one of the offices, e.g., in that of the process engineer. The group meetings are cross-organizational meetings and bring together those people from different human user classes who are specialized in the different facets of a particular problem.

Based on the information flow diagram, several technologies and design alternatives for human-machine interfaces have been considered and discussed in the expert group meetings as possible support tools for the described cooperative work situations. Face-to-face communication is absolutely mandatory. Also, the audio channel, e.g., telephone communication, is very important. The logbook can be used as a multi-human-machine interface between the control room operators and the production master with the maintenance personnel. Large projection screens are not welcome because they are very soon too much overloaded and not adaptable enough. Multimedia technologies can be used, e.g., for integrating the video information from some of the equipment, which is now available in the control room with separate video screens. Otherwise, video observations are rejected as being a spy system.

Display screens for group meetings in different offices and the control room are welcome as multi - human-machine interfaces. They will also be accepted as dedicated human-machine interfaces in a network and for discussions of smaller problems at the phone. The display screens for group meetings can consist of one screen with four to five windows. They allow access to all pictures in the control room, rather than just printouts, as presently available. Different user group representatives may select their most favorite pictures initially. All selected pictures then need to be seriously considered by all meeting members because cooperation rather than egocentric views are required, where each user group representative contributes.

Modifications of control room pictures are foreseeable for the display screens in group meetings. Quick-change and easy-to-use editing facilities may allow selection of important lines or variables from a table, qualitative zooming-in and selection of subareas of component flow diagrams, and maneuvering or selection by slider or text menu through different levels of abstraction. The latter range from physical form, such as scanned-in photos, e.g., from data bases, or just representing separate components (inside of a pump, etc.), to goal hierarchies via multi-flow modeling representations.

The overlapping information of the different user classes is already considered in the log book. It has further to be implemented in the presentation, dialogue, and explanation facilities of the human-machine interfaces, particularly the display screen for group meetings. Thereby, the visual momentum between different windows that relate primarily to different user group representatives has to be supported (Woods, 1984).

All the suggested designs of human-machine interfaces for cooperative work have to consider also face-to-face communication. The social contacts will not be improved if this face-to-face contact disappears. Tele-cooperation is not feasible because the social contract for production will be lost, e.g., the feeling for system quality will disappear. Also, the work climate will deteriorate and, thus, there will be no cooperation.

6   Conclusions

The design and functionalities of human-computer interfaces that actually are human-machine interfaces was presented and discussed in this chapter. The examples behind the cited literature are mainly human-computer interfaces for supervisory control of industrial plants, such as chemical plants, cement plants, and power plants. As Stassen (1989) has pointed out, the situation of a patient in a rehabilitation program can also be compared with these industrial applications from the methodological point of view of supervisory control. Another example that shows the strong need for further improvements in the design of human-machine interfaces is presented with the thorough analysis of the aircraft crash at Gottröra by Mårtensson (1995). It shows that multimodality information presentation, in this case voice and visual output, needs to be consistently supportive, needs to exploit human pattern-recognition capabilities, and should never overload the human users. In summary, it can be stated that human-centered design approaches for human-machine interfaces in supervisory control, such as the ones discussed in this chapter, are indispensable in the future throughout all application domains.

Acknowledgments

The author highly appreciates the stimulation of this work through support of the Aalborg Portland Cement Factory in Aalborg, Denmark, and of two research projects that are currently being carried out at the Laboratory for Human-Machine Systems at the University of Kassel. These projects are supported by the Deutsche Forschungs-gemeinschaft (German Research Foundation), Bonn and by the Commission of the European Union, Brussels, under the BRITE/EURAM Programme in Project 6126 (AMICA: Advanced Man-machine Interfaces for Process Control Applications). The consortium partners of the AMICA Project are as follows: CISE (Italy), ENEL (Italy), FLS Automation A/S (Denmark), University of Kassel (Germany), and Marconi Simula-tion and Training (Scotland). The author is very thankful to all industrial partners as well as to several colleagues in his research group. Material of this chapter has partially been published in previous papers written by the author, as indicated, mainly Johannsen (1995 a, b, c).

References

Alty, J.L. and Bergan, M. (1992). The design of multimedia interfaces for process control. In H.G. Stassen (Ed.). *Analysis, Design and Evaluation of Man-Machine Systems.* (Preprints, The Hague). Oxford: Pergamon Press.

Boutruche, P. and Kärcher, M. (1995). DIADEM - A method for UI development. Presented at CEC Human Comfort and Security Workshop, Brussels.

Cawsey, A. (1992). *Explanation and Interaction,* Cambridge, MA.: MIT Press.

Fabiano, A.S., Lanza, C., Kwaan, J. and Averbukh, E.A. (1993). AMICA Toolkit Software Requirements Internal Report IR1-05, BRITE/EURAM AMICA Project.

Fejes, L., Johannsen, G. and Strätz, G. (1993). A graphical editor and process visualisation system for man-machine interfaces of dynamic systems. *The Visual Computer,* 10: 1-18.

Heuer, J., Ali, S., Hollender, M. and Rauh, J. (1994). Vermittlung von Mentalen Modellen in einer chemischen Destillationskolonne mit Hilfe einer Hypermedia-Bedienoberfläche. Hypermedia in der Aus- und Weiterbildung. Dresdner Symposium zum computerunterstützten Lernen, Dresden.

Heuer, J., Borndorff-Eccarius, S. and Averbukh. E.A. (1993). Task analysis for application B. Internal Report IR 1-04, BRITE/EURAM AMICA-Project. Lab. Human-Machine Systems, University of Kassel.

Johannsen, G. (1990). Design issues of graphics and knowledge support in supervisory control systems. In N. Moray, W.R. Ferrell, and W.B. Rouse, (Eds.). *Robotics, Control and Society.* London: Taylor and Francis, pp. 150-159.

Johannsen, G. (1992). Towards a new quality of automation in complex man-machine systems, *Automatica,* 28: 355-373.

Johannsen, G. (1993). *Mensch-Maschine-Systeme.* Berlin: Springer-Verlag.

Johannsen, G. (1995a). Knowledge-based design of human-machine interfaces. *Control Engineering Practice,* 3: 267-273.

Johannsen, G. (1995b). Computer-supported human-machine interfaces. *Journal of the Japanese Society of Instrument and Control Engineers SICE,* 34: no. 3, 213-220.

Johannsen G. (1995c). Conceptual design of multi-human machine interfaces (invited plenary paper). In 6th *IFAC/IFIP/IFORS/IEA Symposium on Analysis, Design and Evaluation of Man-Machine Systems,* MIT, Cambridge, MA. Oxford: Pergamon Press, pp. 1-12.

Johannsen, G., Ali, S. and Heuer, J. (1995). Human-machine interface design based on user participation and advanced display concepts. Post HCI'95 Conference Seminar on Human-Machine Interface in Process Control, Hieizan, Japan, pp. 33-45.

Johannsen, G. and Alty, J.L. (1991). Knowledge engineering for industrial expert systems. *Automatica,* 27: 97-114.

Johannsen, G. and Averbukh, E.A. (1993). General Man-Machine Interface Organisation, Internal Report, BRITE/EURAM AMICA-Project, Lab. Human-Machine Systems, University of Kassel.

Johannsen, G., Levis, A.H. and Stassen, H.G. (1994). Theoretical problems in man-machine systems and their experimental validation. *Automatica,* 30: 217-231.

Kirwan, B. and Ainsworth, L. K. (Eds.) (1992). *A Guide to Task Analysis.* London: Taylor and Francis.

Kobsa, A. and Wahlster, W. (Eds.) (1989). *User Models in Dialogue Systems.* Berlin: Springer-Verlag.

Mårtensson, L. (1995). The aircraft crash at Gottröra: Experiences of a cockpit crew. Internat. J. *Aviation Psychology,* 5 (3): 305-326.

Rasmussen, J. (1986). *Information Processing and Human-Machine Interaction.* New York: North-Holland.

Reason, J. (1990). *Human Error.* Cambridge, UK: Cambridge University Press.

Sheridan, T.B. (1992). *Telerobotics, Automation, and Human Supervisory Control.* Cambridge, MA: MIT Press.

Sheridan, T.B. and Johannsen, G. (Eds.) (1976). *Monitoring Behavior and Supervisory Control.* New York: Plenum Press.

Stassen, H.G. (1989). The rehabilitation of severely disabled persons. In W. B. Rouse (Ed.), *Advances in Man-Machine Systems Research,* 5: 153-227.

Stassen, H.G., Johannsen, G. and Moray, N. (1990). Internal representation, internal model, human performance model and mental workload. *Automatica*, 26, 811-820.

Stassen, H.G. and Smets, G.J.F. (1995). Telemanipulation and telepresence. In: *6th IFAC/IFIP/ IFORS/IEA Symposium on Analysis, Design and Evaluation of Man-Machine Systems*. MIT, Cambridge, MA, Oxford: Pergamon Press, pp. 13-23.

Sundström, G.A. et al. (1991). Process tracing of decision making: An approach for analysis of human-machine interactions in dynamic environments. *International Journal of Man-Machine Studies*, 35: 843-858.

Sundström, G.A. and Salvador, A.C. (1995). Integrating field work in system design: A methodology and two case studies. *IEEE Transactions on Systems, Man, Cybernetics*, 25: 385-399.

Thurman, D.A. and Mitchell, C.M. (1995). A design methodology for operator displays of highly automated supervisory control systems. In *6th IFAC/IFIP/IFORS/IEA Symposium on Analysis, Design and Evaluation of Man-Machine Systems*. MIT, Cambridge, MA., Oxford: Pergamon Press, pp. 821-826.

van Paassen, R. (1995). New visualisation techniques for industrial process control. In *6th IFAC/IFIP/IFORS/IEA Symposium on Analysis, Design and Evaluation of Man-Machine Systems*. MIT, Cambridge, MA. Oxford: Pergamon Press, pp. 457-462.

Vicente, K.J., Christoffersen, K. and Pereklita, R. (1995). Supporting operator problem solving through ecological interface design. *IEEE Transactions Systems, Man, Cybernetics*, 25: 529-545.

Woods, D.D. (1984). Visual momentum: A concept to improve the cognitive coupling of person and computer. *International Journal on Man-Machine Studies*, 21: 229-244.

Zinser, K. (1993). Integrated multi-media and visualization techniques for process S&C. In *Proceedings of IEEE International Conference on Systems, Man and Cybernetics*. Le Touquet, pp. 367-372.

Chapter 22

Models of Models of . . . Mental Models

Neville Moray

1   Introduction

In the last two decades, as the development first of information processing models and then cognitive psychology and cognitive science ensured the final demise of the Ur-behaviorism of the 1930s and 1940s, the notion of "mental model" has become more and more common. At the same time, the literature in the field is extremely confusing. To someone entering the field there appear to be several quite distinct usages with little or no connection between them. Consider, for example, the difference among Johnson & Laird (1983), Gentner & Stevens (1983), and Bainbridge (1991), and indeed Sheridan & Stassen (1979). The conceptual confusion has been clearly demonstrated in two reviews by Rutherford & Wilson (1991) and Wilson and Rutherford (1989). However, closer inspection reveals more coherence in the field than at first sight, and the purpose of this chapter is to provide a guide to the literature and to place the entire corpus in the context of a single unifying formalism.

2   A brief history of mental models

2.1   CRAIK: EPISTEMOLOGY

Historically one of the earliest modern uses of the notion of a mental model was by Craik (1943), who suggested that knowledge consists of a model of the world formed by humans in their nervous systems. This was a remarkable suggestion at a time when Behaviorism was still dominating psychology, at least in the United States. It is probably significant that the notion of a mental model came from a European psychologist, and one who in many ways can be seen to have independently invented the major concepts of Wiener's *Cybernetics*. Thus for Craik, to talk of a mental model is to talk of the way in which our knowledge of the world is represented in the head. Note that Craik's notion requires the mental model to be part of long-term memory.

Although Craik was a psychologist, he was deeply interested in human-machine systems and closed loop skills. This is interesting, and perhaps historically significant, since another main tradition of talk about mental models comes not from psychology, but from the control engineering community. Indeed, Henk Stassen used the notion of a mental model many years ago, both in a formal sense in connection with Optimal Control Theory (OCT) and informally in discussing mental workload (Sheridan and Stassen, 1979; Stassen, 1985, 1988). These two traditions, psychological and engineering, have developed almost independently of each other as far as mental models are concerned. For example, in the major books about mental models, such as those by Johnson-Laird (1983) and Gentner & Stevens (1983), there are no mentions of control theory. Indeed, it is somewhat ironic that, on the whole, control engineers were happily invoking the notion of mental models some 30 years ago, at a time when it was almost impossible to discuss them in the psychological community.

2.2   WHOSE MODEL? NORMAN'S TAXONOMY

In a paper that has become a classical reference, Norman (1983) distinguished three uses of the term *mental model*.   First, one may talk of the system designer's mental model of the system that is being designed and that will be realized in the future through construction or manufacture, or, nowadays, through the development of a piece of software and its implementation on one or more computer systems. Let us call this a Type 1 model.

Second, one may speak of the mental model possessed by a user or operator. In this usage we are speaking of the more or less imperfect knowledge or understanding that a person has of a system or device that is being operated, and the environment in which the person lives and works. Thus one may speak of a user' mental model of a pocket calculator, or a nuclear power plant operator's mental model of the dynamics of the reactor and steam generator, or even of an economist's mental model of an economy or a politician's mental model of society. This is a Type 2 mental model, and it is this use of the term with which the present chapter is concerned. It is very close to Craik's original notion.

Finally, one may speak of a Type 3 mental model, the mental model that a researcher has of the nature of the mental model that the person, user, or operator under investigation possesses. This is a mental model in the mind of an investigator of a Type 2 mental model possessed by someone else. (One assumes that a good designer includes a Type 3 model as part of his or her Type 1 model, at least if the designer has listened to good advice from human factors professionals!)

The notion of mental model with that we will be concerned, then, is that of *the more or less imperfect knowledge that a person has of his or her functional environment* - the environment, in a broad sense of that word, with which moment-to-moment interaction is occurring, from which information is being received, with respect to which decisions are being made, and upon which the person is acting.

The problem is that when one reads the literature, one becomes uncertain whether there is a substantive content of the notion of mental model. It has been noted by various writers that while mental model is certainly a useful heuristic, and one curiously more acceptable for many years by engineers than by psychologists, it is often not clear whether it is more than a shorthand or colorful way of referring to the knowledge that a person possesses. If "she has a mental model of X" means no more than "she knows about X," or "she understands X," it is not immediately obvious that the concept is worth adding to the functional vocabulary of psychologists and engineers. Why not simply direct our efforts to understanding how people know things? It will be one aim of this paper to claim that there is a particular and quite precise reason for using the term mental model that captures an important aspect of human functioning. The latter is perhaps a subset of knowledge, but a very important one for describing, understanding, and predicting human behavior not in over simplified laboratory tasks but in the world of everyday life, be it engineering, psychology, sociology or politics.

2.3   ROLE OF MENTAL MODELS IN COGNITION: SHORT- OR LONG-TERM
         KNOWLEDGE?

Another source of confusion lies in the fact that some writers seem to think of a mental model as being the content of "short term," or "working" memory. Such a mental model is constructed from the background knowledge that an operator has of a system or task, and consists of just those aspects that are needed to solve a particular problem over a short time span. Examples of this use can be found in Johnson-Laird (1983) and Wilson & Rutherford (1989). On the other hand, there are many writers, including Bainbridge (1991), Craik (1943), and the present writer, who more typically use the term *mental model* to refer to certain contents of long-term memory, and think of the contents of working memory being constructed from moment to moment from the interaction of the contents of long-term memory, including mental models among other such contents, and information obtained anew from the environment.

As stated earlier, a particularly good discussion of the problems that arise due to these conflicting uses will be found in two papers by Rutherford and Wilson (1991) and Wilson & Rutherford (1989). There are even usages that seem to lie between these two, such as the recent model of situation awareness proposed by Endsley (1995). In the present chapter, as will become apparent, the proposed unified formalism will mainly apply to long-term representation of system knowledge, although in so far as the contents of such mental models can be transferred to working memory for particular purposes, the contents of the user's mental model can be made available, along with other information, to working memory.

2.4   WHAT KIND OF TASK?

A major source of difficulty when studying the literature on mental models lies in the differences between the kinds of tasks that have been used by researchers who use the concept. Perhaps the extremes are exemplified by Johnson-Laird's experiments on reasoning at one end of the spectrum of use, and by the application of the notion of mental model to operators in large industrial plants by Bainbridge (1991) and De Keyser (1986) at the other. In between lies the research on understanding causality collected by Gentner & Stevens (1983) and that on perceptual-motor tasks (Jagancinski & Miller, 1978), control theory (Stassen, 1985,1988; Sheridan & Stassen, 1979), situation awareness (Endsley, 1995), and the application of mental models in human - computer interaction (Green, 1990; Schiele & Green, in press).

Newcomers to the field, and even some of those who have been working in the field for many years, frequently find it difficult to detect any common theme or meaning to the notion of mental model when looking at the very disparate investigations where the concept has been used. However, close examination of these tasks reveals a systematic pattern.

There seem to be five main uses of *mental model,* which can be characterized by the type of task, the degree to which the human interacts with the task and the environment, the temporal dynamics of the task, and the extent to which the environment interacts with the system (the latter comprising the human and the task or plant with which he or she interacts). Table 1 provides a summary.

In Figure 1 three aspects of a work situation are distinguished graphically. The task is the particular set of operations that is defined by the work context as containing all the specifically relevant information. Thus in Johnson-Laird's work, the task is the reasoning problem that is set to the subjects in the laboratory, and may consist of cards such as used by Wason, verbal descriptions written on a sheet of paper, etc.

In the Genter & Stevens papers the task is to manipulate the small devices, and again the environment plays a negligible role. For DeKeyser the task is the work setting of a continuous casting steel mill, with the control room, the massive machinery used to cast molten steel, the rollers that form the molten steel into a sheet or ribbon, etc. In Stassen's work it might be the bridge of a super tanker where the helmsman controls the course of the vessel. The environment is the supposedly "irrelevant" or contingent aspects of the world in which the task occurs. For Johnson-Laird it would be the physical characteristics of the laboratory, for Stassen the state of the sea, and for DeKeyser the noises and temperature of the control room, unexpected fluctuations in the electrical supply to the furnaces, or chemical composition of the raw materials, etc.

In Figure 1 the arrows with solid heads indicate major flows of information or causality. The arrows with shaded heads indicate weaker effects. Thus in the work of Schiele & Green there is a strong influence of the human on the task, and a rather weaker effect of variations of the task on the human. In the case of perceptual-motor control, variations in the evironment affect both the task and the hman. For example, an automobile or airplane may be subject to wind gust disturbance, and the human may be affected by bodily vibration,which results in accidental movements of the controls.

Researcher	Type of Task	Temporal dynamics	Human to system effects	System to human effects	Environment disturbance effects
Johnson-Laird	Logical Reasoning	Negligible	Negligible	Negligible	Negligible
Gentner & Stevens	Causal analysis of simple systems	Negligible	Slight	Slight	Negligible
Schiele, F., Green	HCI - drawing	Negligible	Strong	Slight	Negligible
Jagacinski & Miller; Jex	Perceptual-motor skill	Strong but short term	Strong	Strong	Slight
Young; Stassen	Complex dynamic skills	Strong	Strong	Strong	Considerable
Bainbridge, De Keyser	Industrial Human-machine Systems	Very strong & long	Very Strong	Very Strong	Very strong Strong & long

Table 1. Relation between human, task, and environment in different mental model studies.

In all cases there is a self-directed loop on the left hand side of the human, which represents the use by the human of his or her mental model for internal reflection, calculation, planning, etc., the central concept with which we are concerned. It is left as an exercise for the reader to place the ideas of other mental modelers in appropriate places in Figure 1.

The many uses of the concept are confusing because they seem to apply to very different fields and very different kinds of tasks. The key to understanding the literature is to examine the relative roles of human, task and environment in each research paradigm.

2.5 JOHNSON-LAIRD: "LOGICAL" REASONING

In the justly influential use of the mental model concept by Johnson-Laird (1983), the task for the human is "logical reasoning." This description of it must be put in quotes, since it is precisely the contention of Johnson-Laird that logical reasoning is not how humans solve these tasks. Rather, they use "mental models" which are some form of internal representation of the information presented to them, which represents alternative possibilities justified by the information. The information required to carry out the task is presented as a verbal problem, usually written or graphically drawn on cards or paper. Nothing the person does influences the information provided (the task), nor does the environment act either on the state of the task or the state of the person. Indeed, there is, in principle, nothing to prevent the person from thinking about the problem for may days before providing a verbal or written answer. The information becomes a mental model because it is turned into an internal representation, and it is that representation that is manipulated by the person to arrive at the answer.

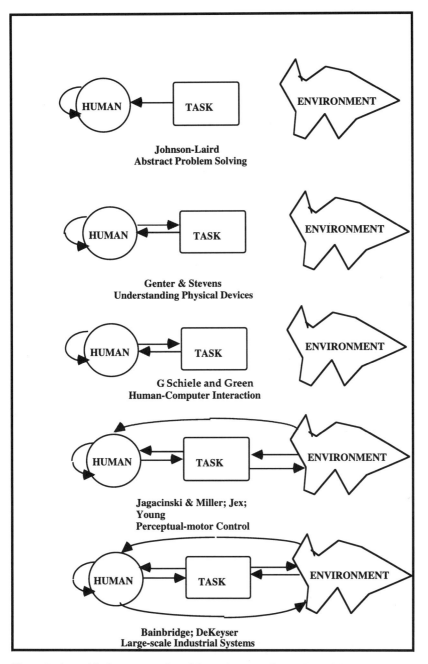

Figure 1. A graphical representation of the main types of mental model.

2.6   GENTNER & STEVENS: NAIVE PHYSICS: SIMPLE QUALITATIVE CAUSALITY

Gentner & Stevens (1983) edited a series of studies of how people understand the causal processes in very simple physical devices. These include doorbells, toys, pocket calculators, and abstract simple mechanical problems, such as the path taken by a projectile or a weight on the end of a string if it is swung in a circle and then released. Their purpose was to develop a Type 3 model of the Type 2 models that people have of causality, with a view to modeling the "naive physics" that people use to understand the world, and developing a "qualitative physics" of human understanding of real systems (Bobrow, 1985). In such tasks there is some interaction with the devices that people are trying to understand. Switches are moved, buttons are pressed, and data are entered into calculators. Conceptually at least, stones are swung in circles on the end of strings and then released. But the main interaction is that people observe the states and actions of the devices they are trying to understand. The predominant flow of information and control is from the device to the person. The information thus received becomes part of the mental model, and the contents of the mental model are then examined and manipulated to construct a causal account of the functioning of the device. [Note that the word *device,* which is often used, has in ordinary English the implication of a (somewhat) small size and complexity by comparison with a *plant* or a *system.]*

There are no significant dynamics in these devices. That is, left to themselves after they have been put into a new state by the operator, they tend to remain in that state. They do not degrade, heat up indefinitely, oscillate, or explode. Nor is there a significant amount of disturbance from the environment in which the person and the device are located. The temperature of the room does not vary significantly, nor do such changes as occur cause changes in the state of the device. Indeed, in line with classical experimental psychological research paradigms, such environmental impacts would be excluded from the research. People must try to understand the function of the *device itself*, not the device in its context or environment.

2.7   SCHIELE AND GREEN: HUMAN-COMPUTER INTERACTION

There has been an increasing use of the notion of mental model to describe human - computer interaction, as witnessed by the two volumes of papers edited by Ackerman & Tauber (1990) and Tauber & Ackerman (1991).

At first sight the use of mental models in this context seems a little strange. Most of those who talk about such models seem to identify them with the structure of a computer program that would behave in the same way as a human. That is, to a considerable extent, they are a Type 3, not a Type 2, model in terms of the taxonomy that was introduced at the beginning of this paper, and are an example of artificial intelligence or cognitive science rather than psychology. For example, Schiele and Green describe a Task-Action Grammar (TAG) of the computer program MacDraw that they seem to claim is a person's mental model of how to use MacDraw. Clearly, if such is the case, the main flow of information and control is from the model to the task. The user of MacDraw decides to draw something (a triangle, a circle) and performs certain actions on the computer system using a mouse or keyboard. The system reacts and something appears on the screen.

Schiele and Green want to say that what allows the user to choose the appropriate sequence of actions is his or her mental model. In accordance with current paradigms of psychology they "validate" their (Type 3) model of the user's mental model by writing a computer program that can perform the same actions in the appropriate sequence. It seems a little strange to think of an algorithm, a procedure that guarantees effective results, as being a model of something. However, if one thinks of the algorithmic mental model as a *script* in Minsky's sense of the word (Minsky, 1975), and hence as a program that contains default values that may not always be set correctly, and therefore can lead to unwanted results, the notion of a program as a mental model may be justified. There may be some input from the task to modify the model, but that is not the main way in which the mental model operates. It is a mental model that generates action rather than one that models the task. Dynamics are negligible, as are environmental effects. The drawing that appears on the screen when MacDraw is used does not change with the passage of time, and there are no unexpected disturbances from the environment to corrupt the image. (It is, of course, possible to imagine a power failure that will cause an inadvertent shutdown of the computer with a loss of the image, but, in the same sense that Gentner and Stevens would exclude arbitrary interference from the environment, so HCI researchers are not really interested in such random interference.)

If the mental model in HCI is thus a generator of action, instantiated for the researchers in a computer program or algorithm, and if the main purpose of the model is to generate action in the device, where does the model come from in the first place? How does it get into the head? Not many people learn to use software merely by playing randomly with the keyboard or mouse to see what happens after loading a new program. Rather, the instructions in the user's manual, and informal or formal instruction from other users, build an initial model in the head that is sufficient for the user to begin to interact with the system. In doing so, and in observing the results, and in the light of further input from formal or informal instruction, the mental model increases in detail, and the user becomes more and more expert. Thus we do see an arrow in Figure 1 from the task to the human, albeit, since the researchers put the main emphasis on action generation, an arrow that is less emphasized than the outgoing arrow. The arrow from the task to the human, in fact, represents feedback rather than primary information. The task tends to be a sink rather than a source of information except for this feedback.

2.8 JAGACINSKI AND MILLER: CLOSED AND OPEN-LOOP CONTROL WITH SHORT-TERM INTERACTION

In all the applications of the notion of mental model that have been discussed so far in this section, there is probably at least an implicit notion that the use of that model if not its detailed content, is conscious. In the application of the notion to closed- and open-loop perceptual-motor skills by control theorists, such is not the case. As we shall see later, the notion seems to be that during practice at a perceptual-motor task, the human operator, behaving as a component in a closed-loop control system, constructs an appropriate transfer function to achieve control of the plant, so that the overall human-machine system performance is stable and effective. It is the transfer function that is the operator's mental model (Jagancinski & Miller, 1978). As practice proceeds the operator is able to develop also a model of the future behavior of input from the environment (which disturbs the plant), and of the future behavior of the plant (in response to the human's control actions). This results finally in a series of mental models that allow open-loop, predictive behavior.

In this kind of research we find for the first time a strong dynamic coupling to the environment external to the task. Changes in the environment disturb the task and the task changes the environment (at least in the sense that, typically, a perceptual-motor task moves the operator and his system to a new "location" in the environment, whether the task is flying an aircraft in real space, tracking an abstract variable in conceptual space, or moving through a so-called virtual reality). There are also temporal dynamics, and usually there is tight coupling of the actions of the human to the state of the system. (It is not advisable to take one's hands off the steering wheel for an extended period of time.) There may also be direct actions on the operator by the environment, rather than through the task (sudden accelerations of the body, changes in temperature or humidity, etc. as a function of location in the environment).

2.9 YOUNG AND MCRUER: CLASSICAL CONTROL THEORY

In classical control theory, as epitomized by the work of Tustin, McRuer, and others, we can identify the mental model as the embodiment of the human operator transfer function. That is, we consider the human as an element in a closed- or open-loop control system, including both the human as a controller and the physical plant as the controlled element, and determine the input-output functions of each element, usually expressed in Laplace transform notation. The task is to control some dynamic system such as a vehicle, and the environment is the road, wind shear, etc. The initial notion that there was a unique human operator transfer function was quickly found to be wrong, and instead it was found that human controllers adapted their behavior in such a way as to match the requirements of the controlled element, the physical plant. It turned out that the human is able with remarkable accuracy to chose a transfer function that "equalizes," or matches in the inverse sense, the plant transfer function.

There is considerable evidence that, over a wide range of plant transfer functions, the human adopts a transfer function so as to provide an adequate gain and phase margin, so as to make the overall joint transfer function of human and machine, in open-loop terms, close to a pure gain, a pure delay, and a single integration. In this case the joint transfer function approximates, in Laplace notation, ke^{-ts}/s.. One might say, in such a situation, that the human operator has a mental model of the controlled element, or that the chosen human operator transfer function is chosen precisely as an inverse model of the controlled element transfer function (hence "equalization function," McRuer & Krendel, 1957, 1974).

In addition, both McRuer & Krendel (1957) and Young (1969) have made use of the notion of mental models in a qualitative description of the acquisition of control skills in tasks such as vehicular control. The former described the acquisition of open-loop skills as the development of a series of models in the Successive Organization of Perception (SOP) (McRuer & Krendel, 1957). The operator begins by controlling a dynamic system, such as vehicle in closed-loop, error-compensating mode in a statistically stable (usually zero-mean Gaussian) environment subject to random perturbations. The experience of an environment means that statistically stable patterns in the input begin to be detected by the operator, and as this happens the latter begins to use the regularities that are detected in input and plant response as a basis for bursts of open-loop predictive control. As time passes this results in the development of a series of motor programs, each triggered by the occurrence of certain recognizable subsets of input that can be used to predict actions required in the future, and these are a series of models of the environment and models of required action. Only if the operator's output leads to unusually large error signals does the latter revert to closed-loop control.

This kind of model was taken even further in a classic description of skills in an everyday environment by Young (1969), who described the human operator as an adaptive controller who made use of four models: of input, of the response of the controlled element to operator actions, of the state of the environment (weather, icy roads), and of goals and values. All four were built up (not necessarily consciously) as in the SOP model, and together comprise a fairly complete set of models that allow predictive, predominantly open-loop behavior in real-time systems with a human operator exercising manual control.

2.10   OPTIMAL CONTROL

In the Optimal Control Theory (OCT) of human-machine systems, the use of the concept of an operator mental model is much more explicit. Initially used to describe the human operator by Kleinman et al., (1970), this is of course an area to which Henk Stassen has contributed notably over many years, applying OCT to the control of ships, to the design of prostheses, to the management of hospitals, theories of mental workload, and many other problems. The critical requirement of a good optimal controller is the Kalman filter, which provides a best estimate, in a statistical sense, of the true system state on the basis of noisy, delayed information received through sensors that measure system state variables. The Kalman filter contains a model of the plant dynamics, and it is by using this model that compensation for the delays and noise can provide a best estimate for actions in the light of a clearly specified decision criterion. Many experiments have shown that a well-practiced human, when behaving as part of a closed-loop control system, shows behavior that is almost indistinguishable from that shown by an optimal controller. On the basis of such data one might wish to say that the person behaves "as if" he or she has a mental model of the plant that is being controlled.

2.11   ENDSLEY: SITUATION AWARENESS

In the somewhat confused literature of the burgeoning field of "situational awareness", there is one clearly formulated qualitative model, that of Endsley (1995). She discusses how people keep track of dynamic changes in their environments, and how incoming environmental information interacts with the operators' existing knowledge of the recent and long-term past, and with their choice of action. Although it is not her concern to discuss the nature of mental models, her model of how people keep track of their "situation" makes use of the notion of mental models, and her Figure 4 bears a very close resemblance to Figure 2 in this chapter.

2.12   INDUSTRIAL HUMAN-MACHINE SYSTEMS (LARGE SYSTEMS WITH ENVIRONMENTAL DISTURBANCE AND RICH INTERACTION)

To those in the field of human factors this is, *par excellence*, where to apply the notion of mental model. Operators of power plants, nuclear or otherwise, of chemical factories, of petroleum distillation plants, of economies, etc. are all dealing with immensely complex systems. There may be very many degrees of freedom (nuclear power plants have on the order of 45 degrees of freedom), hundreds or even thousands of displays and controls, and computer-driven automation with hundreds of pages of displays. Plants may run for months or even years at a time without shutting down, in the face of major disturbance inputs from the environment. They are far too large for an operator to be able to track the exact values of all state variables from moment to moment, they may have enormous time lags and phase lags in response to inputs, and they may have many levels of automation from manual control to autonomous robots. A disturbance caused by the environment may result in the loss of a major component in the plant. A disturbance may propagate through the plant, generating other transients in its wake. There is tight coupling between the environment and the plant, and even tight coupling between the environment and the operator. There is, one hopes, tight coupling between the operator and the plant, but abnormal plant states may be generated by incorrect actions by the operator and may persist, whatever their origin, for hours, days, or months (Perrow, 1984).

Clearly even an experienced operators cannot be fully up to date in their knowledge of plant state and the desirable control actions in such large and complex systems. How then can they manage to act efficiently? The answer, acceptable alike to engineers and human factors practitioners, is that they have a mental model of the plant, and this model allows them sufficient understanding of the plant for them to control it.

3   Canonical mental model

There is a fairly obvious sense in which this last application is the most complete and extensive form of the mental model concept. Potentially it contains all the other uses of the concept as subsets. Looking at Figure 1, we see in all the domains of application a self-referential loop within the human. Humans can, as it were, take their knowledge, walk away from the task, and use the knowledge to think about the task. In such a case the task, device, plant, or system is not taken with the person. What is taken is a representation of the task device, plant, or system, that is, a model. The human may make use of this representation to act on the task, and to understand inputs from the task, including using these to modify the input or output properties of the mental model (see Young, 1969). The plant may change its state spontaneously or in response to disturbances from the environment and may in turn act on the environment. The environment may act directly on the human to modify the mental model.

We can now see more clearly the relation between the various uses of the concept. In the Johnson-Laird experiments, the mental model may be almost a perfect representation of the problem, in that it includes all there is to be known about the task. The person takes information and forms a model with which he or she then operates, but does not significantly change the environment, nor does the latter change spontaneously. The model may not even be permanent, and may exist only in working memory during the time the person is thinking about the problem, although background knowledge certainly is retained.

That is not true of any of the other instances of mental models. In these latter we see an increasingly difficult situation for the human, in which it is less and less likely that the internal representation is a complete representation (model) of the task and its environment at any time following its construction. Indeed, as context is allowed to influence the task and we move from the laboratory with its table-top experiments to massive industrial plants or even political economies, the range of required knowledge comes increasingly to approach the domain of the natural world. Even if internal representation is through symbols (which may or may not be the case, as we shall see) "to imagine a language is to imagine a way of life" (Wittgenstein, 1953).

The canonical form of a mental model, as indeed of any model, is a homomorphic mapping from one domain to another, resulting in an " imperfect" representation of the thing modeled. In this case one domain is the world external to the human, including the task and perhaps its environment, and the other domain is the mind, brain, or head. (For our purposes there is no need to consider philosophical niceties. Although we talk of mental models, we consider that whatever the representation – analogue, digital, propositional, imagery, etc. – they are instantiated in some way as events in the physiology of the nervous system, albeit their owner is not aware of those events but merely of the model.)

4   An inclusive model of mental models

Given the above approach and the outline of paradigms in Figure 1, let us examine the set of mappings that make up the human mental model. This is shown in Figure 2.

The "operator" or more generally the "person" is assumed to have had a prolonged period of experience of the task and the environment. Such experience may consist of specific training, the result of apprenticeship and day-to-day operation of a task, or may be a carryover from general knowledge of the world (Rasmussen et al., 1994). That experience results in a mapping of features and properties (relations, dynamics, entities) from the task and environment into the long-term memory of the person. It is a reducing, many-to-one mapping, which results in memory containing a simplified version of reality (the properties of the task and its environment). We shall not consider the question of how the model is constructed: that is a question for learning theory (see, e.g., the suggestions of Holland et al., 1986).

5   Some model theory

Once the model has been formed by mapping into long-term memory, it can be used in response to the demands of the task facing the human, and this can account for several important aspects of human operator performance in the several paradigms.

1. The model can be used as a base for further mappings, thus making models of models with differing amounts of detail (or abstraction) that represent parts and wholes, etc. This can be formally represented in lattice and mapping formalisms and corresponds, e.g., to the development of abstraction hierarchies and part-whole hierarchies (Rasmussen, 1986; Rasmussen et al., 1994; Moray, 1987, 1989).

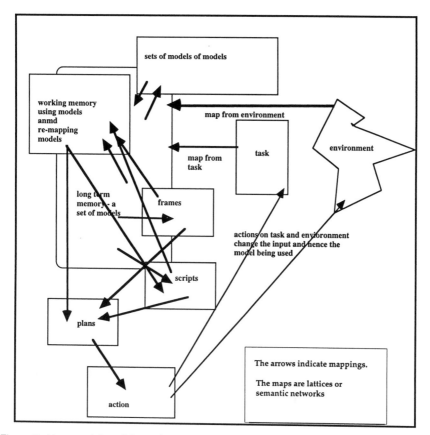

Figure 2. How mental models are formed and used.

2. The model can produce, by such mappings, both models of knowledge (hypotheses, facts, and expectancies about the task and the world), and also plans for action – models for what to do. This is particularly clear in Young's discussion of the development of skill (Young, 1969) and in McRuer & Krendel (1957). These models reside in long-term memory.

3. The long-term models waiting to be activated by task demands and intentions are sometimes called *schemas* (more correctly, *schemata*) by other writers. The long-term models that pop up automatically when activated by a task or an environment are called *frames* by Shank &Abelson (1977), and the long-term action orientated models that pop up automatically are called *scripts* by some writers (Minsky, 1975). These models are defaults, the basic background assumptions by the human about the task, and can be modified or overwritten by the person who consciously thinks about the problem, thus in turn calling up or constructing other models. Being defaults they are, by definition, homomorphs not isomorphs of the world, except in very special cases.

4. The models can be loaded into working memory, where they become dynamic, short-lived running models, and it is probably at this point that people are aware of their contents and where they are consciously constructed, modified, and explored.

From this point of view, Johnson-Laird's subjects use mappings from their long-term knowledge about the world and a mapping from the instantiation of the logical puzzle in the environment to form short-lived models in working memory that they then manipulate for problem solving. Schiele and Green's computer operators use their models of software to generate action sequences, having mapped a subset of the properties of the software into a model that might be called in some sense a *script* for using a word processor or graphics package, and whose properties can be, for Schiele and Green, represented as a task action grammar (Schiele & Green, in preparation).

Operators in very complex dynamic industrial systems, such as nuclear or fossil power plants, oil and chemical refineries, etc., normally operate at a high level of abstraction (Rasmussen, 1986) because only thus can they handle the workload of very-high-degree-of- freedom systems with many components. They therefore use a model of their model of a model of . . . . Furthermore, while the relation between the models and submodels is hierarchical, the overall knowledge structure, at least of causality, is neither a tree nor a heterarchical network but a lattice (Ashby, 1956; Moray, 1987, 1989).

The appropriate formalisms for discussing both models and relations are lattice theory and mapping theory (Ashby, 1953; Moray, 1987, 1989). An appropriate mapping can account for the fact that two pieces of knowledge that are quite disparate can interact. One does not necessarily have to retrace a path from a deep level of a tree structure up to a higher node. The structure of each model is a lattice of relations over propositions, images, plans, programs, etc. That is, what has often been called a *semantic network* covers both knowledge and action, and is a lattice, with all that implies for ordering relations between things that a person knows. On the other hand, there may be aspects of knowledge (nodes) that are not directly comparable or reachable from each other on a lattice. In that case, it should be quite literally impossible to think about them in relation to each other. However, since a person may have several different models (mappings) of a given task and environment, some of the mappings may contain relations that others do not. When people change their minds about how to think about a problem, or what to think about a problem, or at what level of abstraction to think about a problem, etc., they jump from one model to another. In such a case, it may be necessary for the operator to construct a new mapping from the experiential data, or to move from one existing map to another, on which the knowledge *is* related. Only by making the remapping and then exploring the new lattice will the person become aware that the relation is already known. ("Oh, I see now that if you think about it in *that* way there is *obviously* a relationship!")

Models are highly dynamic and can be modified in real time by the flow of information arriving from the task and environment, from actions flowing out to the task and environment, and by remappings from one existing model to another. An interesting question is the mechanism by which mappings are made from images to propositional descriptions and vice versa. A person can describe a mental image verbally and then make use of the propositional description. This indeed seems to be how Johnson-Laird thinks of his subjects performing their problem solving tasks. And most people – though not all – can conjure up an image in response to a verbal description.

5. Since thinking is based on a series of models, homomorphic mappings, and not a veridical isomorphic representation of the world, people make errors. By definition the models do not usually contain all the necessary information to represent unequivocally all properties of the task and environment, and internal remappings will further distort the representation. In the case of Johnson-Laird paradigms, the mapping may be isomorphic with respect to those characteristics of the task and environment that are relevant to the problem to be solved, since the experimental setting so clearly defines the relevant information, and so reduces the ambiguities in the task and environment. As we move down the paradigms in Figure 1, this becomes less and less likely.

6   Explanatory role of the formalism

Three assertions are being made in this chapter. The first is that the notion of mental model can be given a well-defined meaning that can unite all the existing mental model paradigms in the literature. The second, following Ashby (1953), is that the canonical formalism for discussing models (of any kind) is *mapping theory*. The third is that the domains and ranges of mappings are *lattices* whose nodes and arcs represent a hierarchy of knowledge possessed by humans about their tasks and environments.

It is clear that the proposed approach provides a formal way to describe what happens as operators learn and use abstraction hierarchies and part-whole relationships in the sense of Rasmussen (1986). Furthermore, the possibility of comparing items on different models by comparing the mappings and exploring the relationships on the lattices can allow for the discovery of "serendipitous" ideas. Indeed, in a sense it is only on a lattice, and not on a tree, that serendipitous ideas can be represented. (Trees do not allow the arbitrary connection of knowledge on different branches and different levels of abstraction.) Fourth, the large effects of individual differences in the way people behave, and the errors they make, when working in complex industrial systems is clearly explained by the fact that any individual can construct a variety of mappings unique to that individual, and linked to idiosyncratic experience. Furthermore, one would expect increasing individual variability as systems become more complex, and less individual variation as extensive experience minimizes the degree of disparity in experience due to the effect of the law of large numbers. Fifth, what has been called *opportunistic problem solving* can also be seen as a result of exploring the implications of various mappings and connections on the lattices of knowledge. Finally, by considering the implications of Lattice theory, it is possible to provide a detailed account of how operators understand different kinds of causality when interacting with complex human-machine systems (Moray, 1989), and hence to provide a way of developing a formal account of mental models in real tasks.

7   Conclusions

The notion of a mental model has been applied to many paradigms. Indeed, the great variety of uses has made it difficult to see what underlies the widespread intuition by researchers and designers that such a notion is needed to account for human behavior. It is particularly interesting to read Henk Stassen's account of the development of engineers' models of the human operator (Stassen, 1985, 1988). In these papers he shows how there has been an evolution from the strong quantitative modeling of tightly coupled skill-based manual control, through qualitative modeling of rule-based behavior. Wonham (1976) (and one may add Conant & Ashby 1970) have, as Henk points out, shown that the possession of a good "internal representation" is necessary but not sufficient for control. When we deal with situations in which the operator is tightly coupled to a dynamic environment in tasks with a high bandwidth, that is, from about 0.1 Hz up to the limit of effective manual control at about 2 Hz for a zero-order control task, the meaning of the operator's mental model is quite clear. The operator's internal representation plays the role that the Kalman filter model plays in optimal control. But as we move to more complex and lower bandwidth systems, such as those typical of supervisory process control, and as tasks become less stationary in a statistical sense and qualitatively more varied (Henk cites startup and fault management procedures), the strict quantitative meaning of *operator mental model* becomes inapplicable. Stassen describes a kind of evolution, both of engineering models of human performance (Pew & Baron, 1982), and also of the notion of what a mental model or internal representation must be as we move from skill-based to rule-based to knowledge-based behavior.

In a sense the attempt that has been made in this chapter to unify the literature on mental models is almost analogous to Henk's account. The difference seems to be that the rigorous definition of an operator's mental model that we find in the control-theory applications falls off in two directions. On the one hand, as we move toward supervisory control situations (Bainbridge, De Keyser), we move into the realm that Henk identified as ill-defined tasks and ill- defined environments, system complexity that is too great for the operator to comprehend, and very long time constants that cannot be computed effectively by humans. HCI applications are clearly rule-based (grammar-like) models. But in the other direction, we also find that with a loss of coupling to the environment the notion of the mental model becomes less well defined. Again we see a progression from skill-based to rule-based behavior, or at least rule-like formulation of knowledge in the qualitative physics of the Gentner and Stevens domain, and the ultimate knowledge-based paradigm of the logical reasoning tasks in the work of Johnson-Laird and similar workers, where there is nothing but knowledge of and no overt action on, the world.

Although there has not been a temporal evolution of the use of the concept of mental model, as Henk described the temporal evolution of models of performance, one can see many similarities. In these papers, as so often in his career of working on complex real-world systems, Henk's insights provide a valuable framework for approaches to other domains.

In this chapter the author has tried to show how the various applications of the concept of a mental model can be unified and to indicate directions for future development. Since it turns out that engineering models of human operator performance are so central to understanding the nature of mental models, I would like to think that such an effort is particularly fitting in the context of a *Liber Amicorum* dedicated to Henk Stassen, since Henk has, throughout his career, played an outstanding role in synthesizing the fields of engineering and psychology, with a view not merely to develop theory and design human-machine systems, but to improve the human condition. He has been a model for many of us.

8   Acknowledgment

I would like to thank Erik Hollnagel for helpful comments on an early draft of this chapter.

References

Ackerman, D. and Tauber, M.J. (Eds.) (1990). *Mental Models and Human-Computer Interaction I.* Amsterdam: Elsevier Science Publishers B.V. (North Holland).
Ashby, W.R. (1956). *Introduction to Cybernetics,.* London: Chapman and Hall.
Bainbridge, L. (1991). Mental models in cognitive skill. In A.Rutherford and Y.Rogers (Eds.), *Models in the Mind,* NewYork: Academic Press.
Bobrow, D.G. (Ed.) (1985). *Qualitative Reasoning About Physical Systems.* Cambridge, MA: MIT Press.
Conant, R.C. and Ashby, W.R. (1970). Every good regulator of a system must be a model of that system. *International Journal of System Science,* 1: 89-97.
Craik, K. (1943) *The Nature of Explanation.* Cambridge: Cambridge University Press.
De Keyser, V. (1986). Cognitive development of process indutry operators. In, J. Pasmussen, K. Duncan, and J. LePlat, (Eds.) *New Technology and Human Error.* Chichester: John Wiley.
Endsley, M.R. (1995). Toward a theory of situation awareness in dynamic systems. *Human Factors, 37,* 65-84.
Genter, D. and Stevens, A.L. (1983). *Mental Models.* Hillsdale, NJ: Lawrence Erlbaum..
Green , T.R.G. (1990) Limited theories as a framework for human-computer interaction. In D. Ackerman and M.J. Tauber, (Eds.) *Mental Models and Human-Computer Interaction I.* Amsterdam: Elsevier Science Publishers B.V.
Holland, J.H., Holyoak, K.J., Nisbett, R.E. and Thagard, P.R. (1986). *Induction.* Cambridge, MA: MIT Press.
Jagancinski, R.J. and Miller, R.A., (1978). Describing the human operator's internal model of a dynamic system. *Human Factors,* 20(4), 425-433.
Johnson-Laird, P.N. (1983). *Mental Models.* Cambridge, MA: Harvard University Press.

Kleinman, D.L., Baron, S. and Levison, W. H. (1970). An optimal control model of human response. Part 1: theory and validation. *Automatica,* 6, 357-369.

McRuer, D. and Krendel, E. (1957). Dynamic response of human operators. *WADC Technical Report.* 56-124.

McRuer, D. and Krendel, E. (1974). *Mathematical Models of Pilot Behavior.* NATO AGARD Rep. No. 188. NATO, Brussells.

Minsky, M. (1975). A framework for representing knowledge. In: P.Winston, (Ed.) *The Psychology of Computer Vision.* New York: McGraw-Hill.

Moray, N., (1987). Intelligent aids, mental models, and the theory of machines, *International Journal of Man-machine Studies,* 27, 619-629.

Moray, N. (1989). A lattice theory approach to the structure of mental models. *Philosophical Transactions of the Royal Society of London,* series B, 327, 447-593.

Norman, D. A. (1983). Some observations on mental models. In D. Genter & A.L. Stevens, (Eds.) 1983. *Mental Models.* Hillsdale, N.J.: Lawrence Erlbaum..

Perrow, C. (1984). *Normal Accidents: Living Wwth High-Risk Systems.* Cambridge, MA: MIT Press.

Pew, R.W. and Baron, S. (1982). Perspectives on human performance modelling. *Proceedings IFAC Conference on Analysism Design and Evaluation of Man-Machine Systems.* Baden-Baden. pp 1-13.

Rasmussen, J. (1986). *Information processing and human-machine interaction: an approach to cognitive engineering.* Amsterdam: North-Holland.

Rasmussen, J., Pejtersen, A.M. and Goodstein, L. (1994). *Cognitive Systems Engineering.* New York: Wiley.

Rutherford, A. and Wilson J.R. (1991). Models of mental models: and ergonomist-psychologist dialogue. In , M.J. Tauber & D. Ackermann, (Eds.) (1991). *Mental Models and Human-Computer Interaction 2.* Amsterdam: Elsevier Science Publishers B.V. (North Holland).

Schiele, F. and Green, T.R.G. Using task-action grammars to analyze "the Macintosh style". To appear in H.Thimbleby (Ed.), *Formal Methods in HCI.* Cambridge: Cambridge University Press.

Shank, R. and Abelson, R. (1977). *Scripts, Plans, Goals and Understanding.* Hillsdale, NJ: Lawrence Erlbaum Associates.

Sheridan, T.B. and Stassen, H.G. (1979). Definitions, models and measures. In N.Moray, (Ed.) *Mental Workload.* London: Plenum Press. pp. 219-234.

Stassen, H.G. (1985). Decision demands and task requirements in work environments: what can be learnt from human operator modelling. In E.Hollnagel, G. Mancini, & D.D.Woods (Eds.) *Intelligent Decision Support in Process Environments..* Berlin: Springer-Verlag. pp. 293-306

Stassen, H.G. (1988). Human supervisor modelling: some new developments. In E.Hollnagel, G. Mancini, and D.D.Woods (Eds.) *Cognitive Engineering in Complex Dynamic Worlds.* New York: Academic press. pp. 159-164.

Tauber, M.J. and Ackermann, D. (Eds.) (1991). *Mental Models and Human-Computer Interaction 2.* Amsterdam: Elsevier Science Publishers B.V.

Wilson, J.R. and Rutherford, A. (1989). Mental models: theory and application in human factors. *Human Factors,* 31, 617-633.

Wittgenstein, L. (1953). *Philosophical Investigations.* Oxford: Oxford University Press.

Wonham, M. (1976). Towards an abstract internal model principle. *IEEE Transactions on Systems, Man and Cybernatics,* SMC-6, 241-250.

Young, L. (1969). On adaptive manual control. *IEEE Transactions on Man-Machine Systems,* MMS-10, 292-331.

Chapter 23

Modeling the Human Operator and Human-Machine Interface for Dynamic Reliability Analysis

Anil P. Macwan

Human reliability analysis is an aspect of human-machine systems study that focuses on human error. With increasing complexity of industrial processes, increasing automation, and high reliability of hardware systems, the human operator's role in overall reliability and safety has become important. This can be confirmed by the percentage of industrial accidents that have been attributed to human error. Classical reliability and safety analysis techniques have been found to be limited for addressing the dynamic interaction between the human operator and the plant being controlled. Dynamic techniques have been developed that perform an integrated, dynamic simulation of the operator and plant to generate sequences of events, including hardware failures and human errors. An operator model developed for a dynamic reliability analysis is briefly described in this chapter. Additionally, issues related to modeling the human-machine interface are also discussed.

1   Introduction

Trends of increasing automation and increasing complexity have changed the operator's role from direct, manual control to one of supervisory control (Stassen et al., 1993). This, in turn, has given rise to new, theoretical problems for human-machine system studies (Stassen et al., 1995). These include task allocation, especially between automatic control system and human operators, design of Human-Machine Interfaces (HUMIF), and modeling cognitive behavior of operators. In particular, the last two are significant for safety and reliability analyses that are routinely performed for complex systems, such as power and process plants.

While Human Reliability Analysis (HRA) is an aspect of human-machine studies, it has been extensively developed for application to probabilistic safety assessment, especially of nuclear power plants. An interesting comparison of human-machine system studies and HRA can be found in Stassen et al. (1995). In essence, HRA explicitly focuses on human error and the many techniques available for assessing human error probability, which is subsequently incorporated into the safety assessment. Human error is defined in terms of the human operator not performing a required task within a specified time interval. For example, a human error can be the failure to provide cooling to a pump within T minutes. Both required task and time interval are derived from engineering analysis of the plant. In the present discussion, *plant* refers to the physical plant; whereas *system* refers to the human-machine system consisting of physical plant, human operator, and HUMIF.

Techniques currently used to perform HRA have been recently criticized for their inability to analyze the cognitive behavior of operators (Dougherty, 1990; Macwan & Mosleh, 1994). This limitation is important for two reasons. Firstly, intentional (or cognitive) and not unintentional behavior has been found to be of significance in many industrial accidents. Examples are the Three Mile Island 2 accident (Rubinstein, 1979) and the Davis Besse nuclear power plant accident (US NRC, undated). Secondly, with increasing automation, intentional errors will become considerably more important in their impact on system safety. This can be seen from the changes in interface design where through use of coded keys and buttons, the likelihood of operators inadvertently pushing the incorrect key or button is reduced to almost zero.

Techniques currently used to perform safety and reliability analyses are limited in their handling of the dynamic (i.e., time-dependent, stochastic) behavior of the system.

Dynamic reliability analysis techniques are being developed to overcome this limitation. These techniques have matured with respect to modeling the hardware part of the system. However, they lack good (cognitive) models of operator behavior and do not explicitly focus on the HUMIF as an element of the system. This chapter presents a model of operator behavior that was developed for incorporation in dynamic reliability analysis and issues related to modeling the HUMIF for inclusion in these analyses. A discussion of classical and dynamic reliability analyses is also included to provide some background. The operator model was developed to perform integrated, dynamic accident analysis of nuclear power plants. Discussion about the HUMIF is limited to issues about modeling since current techniques include no model for the HUMIF, or treat it within the plant model in a limited sense. However, the HUMIF has been a factor in many industrial accidents (CCPS, 1994).

2   Classical reliability and safety analyses: Event tree and fault tree

One of the primary tools used in current reliability and safety assessments is the event tree. It is constructed in the form of a forward branching tree and describes event sets that can be associated with random (stochastic) variables or events. A node of the event tree represents the required functioning of component or operator action. Possible paths at each node have a binary structure, representing success and failure. After constructing all possible paths, a success or failed end state is assigned depending on the values of important physical variables. For example, if the temperature exceeds a certain limit, the corresponding path is labeled as having a damage state; otherwise it is labeled with a safe state. Figure 1 provides an example of an event tree.

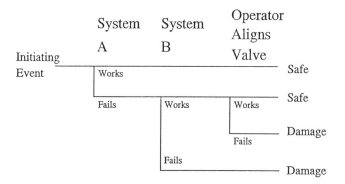

Figure 1.   Event tree

The probabilities of success and failure at each node are obtained by a complementary tool called *fault tree analysis* (Fig. 2). In a fault tree, a top-down deductive logic is used to link the undesired condition (e.g., subsystem not working) to its more proximate subcauses, continuing down to the basic events that are not further classified. The fault tree leads to identification of a logical set of events that could result in a top event, e.g., failure of subsystem. These logical functions are represented as Boolean (AND, OR) gates. Based on the relationship of top event to the basic events (which occur at the bottom most level of the fault tree), the probability of the top event is obtained using Boolean expressions.

In performing the event tree/fault tree analysis, the physical behavior of process parameters is decoupled from the analysis of subsystems and components. Thus, for a given combination of component availabilities and unavailabilities, a separate analysis is

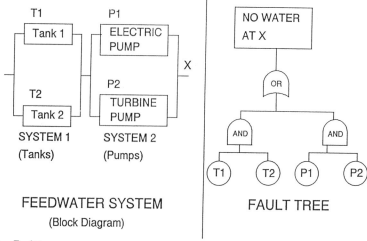

FEEDWATER SYSTEM

(Block Diagram)

FAULT TREE

Figure 2.   Fault tree

conducted to determine the values of process parameters necessary to assign safe/damage states on the event tree. It must be noted that conservative assumptions are made in doing so. For example, when a component is found unavailable but can be subsequently repaired, the conservative assumption assumes it to be always unavailable.

Because of its simplistic treatment of operator errors and hardware failures, "many of the conditions affecting control system actions and operator behavior (e.g., behavior of plant process variables, previous decisions by the operating crew) are not explicitly included in the model" (Siu, 1994). As pointed out by Siu, these "models rarely identify risk significant situations in which the operators turn off needed safety systems, although this was a prime contributor to the TMI-2 accident." Additionally, such treatment of human errors inhibits accurate modeling of consequences of the error to the plant, and the resulting dynamic responses of the operator/plant system.

3   Dynamic reliability analysis

DYnamic Reliability Analysis (DYRA) techniques perform an integrated analysis of stochastic behavior of hardware components and deterministic behavior of process parameter behavior and treat time in an explicit manner. One of the widely used techniques is DYLAM (DYnamic Logical Analytical Methodology) (Cojazzi et al., 1993). DYLAM has been applied to chemical process systems and to a liquid metal breeder reactor. The technique has been described in detail by Cojazzi et al. (1993). A brief overview of DYLAM is presented here.   In addition to DYLAM, other DYRA methodologies have been developed. These include DETAM (Dynamic Event Tree Analysis Methodology) (Siu & Acosta, 1994), ADS (Accident Dynamic Sequence Scheduler) (Hsueh & Mosleh, 1992), and DYLAM-TRETA (Izquierdo & Sanchez-Perea, 1990). So far, these methodologies have been applied primarily to accident sequence analysis for nuclear power plant probabilistic safety assessments. Additionally, DYLAM has been applied for analysis of operating procedures in aviation (Cacciabue & Cojazzi, 1995), and in the nuclear industry (Cacciabue & Cojazzi, 1994).

Initially, an engineering analysis is performed that results in a description of the plant in terms of components, each with a nominal correct state and a number of possible failed states. For example, a valve can have a nominal correct state that is closed but operable, and two failed states, e.g., a failed open and a stuck closed state. Dynamic behavior of process parameters that describe the physics of the plant is described by using a number of mathematical equations. When the plant configuration changes, say, by a pump starting, the equations that describe the physical behavior also change. The DYLAM computer program performs the dynamic simulation. It calculates the course of process variables and generates a set of scenarios from transitions that occur in time. Transitions are allowed at discrete points in time and are described by one or more components changing states. Additionally, when a process parameter crosses a threshold value, a component may change state. For example, when a pressure falls below a certain value, a pump may come on. This allows one to analyze the effect of failure of components on the physical behavior of the plant and vice versa. Event sequences are generated by rules that are specified before the simulation. These rules are meant to control the number of sequences, which can grow rather large.

A DYRA technique such as DYLAM includes the following three major components:

1. *Simulation driver*. The most important component is the simulation driver, or event sequence generator. The main function of this component is to keep track of the transitions that occur in time, and to follow the physical process for each state associated with the transitions.

2. *Plant model*: In order to perform this function, the simulation driver needs a simulation of the plant in question. Here, *plant* is meant to include both the physical processes and hardware components. Physical processes are modeled through differential equations and algorithms to solve these equations, while hardware components are modeled through their failure characteristics (implying a nominal and one or more failed states) and associated failure rates.

3. *Operator model*. Additionally, a model of the operator is needed. It should be noted that the operator models used for DYRA are in a form that is compatible with the dynamic simulation environment. Models from cognitive science are not suitable in their original form for such uses. They need to be modified, or new models need to be developed. Within the DYRA approach, the human operator can be seen in the following way. In one sense, the operator can be one of the causes of transitions of operating status (by turning components ON or OFF). In a broader sense, the operator can be considered as a separate but integral part of the system (in the loop), with its own (cognitive or mental) states, who is constantly interacting with the hardware part. Since intentional behavior of the operator is often of interest, models of operator behavior developed for DYRA often include cognitive modeling. Operator modeling is discussed in Section 4.

4   CAB-SIM: An operator model for DYRA

Various cognitive and semicognitive behavioral models have been developed for inclusion in DYRA techniques. The cognitive models include the COgnitive SImulation MOdel, COSIMO, (Cacciabue et al., 1992), and Cognitive Environment Simulation, CES (Woods et al., 1987), though CES has not been fully implemented in any DYRA. These models are too detailed and are not in a form that can be used to perform practical analysis. Other models include CREW response SIMulation, CREWSIM (Dang & Siu, 1994) and CAuse-based Behavioral Model, CAB, (Macwan et al., 1995a), which are not too detailed, yet are practical. The implementation of the CAB model, CAB-SIM, will be described here.

The cognitive models listed here are too detailed for specific applications, such as analysis of operator errors, and have not been implemented for practical application into safety and reliability assessments. They also lack guidelines in terms of incorporation of their results into such probabilistic frameworks. On the other hand, the cause-based behavioral model (CAB model) has been developed as part of a methodology to analyze and incorporate intentional errors into PSA (Macwan & Mosleh, 1994). The model is inspired by classical models used in human reliability analysis such as THERP (Swain &

Guttmann, 1983) and SLIM (Embrey et al., 1984). However, it goes deeper in modeling the causes of intentional errors. Failure modes of intentional errors are identified on the basis of operating procedures that the operators are trained and required to follow. The model is based on formation of causal links from the error modes to the factors that influence them.

The basic modeling elements of the CAB model are mapping tables that perform the simplified simulation of cognitive processes for each operator-plant interaction. Some of the mapping tables are generic and do not change with the application, while others are application-specific and need to be composed before carrying out the analysis. This is done by interviews with operators and training instructors, and by review of appropriate procedures and training material. The following three cognitive processes are carried out:

1. *Selection of parameters and systems* is done under the assumption that operators do not   monitor all information all the time, but rather pay attention to the parameters of interest, based on their understanding of how the plant is behaving. This is a useful process for complex systems in which the interface contain enormous information. It is also supported by the fact that operators cannot, and do not, look at all available information all the time.

2. Next, the *selected parameters are checked against expected process behavior* (in terms of overall trends and not exact values) to check instrument failure and/or to change the assessment about plant behavior. This process captures the fact that operators do have an expectation about how the plant should behave, given their understanding and the control actions that they perform. The aspect of expectation has been also mentioned in air traffic control (Weston, 1982).

3. The third and perhaps the most important cognitive process, *the generation of a diagnosis about plant behavior. Diagnosis* is defined as assigning one of the many accident types that can occur in a given plant. Within studies on pilot error, diagnosis can be taken to mean situation awareness, which is one of the most important factors affecting pilot behavior.

The results of cognitive processes are subsequently translated into a set of factors, called performance influencing factors, which are then used to generate and quantify possible intentional errors. Examples of influencing factors are time pressure and memory of previous actions. In addition to generating intentional errors, the model also generates expectations, perceptions, and diagnoses. These three variables simulate the so-called mental state of the operator. These are carried to the next operator - plant interaction to account for dependencies due to the effect of current action on subsequent actions. Additionally, phenomena like fixation on a diagnosis or recovery from a misdiagnosis can be studied using such an approach.

The CAB model has been implemented as the computer program CAB-SIM (Macwan et al., 1995b). In developing CAB-SIM, some of the modeling elements of the CAB model were strengthened and improved. An object-oriented approach was used, and the model is flexible so that it can be implemented for different applications. CAB-SIM was used to perform an exhaustive analysis of a simple, hypothetical system consisting of two tanks and two coolant loops that resembles a very simplified model of a nuclear power plant (Macwan et al., 1995a). Additionally, CAB-SIM has also been incorporated into ADS (Accident Dynamic Sequence Scheduler), which is a DYRA technique that includes full simulation of a nuclear power plant risk model for limited conditions (Groen, 1995). The CAB model is also being applied to study intentional deviations from approach-to-landing procedures for civil aviation, using DYLAM (Peek, 1995).

5   Human-machine interface model for DYRA

The human-machine interface (HUMIF) has been recognized as an important influencing factor for plant safety, especially in its influence on the operator-plant (human-machine) interaction. New interfaces are being designed, and include human factors in their design. However, their effect on the dynamics of human-machine interaction and, ultimately, their effect on safety is not given much attention. While the DYRA techniques mentioned earlier have the capability of performing such analyses, these techniques currently don't include the HUMIF in their analyses.

5.1   CONVENTIONAL HUMIF

In addressing the HUMIF, two roles of the conventional HUMIF have been identified. One is that of providing information about the plant being controlled to the operator. The other role is to translate the operator action(s) into appropriate control action(s) that will be carried out on the plant. In modeling the HUMIF for DYRA, the main focus is on how the HUMIF affects the dynamic reliability of the overall system, and one of the important aspects is how it affects operator behavior. Specifically, this means which aspect of operator behavior (intentional or unintentional, decision-making, etc.) is influenced and what questions can research answer?

5.2   OPERATOR DECISION SUPPORT SYSTEMS

However, with increasing automation and increased use of decision support systems, the systems' influence on safety will become important, especially in terms of how they affect operator behavior. DYRA techniques typically do not currently include these systems. From a physical viewpoint, a decision support system can be considered to be part of the HUMIF. However, it can be included in the HUMIF model or it can be modeled as a separate element. In terms of its role in operation and how the operator interacts with it, there are differences from the conventional role of the HUMIF.

- An important issue is whether the decision support system receives information directly from the plant, from the conventional interface, or both. This is particularly so when studying the failure of plant components or interface components. Depending on whether it receives its information from the plant or the HUMIF, its advice might differ. Alternatively, it can receive information both from the plant and the HUMIF, and compare them to check for inconsistencies.

- In modeling the decision support systems, human factors that are important for HUMIF may be no more important.

- The effect of decision support systems is important for cognitive behavior of the operator, and not for unintentional behavior.

- In a conventional HUMIF the operator can influence the plant. This is not true, however, of decision support systems, whose role is only to provide information to the operator.

- It is also important to consider whether it is necessary to model the reasoning process that is designed in the decision support systems, what results are produced by the system, and in what form they are presented to the operator, etc.

Recently, industries such as nuclear power have started to computerize operating procedures that are used during all phases of plant operation. Similar to automation and the use of decision support systems, this will bring new problems and issues. DYRA techniques seem to be a useful tool when analyzing the role and effect of computerized procedures, since in a modeling sense, computerized procedures are similar in nature to decision support systems. These similarities can be exploited while modeling and analyzing. However, differences between the two should also be considered. One difference could be the additional possibility for the computerized procedures to act on the plant. Another one is in the logic (or reasoning) structure of the computerized procedures. How this affects overall reliability, and specifically operator behavior, needs to be studied.

6   Concluding thoughts

Increasing automation and development of operator decision aids are changing the role of the operator. Given the high reliability of hardware components, the contribution of the human operator to the overall risk and reliability has increased significantly. Dynamic reliability analysis techniques were developed to systematically study the effect of process dynamics on hardware components and vice versa. These techniques have matured in terms of modeling and analyzing the system part. However, the models currently used for the human operator are rather simple. Cognitive models that address the reasoning and intention formation behavior of the operator are being developed. These, however, have not been demonstrated for any real application. Evolutionary models that are based on classical human reliability models but are simpler than the cognitive models can be useful for a more realistic portrayal of the operator's role in such analyses.

In addition to the human operator's individual role in overall system performance, the role of the human-machine interface is also important. HUMIF designs are continuously improved, and in some cases, replaced entirely. Dynamic techniques are a useful tool in analyzing the effect of the HUMIF on reliability. In addressing the HUMIF, the role of operator decision aids and computerized operating procedures can also be addressed.

References

Cacciabue, P.C. and Cojazzi, G. (1994). A human factor methodology for safety Assessment based on the DYLAM approach, *Reliability Engineering and System Safety*, 45: 127-138.

Cacciabue, P.C. and Cojazzi, G. (1995). An integrated simulation approach for the analysis of pilot-aeroplane interaction, *Control Engineering Practice*, 3: 257-266.

Cacciabue, P.C., et al. (1992). A cognitive model in a blackboard architecture: synergism of AI and psychology, *Reliability Engineering and System Safety*, 36: 187-197.

Center for Chemical Process Safety (CCPS) (1994). *Guidelines for Preventing Human Error in Process Safety*, New York: American Institute of Chemical Engineers, CCPS.

Cojazzi, G., Cacciabue, P.C. and Parisi, P. (1993). DYLAM-3. *A Dynamic Methodology for Reliability Analysis and Consequences Evaluation in Industrial Plants: Theory and How to use*, Report EUR 15265 EN, European Commission Joint Research Centre, Ispra.

Dang, V.N. and Siu, N.O. (1994). Simulating operator cognition for risk analysis: current models and CREWSIM, Proc. of PSAM-II, March 1994, San Diego, G.A. Apostolakis and J.S. Wu (Eds.) , pp. 066/7-066/13.

Dougherty, E.M. Jr.(1990). Human reliability analysis: where shouldst thou turn?, Guest Editorial, *Reliability Engineering and System Safety*, 29, pp. 283-299.

Embrey, D.E., et al, (1984). SLIM-MAUD: An Approach to Assessing Human Error Probabilities Using Structured Expert Judgment. NUREG/CR-3518, U.S. NRC, Washington, DC.

Groen, F.J. (1995). CAB-SIM: A cause-based operator model for use in dynamic probabilistic safety assessments for nuclear power plants, Report A-713, Laboratory for Measurement and Control, Dept. of Mechanical Engineering and Marine Technology, Delft Univ. of Technology, Delft, The Netherlands.

Hsueh, K.S. and Mosleh, A. (1992). An integrated simulation model for plant/operator behavior in accident conditions, Report UMNE 92-004, Department of Nuclear Engineering, University of Maryland.

Izquierdo, J.M., and Sanchez-Perea, M., (1990). DYLAM-TRETA: An approach to protection systems software analys. In *Advanced Systems Reliability Modeling,* Proc. Ispra course held at ETSI Navales, Madrid, Spain, September 1988.

Macwan, A., and Mosleh, A. (1994). A methodology for modeling errors of commission in probabilistic risk assessment, *Reliability Engineering and System Safety*, 45, pp. 139-157.

Macwan, A.P., Groen, F.J., and Mosleh, A. (1995a). Application of HITLINE methodology to analyze intentional deviations from operating procedures, submitted to Special Issue of *the IEEE Transactions on Systems, Man and Cybernetics* on Human Interaction with Complex Systems.

Macwan, A.P., Groen, F.J., and Wieringa, P.A. (1995b). CAB-SIM: A computer code to analyze errors of commission for probabilistic risk assessment. *Proedings. 6th IFAC Symposium on Man-Machine Systems*, Cambridge, MA.

Peek, F., (1995). Development and implementation of CAB-FLIGHT: A causal pilot model for application of human reliability analysis in civil aviation, Report A-737, Laboratory for Measurement and Control, Dept. of Mechanical Engineering and Marine Technology, Delft Univ. of Technology, Delft, The Netherlands.

Rubinstein, E. (1979). Three Mile Island: How and why the unexpected happened, *IEEE Spectrum* (Special Issue), pp. 33-57.

Siu, N., (1994). Risk assessment for dynamic systems: An overview. *Reliability Engineering and System Safety*, 43: 43-73.

Siu, N., and Acosta, C. (1994). Dynamic event trees in accident sequence analysis: application to steam generator tube rupture. *Reliability Engineering and System Safety*, 41, pp. 135-154.

Stassen, H.G., Andriessen, J.H.M., and Wieringa, P.A. (1993). On the human perception of complex industrial processes, *Proc. 12th Triennial World Congress of IFAC*, Sydney, Australia, 18-23 July, 4: 441-446.

Stassen, H.G., Macwan, A.P., and Wieringa, P.A. (1995). "Man-machine system studies and human reliability analysis", to appear in *Proceedings of Workshop on Human Reliability Models: Theoretical and Practical Challenges,* Stockholm, Sweden, August 1994.

Swain, A.D., Guttmann, H.E. (1983). Handbook of Human Reliability Analysis with Emphasis on Nuclear Power Applications, NUREG/CR-1278, U.S. NRC, Washington, DC.

U.S. NRC, Loss of Man and Auxiliary Feedwater Event at Davis Besse Plant on June 9, 1985, NUREG-1154, undated.

Weston, R.C.W. (1982), Human factors in air traffic control, in *Pilot Error*, Hurst, R. and Hurst, L., (Eds.), pp. 118-135.

Woods, D.D., E.M. Roth, and H.E. Pople, Jr. (1987). *An Artificial Intelligence Based Cognitive Environment Simulation for Human Performance Assessment,* Vol. 3, NUREG/CR-4862, Washington, DC.U.S. NRC.

Chapter 24

Control Models of Design Processes: Understanding and Supporting Design and Designers

William B. Rouse

Design can be conceptualized as a control process that involves outer loops for goal setting and planning, and inner loops for development and assessment. Adoption of such a model enables systematic identification of human abilities and limitations in design processes. This, in turn, provides the basis for specifying aiding, as well as training, concepts for supporting design and designers. This method for the development of design tools is discussed. Two tools are considered, including a discussion of their impact on two companies in particular and on a broad cross section of companies in general.

1 Introduction

Early in the 1980s, several colleagues and the author started to explore how human-machine systems technology transfers from R&D to applications. We perceived that many useful research results are never applied or, at the very least, take a long time to be applied. This caused us to study the "channel" from laboratories to the applied world.

This channel begins with researchers whose goals include gaining fundamental understanding of the abilities and limitations of operators and maintainers within complex human-machine systems. Their goals often include the development and evaluation of training and aiding schemes for supporting operators and maintainers. Despite much excellent R&D in these areas, real applications are few and far between. Why?

This question led to an important insight. While operators and maintainers may be the intended beneficiaries of this R&D, they are not the users! Designers of human-machine systems are the users of research results. If designers are unaware, unable, or unwilling to employ the results of human-machine systems research, this technology does not transfer into application.

This insight led us to focus on the nature of designers, the tasks they perform, and how information can affect these tasks. This, in turn, caused us to investigate the abilities and limitations of designers. The resulting understanding enabled us to create methods and tools for supporting designers in terms of both training and aiding.

Our conclusion is quite simple. In order to improve the jobs and tasks of operators and maintainers, you have to improve the jobs and tasks of designers. In other words, if you help designers, they will be able to help operators and maintainers. This brief chapter reports on our efforts to make this happen.

2 Nature of design

Figure 1 (Rouse, 1991) depicts a control model of relationships among design tasks and issues. The overall steps of the process include recognition of needs or opportunities, formulation of one or more concepts, analysis of requirements, synthesis of one or more solutions, fabrications of solutions, and utilization of solutions.

This process involves several feedback loops. The inner loops involve testing, verification, and demonstration as design and development or manufacturing interact.

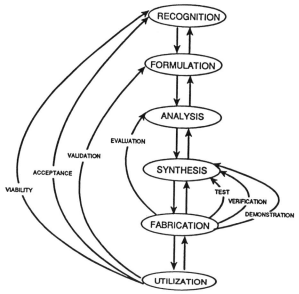

Figure 1.   Design as a control process

Evaluation is a slower loop in that more complete designs are needed to assess the extent to which usability requirements, for example, are satisfied.

The outer and slowest loops include validity, acceptability, and viability. Validity is concerned with whether meeting requirements actually results in the users' problems being solved. Acceptability concerns the extent to which problems are solved in agreeable ways. Viability reflects the "bottom line" of benefits versus costs.

This model provides the context within which designers perform their tasks. Different types of information are relevant at different times. Consequently, it is easy for R&D results to be too early and too late. In either case, these results will be judged to be irrelevant.

Figure 2 (Cody et al., 1995) provides a more detailed representation of designers'

PURPOSE	FUNCTION	FORM
Explore Problem/Need	**Conceptualize Solution Functionality**	**Compose Form of Solution**
Study current requirements (e.g. Statement of Work) -- read and analyze	Review functionality of past designs -- read and analyze	Review forms of past designs -- read and analyze
Study scenarios of operational need -- view and analyze	Synthesize/derive input-output relationships -- create and represent	Synthesize form of solution -- create, visualize and "sketch"
Review requirements for past designs-- read and analyze	Develop model of functionality -- integrate, analyze and test	Prototype/mockup solution -- integrate and fabricate
Explicate performance criteria and attributes -- integrate and decide	Predict performance (exercise model) -- calculate/simulate, analyze and interpret	Measure performance (collect data) -- observe, measure, analyze and interpret

Figure 2.   Archetypal design tasks

tasks. Our studies showed that designers heterarchically move among these tasks as they proceed with the elements of Figure 1. Trajectories through the "space" depicted in Figure 2 can appear almost chaotic as designers attempt to understand issues and trade off competing criteria.

The complexity of design is also affected by the fact that designers seldom work by themselves. Cross-functional design teams are prevalent. Thus, much of the information collection and dissemination occurs verbally. Use of formal sources is the exception rather than the rule. Consequently, design support should include both individual and group aspects.

3 Supporting Design

Systematic analyses of designers' abilities and limitations in the processes represented in Figures 1 and 2 led to the development of a set of functional requirements for design support. Six broad functions include: search, transform, display, explain, tutor, and execute. These functions provide the basis for helping designers to overcome their limitations while also taking advantage of their abilities.

This set of functions implies that an ideal design support system should be able to find information for the designer, transform it in ways that make it more useful, display the information, explain this information, tutor the designer if necessary, and execute models, methods, and tools as desired. These requirements imply an integrated system that provides both aiding and training.

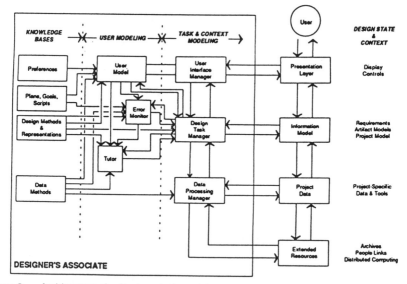

Figure 3.   Architecture of a Designer's Associate

Figure 3 (Cody et al., 1995) illustrates such a system. This *Designer's Associate* provides the types of support that will enable designers, in turn, to better support operators and maintainers. We hasten to note, however, that Figure 3 depicts a vision of what design support could be rather than what is currently possible. In particular, the monolithic nature of such a support system makes it difficult to realize.

Thus far, we have produced two tools that reflect the philosophy and aspirations of Figure 3. The *Product Planning Advisor* supports designers in human-centered design a process of considering and balancing the concerns, values, and perceptions of all the stakeholders in a design effort. This tool assures that many human-related design issues receive attention early in the design process while there are still degrees of freedom to address.

In the process of developing this tool, it quickly became apparent that product planning cannot be pursued independent of business processes. This led to creation of a sibling tool called the *Business Planning Advisor*. This tool focuses on integrating product-related issues with marketing and finance issues. This reflects the basic fact that in order to please customers and users with human-centered products, one has to be able to market and sell these products and finance the whole process. This, in turn, implies that one has to be able to generate sufficient return on investment to attract the necessary resources.

Our experience in developing and selling these methods and tools to over 2000 users in roughly 70 companies has led to the notion that such tools need to be part of an overall suite of software tools for a wide variety of jobs and tasks. Figure 4 (Rouse, 1995) shows how these various tools can fit together into an *Enterprise Support System* (ESS). Note how this architecture depicts a distributed control system. However, in contrast to Figure 1, the complexity of these processes is much greater, as is the realism.

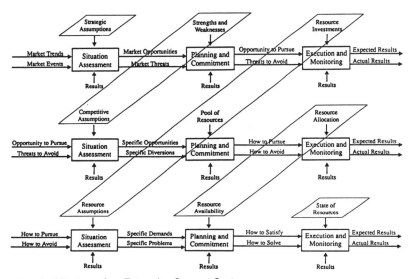

Figure 4.   Architecture of an Enterprise Support System

It is interesting to compare the ESS concept to our earlier notions of a Designer's Associate. While we concluded that a monolithic Designer's Associate is at most a vision, the feasibility of an ESS is much greater due to its modular nature. In fact, the prevalence of computer networking and emerging abilities to move among different software applications are enabling development of fledgling enterprise support systems today. While integration remains a problem, and tends to be the price of the benefits of modularity, progress is surprisingly rapid. Experience using these early systems will quite likely accelerate the maturity of our thinking about such possibilities.

4 Impact on Design

What happens when designers are supported in the ways just outlined? It is too early to provide any evidence regarding the impact of an ESS. We have, however, gained considerable insights into the impact of our design tools. Two cases studies are illustrative.

An automobile manufacturer used the Product Planning Advisor to assess the relative utility of alternative approaches to redesigning one of its engines to reduce emissions. They concluded that the approach in which they had been investing for a year was the worst choice. This was due to negative impacts this choice would have on stakeholders in manufacturing. The human-centered approach, which considers all the humans involved in design success, enabled curtailment of investments in this choice and reallocation of these resources to a much better alternative.

A power systems company employed the Business Planning Advisor to restructure its lines of business and to clarify goals, strategies, and plans. As a result, business teams were formed, leading to more clearly defined roles and responsibilities. This led to many months of losses being transformed into steadily growing profits.

Both of these tools provided these companies with additional benefits. Information was much more efficiently and effectively collected, managed, and disseminated. Team performance was also enhanced by group use of these tools. These companies, like most companies, report that the teams now have shared "mental models" of their markets and businesses.

5 Conclusions

The experiences outlined in this chapter provide support for several strong conclusions. First, design works better if it itself is designed. In other words, if your goal is to better transfer R&D results into application, you should design your design processes with this explicit goal in mind.

Second, design processes are tightly linked with business processes. If you want to improve design, you have to take these links into account. Further, you also may have to improve business processes if you want design improvements to achieve their full potential.

Third, enterprise systems are complex human-machine systems. They also are ubiquitous. There are roughly 100 nuclear power plants and approximately 10,000 airliners in the world. In contrast, there are perhaps 1,000,000 enterprises with sales of at least 10 million dollars. Clearly, the managers and designers in these enterprise systems deserve at least the same investments in studies and improvements as we devote to those who operate and maintain power plants and airplanes. Such investments in enterprise systems will also benefit those who operate and maintain systems such as power plants and airplanes, since the products and systems that they use will benefit from supporting the designers and managers who create them.

References

Cody, W. J., Rouse, W. B. and Boff, K. R. (1995). Designers' associates: Intelligent support for information access and utilization in design, In W.B. Rouse (Ed.). *Human/Technology Interaction in Complex Systems*, Greenwich, CT: JAI Press, Vol. 7, pp. 173-260.

Rouse, W. B. (1991). *Design for Success*. New York: Wiley.

Rouse, W. B. (1995). Enterprise support systems. *Proceedings of the IFAC/IFIP/IFORS/IEA Conference on Human-Machine Systems*. Cambridge, MA: International Federation for Automatic Control, pp. 145-150.

CONCLUDING REMARKS

Chapter 25

A Concluding Perspective on Human-Machine Systems

Henk G. Stassen

The 14th of June 1995 was a very special day. That morning, just before I went to the university, I asked my wife Mia whether she was intending to join me at the symposium dinner of the 14th European Annual Conference on Human Decision Making and Manual Control, being held at the Delft University of Technology. Her reaction was negative. So, I participated that day in the conference and consequently I joined the symposium dinner that evening. Then, a peculiar coincidence happened. At the entrance of the restaurant I met my eldest son and his partner, although they were living at a distance of 150 miles from Delft. I still did not suspect that anything was going on. This changed when I entered the dining room and saw my wife, my other children, and many of my close friends, who were certainly not attending the conference. So it was clear that this was not a coincidence; something had been planned, but I was unaware of the reason.

The confusion would last until the symposium dinner speech given by Neville Moray, when the veil was dropped. Neville passed the baton to Thomas Sheridan and Ton van Lunteren, and they surprised me by announcing this book *Perspectives on the Human Controller,* being written by 28 colleagues and friends, and dedicated to me. I was amazed that I had not suspected any activity going on in our laboratory for such a long time. They ended their speeches with the assignment that I add a final chapter to bring the book to 25 chapters, organized around three major themes. With pleasure I accepted the assignment.

Over more than 30 years I headed a research group in the field of human-machine systems, and as Ton van Lunteren explains in Chapter 2, two major fields were treated: (1) medicine and rehabilitation and (2) human-machine systems in traffic and industry. These two different fields do not seem to be a natural combination, but there are many arguments to defend this lucky choice. Both the fields have many common aspects with reference to the systems approach methodology and complexity, and the interaction between humans and the environment. However, some major differences cannot be neglected. I am therefore pleased to see that the editors of this book have grouped the different contributions into three themes (breaking my second field into two parts): (1) control of body mechanisms, rehabilitation, and the design of aids for the disabled; (2) human control of vehicles and manipulation; and (3) human control of large, complex systems. Clearly these fields support each other. However, I must admit to one reason for my studying the fields collectively, namely, that I never could choose between biomedical engineering and industrial human-machine systems.

As has been elucidated already in Chapter 2, over the past 30 years the research in both the industrial and biomedical fields moved, surprisingly enough, in opposite directions. Starting with the study of manual control problems, the increase in automation and digital control technology forced the shift toward supervisory control of highly complex systems, and hence the shift from monodisciplinary control problems to multidisciplinary human decision making occurred. Or to say it in terms of Rasmussen's three-level human behavior model, the shift from skill-based behavior to rule based and knowledge-based behavior took place (Rasmussen, 1983). In the medical research executed at Delft, the tendency was just the reverse. It started with studies in the very multidisciplinary field of the rehabilitation of disabled with severe motor deficiencies of the upper extremities, and then research turned to more purely medical disciplines, such as neurosurgery and orthopedics. However, when comparing the two fields, it is easy to recognize the commonalities, such as the systems approach methodology and other aspects of the modeling techniques used.

It is also relatively easy to distinguish that one of the most important differences is the influence on human behavior of the performance-shaping factors, such as the psychological and social environment, as well as the emotional state of the human. Note that a process supervisor can be individually trained in a training center, whereas, for example, a single amputee cannot be rehabilitated in a rehabilitation center. In the latter case, one also has to take into account the social and physical environment, i.e., the amputee's partner, family, and friends. Other important differences are the impact of human errors and mistakes with regard to the safety, environment, and social aspects.

Many examples of commonalities, each with its own different problem context and boundary conditions, are discussed in the foregoing chapters. A few are as follows:

- Nonlinear system and control theory is applied to the control of robots and to describing the dynamics of the most perfect robot on Earth: the human upper extremity (chapters 4 and 16).

- The subcritical instability task (Chapter 22), originally developed to measure the mental load of human process operators or pilots, can also be used for the diagnosis of Huntington's chorea, a disease in which muscle function gradually deteriorates. By using eyeball movements as a control input for the subcritical-instability task, one is able to use statistical techniques to detect at an early stage whether or not one suffers from this hereditary disease.

- Three dimensional display technology is making great progress in aeronautical engineering and space control. A 3D display originally developed to manually control the European Robotic Arm (Chapter 16), can be used in environmental dredging technology as well as in minimally invasive surgery (Chapters 17 and 18).

- Evidently the progress made in informatics has led to the development of fault management support systems for the supervision of nuclear power plants and chemical processes (Chapter 20), as well as the design of expert systems for medical diagnoses (Chapter 5), and the training of disabled (Chapter 11).

- The well-known Optimal Control Model (Chapter 13, 19, 20, and 24), has found application in the description of human manual and supervisory control behavior. It can also be used in the description of human control of the musculoskeleton system (Chapter 4).

- Predictive display control concepts are being used more and more in the supervision of complex industrial systems. Many examples can be found in aeronautical engineering and in the navigation of large container vessels and supertankers (Chapter 15). Other examples can be found in medicine, such as in the treatment of quadriplegics (Chapter 12).

- Common features can also be found in human supervisory control in the process industry (Chapter 20), and the treatment team in the rehabilitation of severely physically disabled persons (Chapters 11 and 12).

All examples of analogies that are given are based on a systems theoretic approach, whether it concerns the design of assistive devices for the supervisor (Chapters 13, 15, 16, 17, 20, 21, and 24), for the patient (Chapters 7, 8, 10, and 11), or for the (para)medical disciplines and treatment teams (Chapters 5, 9, 12, and 24). There are also such correspondences between the modeling approaches applied to understand human behavior in the process industry (Chapters 14, 19, 20, 22, and 23) and in medicine (Chapters 4 and 6).

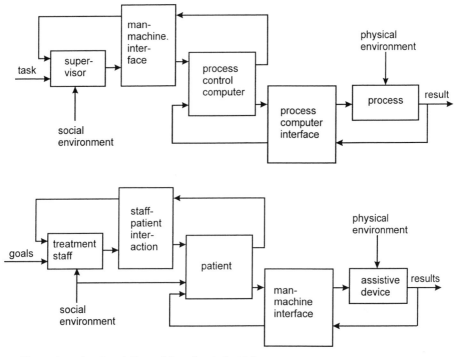

Figure 1.   (a, above) Supervision of an industrial process.
(b, below) Treatment of an amputee.

A few cases will be discussed in more detail for illustration.   Figure 1A is a block diagram of a process supervisor controling an automated industrial process. The diagram shows that the supervisor interacts via the human-machine interface with the controlled process, whereas the process control computer interacts via the process-computer interface with the process to be controlled. A similar diagram can be made in order to describe the rehabilitation process in an amputee. The treatment staff interacts with the patient, whereas the amputee controls the assistive device, i.e., the prosthesis, via a human-machine interface. A remarkable difference is the influence of the social environment, which might influence the process supervisor only slightly, but which may have a strong influence on the behavior of treatment staff and amputee. Moreover, the analogy can also be recognized for the six principal functions that the supervisor (Fig. 2A) or the treatment staff (Fig. 2B) fulfills (Sheridan, 1980), including: monitoring and task interpretation or observation of the amputee; goals definition; process tuning and intervention or treatment; and finally planning and fault management or planning and diagnosis/prognosis. Hence the structure of the interactions is common, but the specific problems require particular solutions.

Another example is the use of predictive displays. A study on the maneuverability of very large crude (oil) carriers (VLCCs), shows that the dynamic behavior of these vessels is extremely difficult to understand and to learn by the helmsman (Veldhuijzen and Stassen, 1977). However, in order to control a system optimally one should have at least an accurate internal representation of the system to be controlled, the task to be achieved, and the statistics of disturbances acting on the system under supervision (Stassen et al., 1990). By modeling the helmsman's behavior, one is able to get insight into the problems the helmsman faces. It was found that a helmsman's behavior can be described by two blocks: an internal representation of the ship, and a decision-making element (Fig. 3A).

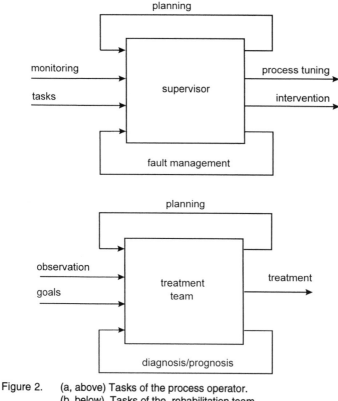

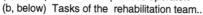

Figure 2.   (a, above) Tasks of the process operator.
                   (b, below) Tasks of the rehabilitation team..

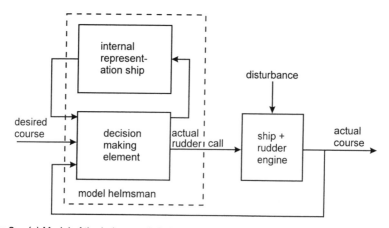

Figure 3.   (a) Model of the helmsman's behavior.

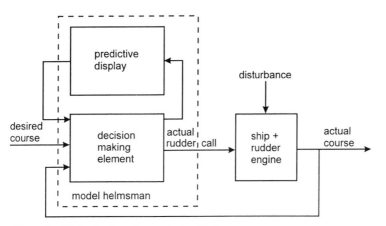

Figure 3. (b) Application of a predictive display.

Hence, by replacing the internal representation with an *internal model*, i.e., a predictive instrument, one is able to create an efficient support system for the helmsman that leads to a major reduction of the mental load of the helmsman.

A similar approach can be followed in the treatment of quadriplegics. By observing the rehabilitation process, the treatment staff can determine the actual state of the patient. Given the goals to be achieved, the staff plans a treatment program and tries to realize it. Note that in process control the goals are usually more or less well defined, but in rehabilitation these goals often are gradually adapted to the actual state of the patient. For example, assume that a quadriplegic is undergoing training to handle a wheelchair but that during the treatment process some activities of his leg muscles regain function. Then the goals will be adapted to the new situation: The patient will also be trained to use crutches. In order to facilitate treatment planning, one can model the rehabilitation process based on measured data of treatment activities and state of the patient. In this way a dynamic treatment model is produced (Fig. 4).

Another, more simple method is to develop a prognosis model, based on the dynamic properties of the state variables of the patient. This entire process can be executed as a running average over the last 50 patients during a treatment period of about 40 weeks. The final result is a treatment staff support system, in fact, a predictive display (Stassen et al., 1980). This support system produced some promising features. First, a uniform treatment protocol became available in The Netherlands and Belgium. Second, a fine data base for scientific research and treatment was made available, and finally, a decrease in treatment period of about 10 weeks was achieved, which resulted in a substantial cost reduction for the treatment and reduced hospitalization of the patient, i.e., an easier integration into the patient's future social environment.

Finally, one last example will be discussed: the similarities and differences between the design of robotic systems and the analysis of motions of the upper extremities. Here robots are discussed only for the execution of special tasks, such as accurate positioning of objects in space, in hostile environments, in minimally invasive surgery, or for inspection tasks in space or under water. The use of robotic systems in automation of production lines will not be considered here (see Sheridan, 1992; Stassen & Smets, 1995). In comparing the robot and upper extremity with one another, it is interesting to note how well the same methodology can be used for both. Both systems are mass-spring systems that are highly nonlinear, and that have to execute accurate positioning tasks.

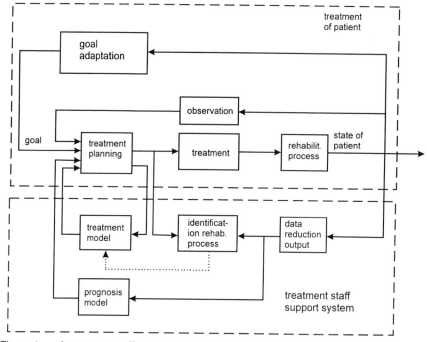

Figure 4.   A treatment staff support system for the rehabilitation of quadriplegics.

As Pronk (1991) and Van der Helm (1991) showed in their doctor's theses, many of the well-known modeling techniques often used in robotics can be applied directly to the description of the dynamic behavior of the upper extremities, such as the finite element method (FEM), the multi-body systems concept, the optimal control theory, and identification methods. Later the theories of fuzzy sets and neural networks will also be used. Hence, the application of well-known theories was a fine start to the development of a dynamic mechanical model of the upper extremity system. However, great difficulties appeared during this development. The human arm has much greater complexity: The number of available sensors is very large, and the number of actuators is also large. Moreover, these actuators are often multi-articular. In addition, because muscles operate only in one direction all muscles work in pairs. The control concept is at least hierarchical, but the number of levels needed for distinguishing them is unclear. Finally, the extent to which one has to go from the macro outlines down to the details in order to understand the system is unknown.

Validation of simulation results by actual measurements is often very difficult, if not impossible. Certainly, multivariate control of nonlinear dynamic robotic systems is still at an early stage, but a start has been made. Much more complex is the development of a shoulder model, which consists of roughly three development stages: (1) a kinematic model, including a description of the thorax, bones, joints, ligaments, and capsules; (2) a dynamic model, including the segmental inertias, muscle geometry, and ligament and muscle dynamics, and (3) the complete model, including muscle control, sensor dynamics, and coordination control by the central nervous system. To implement these models, many special measurement techniques have to be used such as the analyses of cadavers, roentgenograms and magnetic resonance images, 3D video and electromyographic recordings, and palpation methods. Moreover, the use of simulation techniques, such as FEM, control theory, neural networks, and biomechanics is very helpful. In summary, without the experience built up while designing robots, modeling of the highly complex system of the upper extremities would be almost impossible.

Having described the many similarities and differences between the human-machine approach in industrial systems and that in biomedical engineering, it is clear that the systems approach is promising, although it is only a start. It is also clear that many new problems will arise, in particular, with regard to the shift from manual control to supervisory control. To predict the future with precision is impossible, but some tendencies have already been indicated in the literature.

The increase in automation will continue due to laws on energy conservation, environmental pollution, and safety, as well as the need for product quality, whether it involves a chemical plant, traffic, or an intensive care unit. This means that in future the function of the human operator will be that of a supervisor, the behavior of whom will change from skill-based and rule-based behavior to knowledge-based behavior. The human operator will act as a cognitive information processor.

The major question will then be what is the limit to the degree of automation, or to what extent is automation acceptable? After all, the important aspect is what the effect of the automation on the human's supervisory control behavior during malfunctioning of the automated process will be. Can we expect that the supervisor will act adequately during such a situation? Or has all contact and feeling been lost for the fully automated plant? In fact, it goes back to the very fundamental problem of task allocation between human and machine, a problem that was defined in the early 1980s (Sheridan, 1980) and is depicted in Figure 5. Figure 5 clearly shows the consequences of the increase in automation on the human cognitive functions.

1. The computer offers no assistance: the human must do it all.
2. The computer offers a complete set of action alternatives, and
3.   narrows the selection down to a few, or
4.   suggests one alternative, and
5.   executes that suggestion if the human approves, or
6.   allows the human a restricted time to veto before automatic execution, or
7.   executes automatically, then necessarily informs the human, or
8.   informs the human only if asked, or
9   informs he human only if it, the computer, decides to.
10. The computer decides everything and acts autonomously, ignoring the human.

Figure 5.   Levels of task allocation between the human supervisor and the computer (After Sheridan, 1980, with permission).

A simple and accurate answer to this problem cannot be given at this moment, since it lacks knowledge, and thus models, of human cognitive information processing. As stated earlier (Stassen et Al., 1990), in order to control a plant optimally, knowledge about task, process dynamics, and disturbance statistics is needed. In the same way, the design of a human-machine interface requires, in addition to this knowledge, knowledge about human task interpretation, cognitive information processing, and human-performance shaping factors (Fig. 6). It is the latter knowledge that is missing, and thus a direct solution for the design of interfaces is not available.

Moreover, the change in human tasks requires that the human supervisor receive additional assistance. At the skill-based level requests for demonstrations are now available; similarly, at the rule-based level requests for rules can be answered. However, at the knowledge-based level the supervisor might ask for advice that can be generated by prediction systems, mass/energy flow modeling concepts, expert systems or human operator support systems, and virtual-reality systems (Fig. 7).

Many of these advanced hi-tech systems are presently under development. However, almost no validation studies have been executed, and little practical experience is available. Nevertheless, there is an expectation that these systems are trustworthy. The

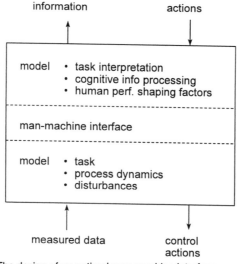

information   actions

| model | • task interpretation
• cognitive info processing
• human perf. shaping factors |

man-machine interface

| model | • task
• process dynamics
• disturbances |

measured data   control
                                      actions

Figure 6.   The design of an optimal man-machine interface.

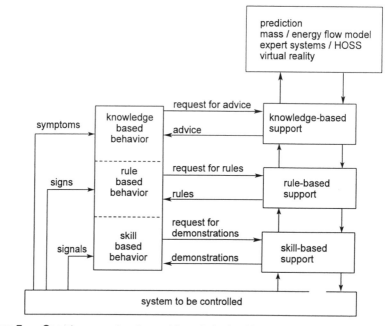

Figure 7.   Operator support systems at three behavioral levels, i.e., skill based, rule
              based, and knowledge based. (Adapted from Rasmussen (1983), with
              permission.)

question remains as to whether the human supervisor believes and trusts the advice, is
willing and able to follow it, and, even more importantly, is able to check the advice.
This means that the advice should at least be transparent. Early optimism is premature,
and much fundamental study is necessary. Moreover, training of the human supervisor
deserves significant attention, since progress in control theory suggests that recently

automated processes are more or less robust for process-parameter variations and disturbances. The human supervisor may be unable to see the consequences of his or her supervisory actions and interventions. Therefore, on-the-job training becomes almost worthless.

Finally, one more direct consequence of progress in automation technology introduced in the late 1980s is the introduction of the *plantwide control* concept (Journal, 1988). Here one tries to integrate the direct process-control systems at the bottom of the hierarchy with the management systems at the top. Unfortunately these two processes are carried out by different disciplines, i.e., the process-control engineer and the information or computer scientist, who each have a different culture and language. So it may be expected that it will take some time before these two disciplines work together.

It should be mentioned that this development leads to the question: Where does supervision and management begin? Or, to say it in another way, who is the responsible authority, and for what, when, and where? This question is of vital importance since it is directly related to the workers' safety, the production process, and the environment. Many incidents and accidents find their origins in vaguely defined authority. This aspect is often underestimated, not only in industry but also in politics. Since safety is a matter of money – how much the industry or society willing to spend for a safe industry and a good health care system – the denationalization of government systems, serving the energy needs of the public, railway and other transport systems, and hospitals – implies there will be some risk when commercial interests take over. With the move towards a society that becomes more and more dependent on technology as well as the tendency for government to withdraw from many public functions, more attention should be paid to human reliability and risk analysis. It should be understood that in at least 80% of the major industrial disasters, human errors were involved.

Conclusions

In summary, progress in automation technology and application of new technological concepts have created many new and very difficult problems. Each technology possesses its own problems, which cannot be solved with existing experience and knowledge. Thus, engineers are making life more difficult. Or, to say it in a more positive way, it is a challenge to work in the field of human-machine systems and to adapt technology to human capabilities and limitations. There are a lot of interesting projects to be executed for many years to come.

In the medical field the human-machine system approach has also shown itself to be a strong tool to integrate different disciplines, but the cultures of medicine and technology are very different. Even with the systems approach as a common language, it takes years before those in one field are able to communicate with and trust those in the other. Integration of the disciplines, however, is a must, since progress in biomedical engineering originates at the boundaries of both fields due to its multidisciplinary character. Happily, the transfer of knowledge from one field to another can be encouraged, a fact that has been demonstrated amply by the examples already discussed.

Future developments in biomedical engineering are difficult to predict, but it is certain that hi-tech solutions will be implemented in diagnostics, magnetic resonance imaging, minimally invasive surgery, and computer-aided surgery, mostly leading to an increase in efficiency and/or a reduction in costs. It may be expected that conventional surgery will move toward automation of certain surgical procedures (e.g., stapling and suturing, drilling holes in bones to fit joint implants), where the surgeon intervenes in a particular procedure only if a particular function is not well performed. There is also a trend toward having surgical consultants available at a distance. The operating theater will more and more be equipped with instrumentation to display variables, images, etc., to position the patient using special controls in an optimal operating position, and to use remote manipulation instrumentation to perform the actual surgery.

It is clear, however, that seldom has the entire surgical process been analyzed, including task allocation among the surgeon, the assisting staff, and the equipment. The moment that new instrumentation becomes available, it tends to be added to existing equipment. The ergonomics of many operating theaters is now far below the standards that can be achieved. Approaching the surgical process as an activity in which the human-machine interfaces (the displays and controls) are well balanced is urgently needed. In this way one is able to see the same problems as faced in industry during the shift from manual to supervisory control.

One may distinguish four steps in surgery: (1) conventional surgery, in which manual tasks are performed with direct perceptive/proprioceptive, tactile, olfactory, and 3D visual sensing; (2) minimally invasive surgery, in which manual tasks are performed with a 2D visual sensing and very poor perceptive/proprioceptive feedback, (3) robotic supported minimally invasive surgery, with only 2D visual feedback; and finally, (4) automatic controlled robotic minimally invasive surgery, in which the surgeon acts as a supervisor.

Important breakthroughs can be expected in the integration of the tasks of the radiologist and surgeon in order to plan the entire surgerical process outside the operating theater, thus resulting in a decrease in the time that the patient is under anesthesia. Imagine the progress that could be made by developing a computer-aided protocol for plastic surgery or a computer aided planning of the positions of the trocars needed for minimally invasive resection of a part of the colon, or by developing diagnostic expert systems to advise medical staff. All of these possibilities are forthcoming. But again, experience teaches us that: new technology will raise new problems.   Two   such problems should at least be mentioned. First, society is becoming more dependent on technology, and the same is true for medicine. However, medical care is much more than just the right application of medical protocols and technology. The human being is a psychosomatic system, so psychosocial aspects are very essential in the care a patient receives. Examples can be found in rehabilitation medicine and geriatrics. The medical community cannot afford to let technology dictate what happens in medicine. The second problem is an ethical one. The biomedical engineer is always creating new possibilities for medical staff. However, it is their decision, in harmony with the patient, to decide whether all possibilities will be used or not. The physician is therefore burdened with the ethical question, should everything be done that is possible?

One final remark remains. In the beginning of this concluding chapter I said that I never could make the decision of whether I wanted work as a human-machine systems engineer in industry or in medicine. Now, 30 years later, I am still undecided. Probably, the mix of both the fields was the perfect scientific environment I needed to cope with the challenges. So, in the next life I think I will do the same.

References

Journal, A. (1988). Plant-wide control. *Benelux Quaterly Journal on Automatic Control.* 29: 3, 100.

Pronk, G.M. (1991). *The Shoulder Girdle. Analysed and Modelled Kinematically.* DUT, Delft, The Netherlands.

Rasmussen, J. (1983). Skills, rules, and knowledge; signals, signs, and symbols, and other distinctions in human performance models. *IEEE Transactions on Systems, Man and* Cybernetics, SMC-13: 257-266.

Sheridan, T.B. (1980). Computer control and human alienation. *Technology Review*, Cambridge, MA, MIT-press, pp. 61-73.

Sheridan, T.B. (1992). *Telerobotics, Automation and Human Supervisory Control.* Cambridge, MA, MIT Press.

Stassen, H.G., Van Lunteren, A., Hoogendoorn, R., Van der Kolk, G.J., Balk, P. Morsink, G. and Schuurman, J.C. (1980). A computer model as an aid in the treatment of patients with injuries of the spinal cord. In *Proceedings of the International Conference on Cybernetics and Systems*, Cambridge, MA, IEEE, pp. 385-390.

Stassen, H.G., Johannsen, G. and Moray, N., (1990). Internal representation, internal model, human performance model and mental load. *Automatica*, 26: 811-820.

Stassen, H.G. and Smets, G.J.F., (1995). Telemanipulation and telepresence. Preprints 6th IFAC/IFIP/IFORS/IEA Symp. on Analysis, Design and Evaluation of Man-Machine Systems, Cambridge, MA, pp 13-23.

Van der Helm, F.C.T.. 1991. *The Shoulder Mechanism, a Dynamic Approach.*. DUT, Delft, The Netherlands.

Veldhuijzen, W. and Stassen, H.G. (1977). The internal model concept: An application to modeling human control of large ships. *Human Factors*, 19: pp. 367-380.

Index